Prosperity, Poverty and Pollution

By the same author

Beyond Marx and Market
Outcomes of a century of economic experimentation

Cluster Publications
Zed Books
1998

Prosperity, Poverty and Pollution

Managing the approaching crisis

Klaus Nürnberger

Cluster Publications
Pietermaritzburg

Zed Books Ltd
London & New York

1999

First published in Africa
in 1999 by
Cluster Publications,
P O Box 2400,
Pietermaritzburg, 3200
South Africa.

ISBN 1-875053-15-8

© Klaus Nürnberger
All rights reserved.

Published in the rest of the world by Zed Books Ltd, 7 Cynthia Street,
London NI 9JF, UK, and Room 400, 175 Fifth Ave, New York, NY, USA in 1999.

Distributed in the USA exclusively by St Martin's Press, Inc., 175 Fifth Ave, New York, NY 10010, USA.

The right of Klaus Nürnberger to be identified as the author of this work has been asserted by him in accordance with the Copyright, Designs and Patents Act, 1988.

A catalogue record of this book is available from the British Library.

ISBN 1 85649 730 5 cased
     1 85649 731 3 limp

Library of Congress Cataloging in Publication Data

Nürnberger, Klaus 1933-
    Prosperity, poverty and pollution: Managing the approaching crisis / Klaus Nürnberger.
    p.   cm.
    Includes bibliographical references and index.
    ISBN 1-85649-730-5 -- ISBN 1-85649-731-3 (pbk)
1. Economics -- moral and ethical aspects. 2. Economic development -- Moral and ethical aspects.
3. Poverty. 4. Pollution -- Economic aspects. 5. Environmental degradation. 6. Social justice.
I. Title
HB72.N87 1999
333.7 -- dc21   98-32145
                                                                                                                   CIP

Cover design: Ad Lib Design, London N19
Editing and Index: Sarah Moses and Arden Strasser
Design, typesetting, layout, and diagrams: Klaus Nürnberger

Printed and bound by
The Natal Witness,
Printing and Publishing Company,
P O Box 362,
Pietermaritzburg, 3200
South Africa.

The financial assistance of the Centre for Science Development (South Africa) towards the publication of this work is gratefully acknowledged. Opinions expressed in this publication, and conclusions arrived at, are those of the author and are not to be attributed to the Centre for Science Development.

To my students and colleagues
over three decades,
the seedbed of insight and commitment

Contents

1. What this book is all about   4

   The approaching crisis 3; a new economic paradigm 6; a supra-disciplinary approach 8; model construction 11; the educational task 12; survey of the book 13; a rational and a prophetic work 15.

Part I: The structure of the global system   19

2. The economic problem in a nutshell   20

   I. Basic elements of the economy 20. II. Economic imbalances 26. III. Comprehensive cost-benefit analyses 31. IV. Ethical considerations 35.

3. Economic discrepancies   39

   I. Geographical centres and peripheries 40; characteristics of centres and peripheries 45. II. Potency distribution in the population 50. III. The relation between potency and need 57. IV. Centres and their offshoots 62.

4. Ecological deterioration   70

   I. The problem 70; causes 72. II. Current dimensions 74: population growth 74, food production 77, resource depletion 79, pollution 84, armed conflict 88; ways out 91.

5. Causes of economic imbalances   97

   I. A model of causation 98. II. Non-economic factors 100: social 100; culture and education 103, poverty 105, population growth 106. III. Factors of production 109. IV: Asymmetrical interaction 116: suction 117, penetration 121, market interaction / comparative advantage 123, the debt burden 127, use and abuse of power 130; the arms trade 132.

Part II: The transformation of collective consciousness   141

6. The social reality of convictions   142

   The rationale of Part II 142. Immanent and transcendent needs 144; truth and conviction 146; interests and ideology 148; popular wisdom 150; privatisation of religion 151; science and religion 152.

7. Activating the resources of faith   156

   I. Faith and fatalism 157; relativity 159; the privatisation of faith 160. II. The inner dynamics of the Christian faith 163. III. Transcending reality 170: time 172, human limitations 173, ambiguity 173, suffering 175, ethnic and social barriers 175, privileges 177; human rights 178; power concentrations 179.

8. Traditionalism and modernity   186

   I. The modernist revolution 186. II. Traditionalism and modernity 195; characteristics of traditionalism 196; of modernity 200; clash between them 204. III. Reactions to conquest 206. IV. Challenges of faith to modernity and traditionalism 214: freedom and mastery 214, responsibility 216, community 217, progeny and profit 218, time 219, truth 221.

9. The dependency syndrome   227

    I. Two analyses 228: Mannoni 228; Freire 232. II. Horizontalising relationships 239; pedagogy of the oppressor 242; pedagogy of the catalyst 243; the missionary 244.

10. Interests and ideology   253

    I. Ideological legitimation 253; ideologisation of economic life 256; of economics 257. II. Faith and ideological distortions 261; acceptance and confrontation 261.

Part III: Towards a new economic paradigm   267

11. The pending paradigm shift in economics   268

    I. Paradigm shifts 270. II. Vision 277: goals 278, cosmology 280; accountability: interdisciplinary 282, methodological 284, anthropological 285, moral 287, universal 288.

12. A more fundamental approach (Market and marginalisation)   297

    I. Need and capacity 297; segmented growth 302; need creation 304. II. Distortion of the peripheral economy 310; rising expectations 310; collapse of production 312; mass unemployment 313; under- and overdevelopment 318. III. The "Asian Tigers" 322.

13. A wider frame of reference (Entropy and evolution)   333

    I. Entropy 334; construction and deconstruction 336; the energy balance of the planet 338; humanity as destructive species 339. II. Evolution and entropy 341; acceleration 342; economic evolution 343; technology as catalyst 346; objections 348.

Part IV: The transformation of social structures   359

14. Agents of change   360

    Levels of power and competence 361; the pivotal role of primary groups 364; the role of convictions 367, motivations 368, institutions 370, the church 371, the nation state 372.

15. Drawing up the agenda   377

    I. A new mindset 377. II. A new economic agenda 383; five priorities 383; a policy model 385; social structures 386; collective consciousness 390; obstacles to change 391.

16. Policies for peripheries   396

    Realistic goals 397; 1. population control 399; 2. self-reliance and sustainability 402; 3. local technology 405; 4. lean and clean state 407; 5. conflict 407; 6. a new mindset 409.

17. Policies for centres   415

    Technology and the centre dynamic as global assets 415; 1. productivity and equity 418; 2. market correction through taxation 420; 3. consumption and declining marginal utility 423; 4. advertising and marketing 425; 5. waste and pollution; 6. arms production 433.

18. Policies for centre-periphery interaction   438

    I. The national level 438: rural development 439; commerce and industry 440; employment 442; development of the marginalised 444. II. The international level 446: the debt crisis 448; international trade 449; international aid 450; isolation / integration 453.

Bibliography   459

Index   477

Diagrams

2-1	Phases of the economic process (model)	21
2-2	Equality of opportunity	27
2-3	Resource increase at the expense of other humans	28
2-4	Four growth processes	30
2-5	Network of costs (model)	34
3-1	GDP per square mile in South Africa	40
3-2	The "Triad" (US, Europe, Japan)	41
3-3	The hierarchy of centres and peripheries (model)	43
3-4	Intensity of agricultural production in Europe	47
3-5	Potency / income distribution	51
3-6	Income distribution in Costa Rica	51
3-7	Global distribution of GNP per capita	53
3-8	Centre-periphery structure of potency distribution (model)	54
3-9	Exponential growth of the economy and the population (model)	55
3-10	The relation between need and income (model)	58
3-11	Three types of need; four income-need relations (model)	60
3-12	Galtung's model (centres and their offshoots)	63
4-1	Economic growth, population growth, and ecological impact (model)	71
4-2	Resource needs and availability over time (model)	72
4-3	Food needs and food production over time (model)	73
4-4	Growth of the world population	74
4-5	Population density per square kilometre	75
5-1	A model of causation (model)	99
5-2	Factors of production (model)	109
6-1	The relation between conviction and ideology (model)	148
6-2	The genesis of popular wisdom (model)	150
8-1	Traditionalism and modernity (model)	204
8-2	Reactions to conquest (model)	209
12-1	The role of money in creating supply and demand	298
12-2	The flow of money, goods and services (model)	299
12-3	Linkages between capacity and need (model)	300
12-4	Affluence and poverty gaps (model)	305
12-5	Widening gaps (model)	313
13-1	The dissipation of energy /entropy (model)	335
13-2	The evolution of reality (model)	343
15-1	The (new) economic agenda (model)	385
15-2	The basic policy model (model)	389
15-3	Obstacles to change (model)	392

Preface

This study has a long history. My interest in the subject of economic development was aroused in the 1950s during first degree studies in Agricultural Economics and subsequent involvement in rural development planning. Having switched to Theology, my interest in economic development was rekindled in 1968 when I became pastor of a scattered parish in a remote and poverty-stricken rural area in South Africa, and subsequently in the hothouse of the black townships around Johannesburg.

The foundations for the present design were laid when I was invited to lecture in Berlin in 1974 during my first sabbatical. Here I was confronted with Marxism, dependency theory and the limits to growth debate. I struggled to find connections between two truncated disciplines which I had carried in my baggage: an existentialist theology, and a "value-free" economics. It was a kind of theology which was unable to relate to the economic dimension of life and a kind of economics which was unable to relate to convictions, values and norms. Four years later I finalised my first major study in the field, entitled *The relevance of the Word of God in the process of economic development* (published in German in 1982).

As director of the Missiological Institute, Umphumulo, I conducted two research projects, one on affluence and poverty, another on economic ideologies in a Christian perspective (1978, 1979). Participation in an interdisciplinary research project of the South African Council of Churches on ideologies of change, led by Dr Wolfram Kistner, widened my horizons considerably (Leatt *et al* 1986). In 1981 a visiting lecturership to Sâo Leopoldo, Brazil, opened my eyes to interesting similarities and differences between two semi-peripheral Third World countries, Brazil and South Africa.

I then initiated a course on *Ethics of Economic Life* at the University of South Africa. During a visiting lecturership to Bochum, Germany, in 1982, I wrote a second volume in German entitled *Ethics of the North-South Conflict* (published in 1987). The theory I had developed by then was tested in an empirical study under the auspices of the Human Sciences Research Council, entitled *Power and Beliefs in South Africa* (1988). It dealt with the relation between socio-economic power structures and patterns of collective consciousness in South Africa.

By now the basic structure of the argument had fully matured. In three popular pamphlets I tried to reach the wider public: *The scourge of unemployment in South Africa* (1990), *An economic vision for South Africa* (1994), and *Making ends meet: Personal money management in a Christian perspective* (1995). A number of journal articles, my lectures on Economic Ethics at the University of Natal, another sabbatical in 1994, and a research trip to some of the "Asian Tigers" in 1995 all contributed to the present version.

After a quarter of a century of intermittent work, I believe I should make the outcome of my studies available to the wider public. They appear in two volumes, the first on the lessons learnt from a century of conflict between capitalism and socialism, entitled *Beyond Marx and Market* (1998) and the present one on economic power discrepancies.

I am not an economist, but a theologian. In terms of professional expertise, this may seem to be a disadvantage. But in terms of professional independence and interdisciplinary horizons, I would not change roles with a main line economist. Deeply disturbed by what I learnt about the economy, I became disillusioned with the narrow perspectives of both theology and economics. Together with others, I embarked on a more holistic approach and challenged some of the most cherished economic and theological beliefs. The outcome is bound to be controversial.

I acknowledge the inspiration and assistance I have received from many people over decades with deep appreciation. To mention names would lead to artificial selections. I am most grateful for the invitations extended, the dialogue offered (not least by my students), the editorial work done, and the acceptance of the book by the Publishers into their respective programmes. The financial assistance of the University of Natal Research Fund towards the research programme underlying this study is gratefully acknowledged.

Klaus Nürnberger

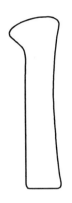

What this book is all about

The approaching crisis

At the turn of the millennium humankind is confronted with a series of seemingly intractable problems which not only imperil its prosperity, but also threaten its long term survival. The most important of these are:
(a) The cumulative destruction of the *natural habitat,* on which all life on earth depends, the wasteful depletion of non-renewable resources, especially fossil fuels, and the overexploitation of renewable resources.
(b) The uncontrolled growth of the *human population.*
(c) Rapidly increasing *discrepancies* in economic capacity between rich and poor sections of humanity, the emergence of large scale unemployment and the growing impoverishment of hundreds of millions of people.
(d) Rapidly growing, and immensely dangerous, powers unleashed by modern *science and technology,* notably in the areas of nuclear, chemical and gene technology.
(e) The deterioration of the moral fibre of society, the breakup of social cohesion and the growth of *conflict potential* - coupled with the development and proliferation of ever more devastating weaponry.

After steady but cumulative cultural advances over at least ten or twelve millennia humanity has reached a phase of *breathtaking acceleration* in the growth of its capacity to populate, dominate, control, exploit, deplete, and - destroy the foundations of its very existence. The decisive enablers were science and technology and the utilisation of fossil fuels.

Fossil fuels are strictly limited and they are rapidly being transformed into waste and pollution. It is plain logic that accelerating growth cannot continue indefinitely on a limited Earth.[1] One cannot help but visualise the thrust of a fountain which shoots into the air in concentrated strength, reaches a peak and collapses into itself in chaotic profusion.[2] Have some sections of humankind already reached the point where economic disintegration begins to snowball? Is it not impossible that the three centuries between the middle of the 18th and the middle of the 21st centuries will go into history as the passing episode of industrialisation, where humankind multiplied recklessly, indulged in a giant spending spree, squandered key resources[3] and plunged coming generations back into misery?[4]

It would be inappropriate, however, to speak about humanity in general. Growth in efficiency, productivity, throughput of resources,[5] accumulation of wealth, and pollution is heavily concentrated in economic centres. Economic activity declines progressively as one moves away from these centres. The outer peripheries are characterised by stagnation, even deterioration. The expectation that the peripheries would be drawn into affluence by the dynamics of the centres may have some validity for the areas nearest to these centres. But as resources are being depleted and the environment polluted, this trend may soon be reversed - long before it has reached the outer peripheries.

If this happens, it will again be the peripheries who suffer first, not the centres. The centres may constitute the peak of a sinking island, as it were, which will be reached by the waves last. Therefore, centre populations are likely to be disturbed least by the whole process. In terms of time, the present generation has a similar advantage over coming generations. Therefore both discrepancies in economic power and ecological deterioration find their common focus in the problem of marginalisation. Thus, marginalisation is the overriding concern tackled in this book.

The current state of popular awareness

Am I alarmist? Maybe. But I am not alone.[6] It has become fashionable to ridicule apocalyptic forebodings. And indeed, cataclysmic global crises, so often predicted in world history, do not seem to materialise. But seen from a long term historical vantage point, revolutionary transformations do happen, and historical processes have been accelerating in recent times. It is no longer good enough to say that all former prophesies of doom have been confounded by history.[7]

Those who have looked at the facts, and who resist the temptation of rationalising them away, agree that intolerable suffering has already engulfed large sections of humanity, that our planet is in grave danger, and that it is high time to wake up. Should history indeed prove the "prophets of doom" wrong, the most concerned among us would be the first to concede their mistake and rejoice - if they were still lucky enough to be alive.

In the meantime, however, the symptoms of a pervasive disease leave absolutely no room for complacency. Catastrophe has already caught up with sufficient numbers of people in the world to substantiate some of the worst predictions. But even if that were not the case, it would not be prudent for a young man to double his tobacco consumption every year just because, as yet, he has not developed lung cancer or emphysema.

In view of the facts, the easy-going optimism of many *liberal* economists is puzzling and disturbing. The assumption that the market, or science and technology, will solve the problems as they come is not very reassuring. With a billion people in absolute poverty, mountains of waste accumulating among the affluent, and pollution levels which endanger life on earth, neither the market nor technology seem to have done a good job in solving the economic problems of humankind so far. On the contrary, they are largely responsible for the genesis of these problems.

Meanwhile, *left-wing* activism has changed from Marxist macro-economics, which has fallen to pieces, to small scale community development. Ideology has made way for romanticism about the symbolic universe of the marginalised. Others concentrate on isolated environmental issues. This is simply not good enough. While the emphasis on community empowerment at grass roots level is important, and while we do not want elephants and tigers to die out, it is the macro-economic context in which grass roots development and ecological sustainability will either flourish or flounder.

It would seem, therefore, that neither liberals nor radicals are currently capable of addressing the issues we are facing. On the other hand the situation is not hopeless.[8] We still have time. We have the means to contain population growth. The productive power of the modern economy is capable of creating moderate but stable levels of wealth for all. We can change from quantitative growth to the enhancement of quality.[9] We can switch to sources of energy which are sustainable and clean. We can reduce waste and pollution to levels that can be absorbed by the natural environment. We can give non-human species a chance to survive and prosper. We can resolve conflicts through institutionalised justice rather than armed conflict. We can shift the emphasis of human culture to the infinite riches of non-material pursuits.[10] We can be released from the solitude and deceit of selfish cravings to a life in community and for community.

To realise these potentials we have to subject current assumptions and convictions to intense scrutiny. At present the liberal-capitalist mindset has triumphed not only over more inhibited and circumspect traditional cultures the world over, but also over socialist and environmentalist lobbies. It has absolutised relative methodological assumptions, such as human control, objectivity and pragmatism. It has elevated to the status of ultimate goals such questionable pursuits as unlimited prosperity and the immediate satisfaction of all individual desires. It has idolised the free play of the market; the unfettered movement of capital across the globe, and the growth of the economy at all costs.[11]

Our dominant culture has become narrow-minded, short-sighted, superficial and self-seeking. These words may sound like unwarranted rhetoric; we shall see that they are not. Elites, whose interests determine the economy, have made themselves comfortable in the express train of progress and few pause to think where the journey might be taking us. Even those who are responsible for the logistics, that is scientists and technicians, prefer to believe that things will sort themselves out in some way or other. And indeed they will; the question is only whether our progeny will like the outcome.

On the positive side, a groundswell of awareness has been growing for at least a quarter of a century. Profoundly disturbed by scenes of world famine on television screens, and alerted by books such as *Silent Spring* and *The Limits to Growth*, a lot of people have become sensitised. Countless studies have been conducted, organisations founded, international agreements reached, substantial sacrifices made.

All this still seems to be insignificant if compared with the vastness of the evolving problem. Much of the existing awareness seems to be on hold. But at least there is hope. Most great developments in history begin small and gain momentum. Once they reach a critical mass they can lead to cataclysmic change. Of course, this principle can work both ways - for a sustainable future or against it. *All depends on whether world citizens with a social conscience can match the determination of influential people working for self-interest.* My ambition is to contribute to the articulation of these concerns and the inspiration of this movement. Therefore the sub-title does not speak of the coming catastrophe but of the emergence of a global responsibility.

The need for a new economic paradigm

If we want to help secure a reasonable, equitable and sustainable degree of prosperity for humankind we need a new economic approach. Any programme of paradigm deconstruction and reconstruction demands criteria and our present effort is no exception. These criteria will emerge as the argument unfolds. But for a preliminary orientation let me enumerate the main assumptions of this book as follows.

(a) The goal of the human enterprise as a whole should be the *comprehensive wellbeing* of the whole of humanity, within the context of the comprehensive wellbeing of its entire social and natural environment. Note that the term "comprehensive wellbeing" does not suggest a utopia where no trade-offs have to be taken into account. It is formulated, rather, as an antithesis against the cancerous pursuit of self-interest at the expense of wider social and natural contexts. It also challenges a reductionist view which concentrates either on the spiritual, the material or the social-structural dimensions of life.

(b) *Ecological* concerns should receive priority over economic concerns, rather than the other way round. The reason is, simply, that an economic growth which is not sustainable is also irresponsible. If we squander the earth's resources now, others have to pay the bill in the future.[12]

(c) The vision of a wholesome economy presupposes a recognition of the *equal dignity* of all human beings, in the context of the hierarchical dignity of the natural world, rather than an attitude which treats both humankind and nature as resources to be exploited at will.

(d) The point of departure for economics, therefore, should be human *need* and the *capacity* to fulfil that need, rather than the concepts of scarcity, supply and demand as used by conventional economics. Supply and demand are derived concepts which conceal rather than reveal the actual economic problems of need and economic capacity. We shall spell out this principle in chapter 12.

(e) It should be recognised that the present economic system is fuelled by *escalating discrepancies* in economic power. The economy is not oscillating around an economic equilibrium, as the terminology of mainline economics suggests. Market equilibrium is just a temporary phase in the process of shifting power relations in the system. Part I will provide the evidence.

(f) Our point of departure, therefore, should be the contrast between *centres and peripheries* of economic power, rather than the market mechanism. The market mechanism is indispensable as an instrument for optimising allocations of resources and distributions of proceeds under any given set of circumstances, but it is incapable of signalling ecological dangers and rectifying economic imbalances.[13]

(g) Concentrations of surplus productive capacity on the local, national and global levels should be directed towards the *economic empowerment of the less privileged*, rather than towards artificially created luxury wants among the affluent.

(h) The means to reach this goal is *responsible participation* by all sections of the population in the economic process, rather than the rapid economic progress of an elite, alleviated by humiliating welfare grants. A spread of purchasing power should not be achieved through handouts but through access to resources and productive capacity.

(i) *The state* should not run the economy but create and defend the space necessary for free initiatives of groups and communities to unfold their potential against dominating economic power. Civil community, state and business should cooperate in safeguarding equitable access to natural and human resources.

(j) Participation in development is to be understood as a process of "interactive learning through action".[14]

(k) Participation has to be achieved on *all levels* of the society. The solipsism, isolation and selfishness of nations and classes is as counterproductive for the human project as those of individuals.

(l) The satisfaction of material needs should be placed into the context of the *balanced satisfaction* of all valid human needs, including psychological, communal, cultural and spiritual needs; material needs should not be allowed to crowd out every other consideration.

The motivations for these propositions are profound and their implications far-reaching. They will become clearer as we go along. The propositions are stated here simply as an "advance warning" that the analyses and conclusions contained in the ensuing argument will take a direction which departs substantially from those of conventional economic wisdom. This brings us to the following question.

A supra-disciplinary approach

A new economic approach cannot be left to the confines of classical economics - just as politics is too important to be left to the politicians, health too important to be left to doctors, and faith too important to be left to the theologians. The problems are vast and complex, and we need to pool our resources.

I do not wish to imply that conventional economics is invalid within its own frame of reference. Nor do I want to deny the professional competence of economists or their impressive achievements.[15] But experts in all fields tend to be dazzled by the immediacy of their particular pursuits and find it difficult to gain a sense of the whole.[16] The attempt to see how single phenomena fit into the giant network of reality, and to formulate a daring overall vision, is indispensable and overdue.

This book is an attempt to contribute to the development of an interdisciplinary approach to the economic and ecological predicaments of our times. The details are not original. Countless people - including prominent economists - have been working on similar ventures for decades and our bibliography gives a glimpse of the wealth of the literature available.[17] Our task is to forge a consistent paradigm which makes economic sense and which takes the complexity of social reality into consideration.

Economic imbalances are caused by social structures and collective mindsets; public policies and personal decisions; obsolete procedures and super-efficiency; factors inherent in economic centres and peripheries, and factors located in the interaction between the two. A systemic view should visualise these dimensions of the problem in their relationships with each other and in the weight of their respective impacts.

This is not an easy task. A dialogue between experts, who are all submerged in their various specialisations, may not be sufficient to do the trick. Because paradigms and conceptual systems vary widely between the social sciences, interdisciplinary studies often resemble a dialogue between the deaf. Most social scientists do not dare to leave the safe methodological confines of their disciplines. The greater their commitment to scientific rigour, the less inclined they are to step out of their predefined models. Those who do are not necessarily heard.[18] Nor have their syntheses found a place in the institutional framework of the academy. As a result they remain outside the routines of established disciplines.

What we need, therefore, is a new conceptual framework which is not just inter-disciplinary, but *supra-disciplinary*. It must allow representatives of all disciplines to suspend, at least temporarily, their professional preoccupations and take a fresh look at the evidence through the eyes of their peers from other disciplines.[19] Only then shall we be able to activate our expertise in a form which is compatible with the whole body of knowledge and accountable to the whole process of world development.

To reach this goal we need to design a paradigm which gleans insights ranging from history to geography, from physics to biology, from sociology to psychology, from political science to cultural anthropology, from education to religious studies. Because of current neglect, the ethical dimension must be accorded particular importance. Ethics is derived partly from the needs of the situation, partly from the ultimate premises of philosophy and religion. To assume that the "depth dimension" of human perception and motivation could be ignored with impunity is one of the fallacies of Western science with disastrous consequences.

Obviously this book can offer nothing but a rudimentary framework. The methodology of designing a scientifically more rigorous approach has to be gleaned from systems analysis.[20] The basic skeleton must then be fleshed out with detailed studies by experts in various fields who are acting in cooperation with each other and who are accountable to each other's fields of expertise.

The competitiveness of the undertaking

It should be obvious by now that this book has an ambitious agenda. How realistic are its goals? Mainline economics presents an impressive and solid edifice whose influence cannot be matched by any rival. Year by year it releases hundreds of thousands of graduates into the decision making processes of the private and the public sectors of society. In theory, economics is nothing but the art of decision making in situations where trade-offs in quantifiable resource allocation have to be faced. It could be a completely neutral tool with no specific agenda of its own. But that is not how it works in practice. In chapters 10 and 11 we shall argue that economics underpins the ideology of the dominant economic system. This ideology, in turn, is upheld by powerful collective interests.

I am realistic enough, therefore, not to expect the current paradigm of liberal economics to be dethroned any time soon. It is also unlikely that many economists will take notice of the views of an outsider. If that is the case, what practical sense does it make to work on a new paradigm?

In the first place, the economy is not just the concern of professional economists, business leaders and politicians. There are also sociologists, ecologists, political parties, administrators, unions, non-governmental organisations, consumer organisations, ecologically concerned citizens, base communities, alert religious groups, leaders of disadvantaged populations - in fact, the whole rest of us who have to make sense of the economy and to take economic decisions.

Representatives of other disciplines and practical decision makers at all levels cannot depend solely on economic experts with their arcane mathematical methodology; they have to engage in analyses and reflections of their own. As mentioned above, economics is too important to be left to economists. Responsibility for the economic and ecological wellbeing of humankind rests with the entire academic community, in fact, with the citizenry at large.

All these people need to make up their minds whether they can accept the dominant paradigm and, if not, which paradigm they want to follow as a basis for their own analyses and judgments. Moreover, to interact with representatives of the dominant paradigm, those who have reservations need sound theory based on empirical evidence. Chances are that they will ride an incoming tide, that history will be on their side. But then they must do their homework now.

In the second place, if the evidence gathered in this book is anything to go by, it is unlikely that the current paradigm is sustainable beyond the middle of the next century. The danger is that humankind will go in the wrong direction just a bit too long. As the Meadows team has argued so persuasively, "overshoot" may lead to collapse.[21] The greatest problem is that the value system underlying this hazardous development legitimates those who benefit from it. Naturally they will tend to push on, even when the victims of the process already begin to pile up. To keep the future open, alternatives must be worked out, tested and propagated long before major catastrophes occur.

This brings us, finally, to the moral argument. It does not count for much in the current economic climate. The ghost of a value-free science still haunts the science of economics. In fact, the assumption that individual self-interest is basic to human behaviour is a profound value judgment with immense consequences. The voices of reason and conscience are drowned by aggressive marketing, suppressed by the competitive pressures on the global scene, and belittled by cynics among the exponents of the art.

Yet these voices must be raised and they must be heard. A humanity which has lost its sense of responsibility has abandoned its birthright. As natural scientists have discovered in the fields of nuclear and gene technology, ethical illiteracy is a luxury we can no longer afford.

The method of model construction

To visualise the economic problem as a whole in its multi-faceted contexts is a formidable task with vast methodological implications. The masses of available data far surpass the capacity of the human brain. Moreover, comprehensiveness involves infinite complexity. There is no way we can avoid complexity if we want to get an impression of the whole. To get an impression of the whole again is necessary if we are to reflect on the direction that we should be taking.

To handle complexity we have to concentrate on understanding, rather than information. The art of scientific research lies in the reduction of complexity by highlighting determinative aspects and relationships among infinite masses of phenomena. In Krugman's words, "it is only through strategic simplification that we can hope to make any sense of the buzzing complexity of the real world."[22] Those aspects which do not stand out, recede into the background. That is the method of model construction.[23] Models are simplified versions of systems. A system is a picture of reality which depicts dynamic relationships deemed to be important for an understanding of a particular dimension of reality, such as the political order.[24] Note, therefore:

(a) A system is always *an extract* taken from reality. By definition, it does not cover the whole of reality. Therefore any system will distort perceptions of reality unless it is seen within the total context. A model is even more abstract than a system: "human mental models must be ludicrously simple compared with the immense, complex, ever-changing universe within which they exist."[25]
(b) There can be an infinite number of models, but not all are equally helpful in understanding reality. A model distorts reality if it elevates minor issues to prominence and neglects major ones. Therefore there must be a critical discussion on the *appropriateness and adequacy* of a model.
(c) The model I offer is deliberately constructed to be *widely applicable*. It can describe phenomena from a rural village to the global economy. Therefore I do not always distinguish between these different levels. The reader will have to judge to which extent the method is successful in shedding light on a particular set of phenomena. A model is nothing but a tool; in many cases it will fit admirably, in some it must be adjusted, in some cases it will not fit at all.
(d) The book is written from the vantage point of a *semi-industrialised Third World country*. It is only natural that I shall refer to the context within which I have done my research. But there is also a deeper justification for this perspective. A country like South Africa has paradigmatic significance because within its borders it mirrors global relationships between rich and poor countries.

The educational task

If responsibility for the future rests with the citizenry as a whole, it is time that the latter be confronted with the facts. Apart from the media, the most important channel of conscientisation is education. It is my conviction that economic and ecological responsibility should become a primary school subject, on which further studies on secondary and tertiary levels should be built. In some rather enlightened societies this is already the case, in most others it is not. In the words of the United Nations, "it is essential to incorporate sustainable development concepts into all levels of education, from basic to tertiary, and for all groups of society."[26] Only in this way can the broad masses of the coming generation be conscientised and empowered to face the future.

To make primary education in these areas possible, colleges of education should begin to offer this field as a major subject. Hoping against hope that this suggestion may find support in the corridors of power, I have constructed this book as a text book. As such it can be used both by ordinary readers and by institutions for tertiary education. For the sake of students and ordinary readers alike I have attempted to use a non-technical language and an easy style. The book should be accessible to any educated person.

For students, in particular, I used didactical tools such as task formulations at the beginning and summaries at the end of chapters; illustrative diagrams; questions prompting revision, application and critique, as well as short sentences and paragraphs. I trust that ordinary readers will also benefit from a lucid presentation and the challenge to react to what is being said.

I know that nobody can be expected to read a book of this size all at once. Therefore the chapters are more or less self-contained and can be read independently of each other. This makes some repetition unavoidable. It goes without saying, however, that everything in the book is linked to everything else and must be seen in its entire context.

The literature is as immense as the area covered. The bibliography concentrates on titles available in the English language. It could keep any student busy for years, yet it only scratches the surface. To enter into a critical dialogue with all the main positions in the debate, even to offer a full documentation, is quite out of the question. Endnotes are only meant to open up the debate and prompt further reading. They are less dense in chapters giving simple information and more dense in chapters which are bound to be controversial.

A survey of the book

The logic of the argument of the book as a whole begins with an *analysis of social structures,* moves to the *transformation of collective consciousness,* then to a *new approach in economics,* and finally to the *transformation of social structures.*

Part I tries to sketch the *contours of the situation.* As in the case of alcoholism, the most fundamental obstacle to finding a solution is the denial that a problem exists. Chapter 2 offers a nursery school presentation of the gist of the *economic problem.* The intention of this chapter is to place, from the outset, the entire argument on appropriate foundations.

Chapter 3 develops a model which demonstrates the *linkages* between economic growth, population growth, growth of marginalisation, growth of the ecological impact and growth of conflict potential in society. Chapter 4 gives an impression of the dimensions of the *ecological* problem: depletion of non-renewable resources, overexploitation of renewable resources and overburdening natural sinks. The provision of food and environmental pollution stand out.

14 | Introduction

Chapter 5 offers a survey of the *causes* of economic imbalances. A model is designed which shows the linkages between factors located *within* poor and rich societies respectively, and factors located in the unbalanced relationships *between* centres and peripheries. It also combines economic with non-economic causes. In doing so, it integrates the valid concerns of major schools of thought in the field: development theory and dependency theory, materialism and idealism, liberal and radical thought.

Part II deals with the fact that discrepancies in economic power are not only caused by social structures and processes, but also by *divergent mindsets*. We begin, in chapter 6, with a brief sketch of the *main components* of collective consciousness and their interaction. Chapter 7 explores the potential of a *particular* conviction, the Christian faith, to act as a critical corrective to detrimental mindsets. The choice of the Christian faith is motivated, apart from the commitment and expertise of the author, by the historical links between biblical assumptions and the emergence of modernity as the dominant mindset of our time.

Chapter 8 takes up the unequal encounter between *traditionalism and modernity* in colonial times, and spell out their socio-economic repercussions. Chapter 9 evaluates the impact of the interaction between *dominant and dependent* collective personality types on colonial and post-colonial developments. Chapter 10 highlights the distortions of perceptions caused by *ideological legitimations* of collective interests. In each of these three cases, the phenomena are confronted with the liberative and reconstructive potentials of the biblical faith.

Part III moves to the level of *academic* pursuits. In chapter 11 we argue for the development and adoption of a *more appropriate paradigm in economics*. Various criteria are proposed: accountability concerning the ultimate goals of the economic enterprise; the preservation of the natural world; the combination of individual freedom with ethical responsibility; inclusive horizons, and so on.

Chapter 12 suggests a more fundamental approach to economics: need and capacity, rather than supply and demand, should form the point of departure for economic theory. This approach is then utilised to explain the emergence and growth of *economic marginalisation* - one of the blind spots of the main line approach. Chapter 13 makes an even more drastic proposal: to gain wider horizons, the theories of *entropy and evolution,* gleaned from the natural sciences, are to be applied to the economic and ecological realms. Such a cross-disciplinary methodology yields startling insights.

Part IV indicates *the direction* in which humankind must move if it wants to avoid global catastrophe. Chapters 14 and 15 prepare the ground for detailed suggestions contained in the last three chapters. Chapter 14 reflects on *potential agents of change*. Chapter 15 spells out the new kind of *mindset* we need and draws up the *agenda*. Chapters 16, 17 and 18 are devoted to *public policy* for centres, peripheries and the interaction between centres and peripheries respectively. The original manuscript also contained a chapter on responsible money management at the *personal* level. In the mean time this material has been published as a separate booklet.[27]

The rational and the prophetic character of this book

I have done my best to avoid the language of emotions, slogans and propaganda.[28] The book is a serious attempt to offer hard-nosed analysis and tight argument. Yet in spite of its "secular", "empirical" and "rational" character, it is essentially a call to *go beyond*. We are all entangled in the network of daily necessities, demands and desires. We are socialised into a framework of unquestioned assumptions, values and conventions. We are stuck in our professional specialisations. We are conditioned by the erratic flow of social life. We are bombarded with short-lived impressions which call for immediate attention. Atrocious scenes on our television screens first numb us, then begin to leave us cold. Because change is costly, it is always easier to let things go and hope for the best.

Yet we have the capacity to go beyond all that - beyond our immediate experiences, needs and desires, beyond what our communities take for granted, beyond what primary groups expect from us, beyond our areas of work and residence, beyond the present moment, beyond our life times. The capacity to go beyond immediate experience belongs to the dignity of the human being as opposed to other creatures. It finds expression in religion, philosophy and science. It opens our eyes for a greater reality out there into which we are embedded. There is a long history of which our little yesterdays and our little tomorrows form but a tiny stretch and to which we are accountable. Where we become submerged under the avalanche of immediate impressions and instinctual responses we lose our humanity.

So this book is a challenge to realise our sovereignty, and the responsibility that goes with it. This is not easy. It goes against the grain of human nature. Yet to go beyond is necessary. We are not the only ones who have a right to live on this singular planet. There are contemporaries in grinding poverty. There are future generations who must be given a chance to enjoy what we are enjoying now. There are nonhuman species which are pushed into oblivion by our mindlessness and greed. In the long run our own worth will be determined by the degree to which we are capable of recognising and defending the dignity of all the creatures of God, present and future.

As a Christian theologian I explore the potentials of the Christian faith to develop a vision which could lead us beyond given constraints and predicaments. This also implies a critical and reconstructive stance over against my own faith community. But my personal commitment is not intended to exclude anybody. My argument follows the phenomenological method. It is designed to make sense to all readers with an open mind, regardless of their religious affiliations. The social and economic challenges facing Christianity today are the challenges facing humankind as a whole.

This certainly does not imply that convictions are immaterial. On the contrary, I am persuaded that they are absolutely decisive if people are to be liberated and motivated to change course. The current selfish, shallow, short-winded and absent-minded mentality is not a basis from which we could face the challenges of the future.[29] I hope that representatives of other faiths will be challenged to reflect on the potentials and weaknesses of their own religious heritage as well.[30]

Notes

[1] Earth with a capital letter denotes the thin life-sustaining layer composed of earth, sea and air surrounding the bulk of the planet.
[2] See the haunting illustration of the "space ship" which resembles such an island in Kubalkova & Cruickshank 1981:220.
[3] "On the scale of human history, the era of fossil fuels will be a short blip, either because of their source limits or their sink limits" (Meadows et al 1992:73f).
[4] Note that the different scenarios on material standards of living presented by Meadows et al 1992:11 all peak before the middle of the 21st Century after which they fall more or less dramatically. None shows continuous growth.
[5] For an explanation of the term "throughput" see chapter 2, section I.
[6] Meadows, Meadows & Randers 1992; Harrison 1992; Douthwaite 1992; Daly & Cobb 1989; Schumacher 1973; Myrdal 1957; especially various annual editions of Brown: "The state of the world" and Brown 1981.
[7] Samuelson & Nordhaus 1989:853.
[8] "These conclusions constitute a conditional warning, not a dire prediction. They offer a living choice, not a death sentence." (Meadows et al 1992:xvi).
[9] Ibid xix.
[10] Not only Karl Marx, but also Keynes, one of the most celebrated economists of all times, believed that technology would soon allow us to reduce the economic problem to secondary importance so that "the arena of the heart and head will be occupied ... by our real problems - the problems of life and of human relations, of creation and behaviour and religion." Meadows et al 1992:233.
[11] Somewhat bitter, Daly & Cobb remark: "Moral concern is 'unscientific'. Statement of fact is 'alarmist'" (1989:2).
[12] Schumacher 1974 has drawn our attention to the fact that depletion of resources is living off one's capital.
[13] The centre-periphery model provides powerful tools of analysis which should not be dismissed simply because they were used by some Marxists and dependency theorists whose theories were flawed. There are much greater flaws in neo-classical theory! Its history goes as far back as von Thünen (1826), was used by main line sociologists (Behrendt 1962), peace researchers (Galtung 1973), economic geographers (Smith 1977, 1979); recently it was "rediscovered" by the American economist Paul Krugman

(1993).

[14] I owe this formulation to a paper delivered at the ICEA 7th World Conference in Jomtien, Thailand, August 1995 by Prof Prawase Wasi, Bangkok.

[15] Cf Daly & Cobb 1989:21 who face a similar dilemma.

[16] Cf Daly & Cobb 1989:32ff.

[17] Cf Heller, Walter H: "What's right with economics" *American Economic Review* 65/1975 1-26.

[18] "There are excellent economic geographers out there ... however, these people are almost uniformly peripheral to the economics profession" (Krugman 1993:3).

[19] A very good example is Krugman's (re)discovery of the relevance of geography to the theory of international trade and economic concentration. " ...the analysis of international trade makes virtually no use of insights from economic geography or location theory" (1993:2).

[20] See Meadows et al 1992:3, Capra 1982:285ff, Checkland 1981; Mesarovic & Pestel 1974, Laszlo 1972. An overview and comparison of methods is found in Tudor & Tudor 1995.

[21] Meadows et al 1992.

[22] Krugman 1993:2.

[23] We could also speak of theories (McKenzie & Tullock 1985:9ff). See the devastating critique of model building by J H Raisin in the *Journal of Economic History* vol L/1990:261: "By the time the gigantic flaws in these models are removed, I doubt there will be any problem or urgent need for models." The point is taken, but the construction of models is absolutely indispensible for understanding complex networks of relationships. See Meadows, Meadows & Randers 1992:105.

[24] Cf Capra 1982:286.

[25] Meadows et al 1992:105.

[26] United Nations 1993:173.

[27] *Making ends meet: Personal money management in a Christian perspective.* Pietermaritzburg: Encounter Publications, 1995.

[28] The celebrated economist Keynes has remarked that "words ought to be a little wild, for they are the assault of thoughts upon the unthinking" (quoted by Daly & Cobb 1989:1).

[29] Note the similar approach of Daly and Cobb (1989:374).

[30] An example is Schumacher's advocacy of "Buddhist economics". (Brown 1981:126/138; also in Schumacher's *Small is beautiful*).

PART I

The structure of the global system

The economic problem in a nutshell

What is the task of this chapter?

As mentioned in the introduction, our approach to economic and ecological problems differs substantially from that of conventional economics. This chapter serves as a first entry into the ensuing analyses. We begin with the most obvious *phases of the economic enterprise,* namely extraction, transformation, distribution and consumption. These phases will be depicted as "throughput" of material from resource base to waste. A model follows which explains the emergence of *discrepancies in prosperity* between various population groups and the danger of ecological destruction. After highlighting the importance of comprehensive *cost-benefit analyses*, the chapter closes with some *ethical* considerations.

Section I: Basic elements of the economy

Human beings are creatures that have material needs. They need space, time and energy to exist. They need protection, food, rest, clothing, and shelter. The availability of these resources cannot be taken for granted; they must be secured.

Animals obtain what they need to survive and prosper by utilising what they find in nature. Our ancient ancestors were not much different in this respect. In contrast to animals, however, humans possess an intelligence and a peculiar body - free hands and a thumb - which enable them to acquire productive powers and to unlock dimensions of satisfaction which go beyond their immediate survival needs. They can work; they can develop tools to acquire a greater quantity, variety and quality of goods and services; they can cooperate in organisations designed to achieve their collective goals. That is what the economy is all about.

In its most elementary form we can define an economy as the process in which humans use tools to transform raw materials into commodities suitable for human consumption. This process has a few clearly distinguishable aspects which can be depicted as follows:

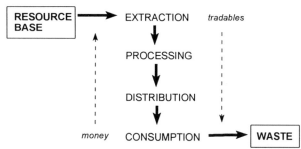

Diagram 2-1
Phases of the
economic process

(a) The point of departure for the economic process is the *resource base*. The resource base provides us with the raw materials which are transformed into consumer goods. Examples are crude oil, iron ore and the fertility of agricultural land.
(b) While for humans the goal of the economic process is consumption of goods and services, for nature as a whole the final stage is *waste*.
(c) Between the resource base and waste lie four identifiable stages. The first is the *extraction* of raw materials from the resource base. We must get the crude oil out of the ground.
(d) The second is *processing,* in which the raw materials are transformed into usable commodities. Crude oil is changed into petrol or plastics. This stage is called production.
(e) The third stage is the *distribution* of the goods and services, which have been produced, to satisfy the different needs of different people.
(f) The fourth stage is *consumption*. Consumption is nothing but the satisfaction of needs.

Economists often maintain that wealth is created through production. But humans cannot create anything out of nothing.[1] They can only transform what already exists in this world. And this transformation consists of breaking down parts of reality and putting some of the components together in new constructs which have utility for humans, such as cars or steaks. The rest is discarded. What is utilised, also ends up in waste, such as scrap or excretions. Even humans, who utilise these constructs, end up in decay. On the whole, therefore, the economic process is a process of destruction, although its intermediate stages seem to be constructive. This can only continue as long as nature is able to construct more than what is destroyed by humans.[2]

Effort, capital and money

The intermediate stages involve human effort. Human effort has physical and mental aspects. The *physical* aspect, usually called labour, is constituted by the utilisation of human energy. The *mental* aspect consists of initiative, the application of technological expertise and organisation.

To enhance their efforts humans use tools. We call these tools capital. Capital consists of all non-human factors which help human beings in extraction, processing, distribution, and even consumption.[3] Examples are knives, ox wagons, fertilisers, stone crushers, computers, assembly plants, railway lines, telephones, chain stores and restaurants. Usually capital is expressed in terms of the money invested in an enterprise.

Both human effort and capital are applied to achieve utility. When a woman uses a hoe to cultivate a field she does so because she expects a crop. In a more developed economy, however, the investment of effort and capital does not lead to immediate utility, but to a symbol of value called money. Money acts as a means of exchange. We do not work to make a shirt or a car, but to earn an income with which we can, among many other things, buy a shirt or a car made by others elsewhere on earth.

Money greatly enhances the versatility of the economic system because it makes the division of labour possible. Those who have specialised in the production of goods and services can do so much more efficiently. When they sell their products they can obtain other products which have also been produced more efficiently by specialists. There is also a "division of labour" between countries, especially rich and poor countries.[4]

While goods and services flow from extraction, processing and distribution to consumption, money flows in the opposite direction from consumers to extractors through all the intermediary stages. I pay the agent (distributor) for a car, the distributor pays the car factory (producer), the car factory pays the supplier of steel, which is, through various further stages, acquired from a mine (extractor). But the money which ends up in the pocket of the extractor is only a fraction of the money the consumer pays. The rest is used to pay the contributors of initiative, expertise, organisation and capital. And a lot of it is lost on the way through inefficiencies.

Economic efficiency

On the basis of this analysis we can distinguish a number of essential prerequisites for the satisfactory functioning of the economic process. Failures occurring on any of these levels will lead to shortages and imbalances:

(a) The capacity of *the resource base* of a society must be adequate to supply the raw materials needed to produce the goods and services which are to satisfy the needs of the consumer population. Obviously this also applies for humanity as a whole, including future generations. A shortage of resources can be interpreted in two ways: either the needs of the population are too high for the resource base, or the resource base is too limited to cater for the needs of the population. In other words, there can be too many people, or they can be living beyond their means.

The *needs* of the population depend on the number of people in that population at any given time and the quantity and quality of the goods and services they want to consume. However, to come to a realistic assessment of needs, one also has to consider the cumulative effect of the needs of successive generations over time. The *capacity* of the resource base depends on the quantity and quality of the raw materials it yields, their accessibility, their diversity and their sustainability. Of course, what is lacking can be imported from elsewhere if the economy has other resources which it can export to pay for the imports. However, one has to distinguish between resources which can be regenerated, for instance forests, and resources which are being exhausted, for instance crude oil.

Examples: Japan, Switzerland and Singapore are countries which have far too few raw materials for their economic needs, but their production capacity is so advanced that it is able not only to cater for local needs but also to export consumer goods and services to other countries. With the proceeds they can import the raw materials they do not possess. Of course, this presupposes that there are other countries that have more raw materials than they need, or resources which they cannot process themselves. This leads to the next point.

(b) The *intermediate stages* - extraction, production, distribution and consumption - must function in a way which ensures the transformation of raw materials into goods and services that actually satisfy the needs of the entire population in an efficient, adequate and balanced way.

Whether the *total volume* of production is sufficient for the fulfilment of the needs of the population, depends on the efficiency of labour, initiative, technological expertise and organisation - as well as on the quantity and quality of the tools (capital) used. Whether *genuine needs* are satisfied depends on the appropriateness of the goods and services that are produced. This again depends on the human needs they are to satisfy and the definition of those needs by the population. Whether the needs of the *entire population* will be satisfied in a balanced way depends on the equity of the distribution of these goods and services between the different population groups and their needs.

Example: In many oil-rich countries the local population does not have the means to extract the oil on its own, let alone transform it into a variety of consumer goods. Fortunately for them, oil is a scarce and expensive raw material on world markets. So companies from developed countries come and extract this valuable resource and pay them for the privilege. This again makes it possible for the oil countries to import consumer goods, thus largely bypassing the stages of extraction and production. What will happen to their economy when their oil reserves have been depleted is anybody's guess.

(c) The *distribution* of the products within the population depends on three factors: (i) The level of active *participation* of the population in the economic process, that is the spread of contributions people make to the economic process: resource exploitation, production, marketing, management, expertise, organisation, capital, and so on. (ii) The *quality* of the input of different groups.[5] (iii) The degree to which their *reward* is commensurate with their contribution.

Example: To make distribution more equitable we have to raise the participation level and the quantity and quality of the input of less productive groups. We also have to ensure that their remuneration is appropriate in terms of their contribution. Because processing becomes ever more complex and uses more capital, it demands less labour and greater sophistication. This means that both the demand and the remuneration of skilled labour has gone up, and the demand and remuneration for unskilled labour has gone down. While there are shortages for specialised skills and professionals, the unskilled labour force is increasingly replaced by machines and becomes redundant. Those who have no contribution to make to the economic process also have no income for buying goods and services. They are "marginalised", that is, they are pushed to the fringes of the economic system.

(d) *Trade,* especially international trade, has become very important in modern times.[6] The world economy is gradually moving towards a single integrated market. This allows countries to concentrate on their strengths, say raw material, abundant labour, or technological sophistication, and to leave aspects in which they are weak to others. But some national economies are weak and others are strong almost across the board, and when more powerful partners interact with weaker partners the outcome is usually to the advantage of the more powerful partners.[7] Reasons for this include: (i) The goods and services which different countries have to offer on the international markets are *not equal in value*. (ii) More developed countries can *outcompete* less developed countries in processing and distribution and destroy the markets of the latter.[8] (iii) Less developed countries are *dependent* on the capital, technology and skills from more developed countries. This can lead to crippling debt burdens.

Example: Processed goods are more expensive than raw materials because the remuneration for human effort and capital used in the production is added to the value of the article produced. During the last few decades prices of raw materials, such as copper or sisal, have been falling steadily relative to the prices of processed goods such as tractors or telephones. Processed goods have become more expensive because technology has become more sophisticated and the remuneration for labour has risen constantly in industrial countries. At the same time raw materials have become cheaper because industrialised economies have developed alternative materials and are increasingly less dependent on imports.

(e) Economic processes presuppose *cooperation*. An isolated individual, or a mass of isolated individuals, could hardly stay alive in this world, let alone attain higher degrees of prosperity. Cooperation depends on two things. First, individuals must be free to develop their particular gifts. This is called specialisation. Second, these specialised functions must be integrated optimally in an institutionalised process.

Example: The individualist selfishness of Western societies is often contrasted with the community spirit found, for instance, in traditional African societies. But because traditionalist societies often imprison individual potentials in predefined roles and statuses, their economies cannot develop beyond subsistence level. Paradoxically there is a far higher degree of social cooperation and integration in industrialised countries than in tribal contexts.

(f) Finally we have to consider two negative effects which the entire process has on the *natural environment* of an economy, and ultimately on the economy itself. The economic process uses up the potential of the resource base. Some of this potential can be renewed, some recycled, some is simply lost. In addition, the economic process produces toxins and wastes which pollute and ultimately destroy the natural environment. It is clear that the impact of these two negative consequences is determined by the size of the population, the volume of the throughput of the economy from extraction to consumption, and how clean the productive process is.

Examples: Oil and gas cannot be restored at all. Every barrel of crude oil that has been used, is gone for ever. The fertility of the soil can be partially restored, though only at great expense and over long periods of time. Quick fixes may cause further deterioration, as in the case of artificial fertilisers.

> *Revision: Summarise the main stages of the economic process and indicate areas where it can go wrong.*
> *Application: Where do you thing the weak points are in the economy of your country? What could be done to overcome them?*
> *Critique: (a) "People are poor because the rich have taken more than their rightful share from the abundance of nature and society." (b) Industrial societies are affluent because they have kept their numbers down, worked hard and used their brains to develop technologies. Why can the poor not do the same?" Respond to these statements.*

Section II: The emergence of economic imbalances

We have seen that creatures depend on a natural resource base for their survival and prosperity and that this resource base must be commensurate with their needs. This fact can also be observed in non-human nature. Some birds delimit their hunting grounds for the day against competitors by means of their morning songs. Some small fish are able to chase away larger fish who trespass upon their hunting ground. They do so not by virtue of their greater power, but by virtue of their greater "authority". Normally animals will take only what they need and these needs remain relatively constant within a species. In other words, distribution in nature is governed by a reasonable measure of sufficiency and equality.

Theoretically such a "natural balance" could also come about among humans. Say each family received one plot which is equal in size to all others. Whether a family were more prosperous than another would then depend mainly on their willingness to exert their energies and cultivate their ground. In other words, the old ideal of equality of opportunity would be realised. In ancient Israel, for instance, every family was supposed to possess its own piece of land as a permanent "inheritance", ostensibly allocated far back in history by casting lots. If the land of a family was lost due to misfortune or lack of prudence, it had to be returned to the family after the 49th year, the year of the Jubilee.[9] Assuming that there were neither population growth nor natural disasters, the material adequacy of the life of each family was secured by the legal system (see diagram 2-2).

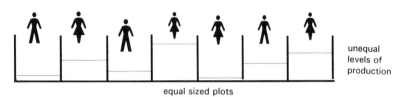

Diagram 2-2
Equality of opportunity

If we had organised our societies according to this principle, I would probably not have written this book. The fact is, however, that humans utilise human intelligence and the versatility of the human body to acquire a level of power which goes beyond the immediate needs for healthy collective survival. These enhancing activities can be directed in either or both of two directions:

(a) Humans can develop and utilise more and more technology and become more and more efficient in their exploitation of nature.[10] If greater efficiency would make them utilise fewer resources to gain the same output, this would be highly desirable. In fact, technology and greater efficiency leads to a deepening of the exploitation of nature. Power saws and dragnets are good examples. Obviously this happens at the expense of *other creatures*. The impact of human beings by far surpasses the impact other animals have on nature.

(b) Secondly, humans can increase their resources at the expense of *other humans*.[11] If elites exploit the resource base of lower classes we call this feudalism. The extension of control over geographical resource bases at the expense of the local inhabitants we call imperialism. However, power struggles for resources do not only occur between nations and peoples, but wherever there is economic interaction.[12] It is self-evident that those whose resource base is being curtailed have to be satisfied with less than their rightful share.[13]

We may add that an exploitation of nature beyond the threshold which nature can absorb without permanent damage, also goes at the expense of other humans, in this case future generations.[14] So a group lives either at the expense of another group, or the current population thrives at the expense of the future population (see diagram 2-3).

Diagram 2-3
Humans increase their resources at the expense of other humans

Obviously such predatory behaviour generates defensive pressures from the side of those whose resource base is being threatened or curtailed. Therefore the effort of building and maintaining control over resources, whether in the form of a feudal system or an empire, necessitates the investment of human effort and capital into means of conquest and defence, which surpass the normal task of securing the resource base itself. This not only implies a further impact on nature, but also on the wellbeing of the conquered. Because empires and feudal systems are built not on benevolence but on self-interest, rulers usually transfer the costs of the empire to the subjugated, which again increases the pressure from the oppressed. It is a typical vicious circle.

The two ways of increasing one's potential, that is technological advance and expansion of control, can be combined. The accumulation of productive potential in feudal elites and imperial centres usually leads to a great upsurge of cultural and technological progress. Both technological development and capital accumulation again enhance the possibilities of the self-interested use of power. There are also structural mechanisms at work in the interaction between stronger and weaker groups which contribute to an escalation of discrepancies in economic potential. The cumulative growth of these discrepancies causes more underdog pressure and necessitates further investments in the maintenance and security apparatus of the feudal or imperial system. Both the development of technology and the security system again increase the impact on nature.

Chapter 2 - The economic problem in a nutshell | 29

The "protective walls" which guard the interests of the elites obviously have changed their forms profoundly in modern times. The conflicts and coalitions of tribes, feudal hierarchies and nation-based empires have made way for the conflicts and coalitions of multinational corporations. Their "armaments" are productivity, competitiveness, flexibility, information, communication, transport and leverage in the corridors of power, rather than force of arms. It is no longer merely a matter of states utilising the economic potentials under their jurisdiction, but also of economic giants utilising the power of states.[15]

The results of the new power structures are no less grandiose in the centres of power, and no less crippling in the peripheries, than those of the old feudal and imperial structures have been. Centre nations have obviated internal conflict through democratisation, they have become militarily impregnable and they have consolidated relations between themselves in blocs of mutual dependence. As a result, warfare has ceased to be the most lucrative option for the centre - except where a centre nation decides to teach a peripheral country a lesson, for instance in the war against Iraq in 1991.

The new situation does not prevent states from building up their armaments.[16] Among and within poor nations, old style struggles for power, privilege and prestige are still being fought on a devastatingly large scale, often clothed in surrogate motives such as ethnic solidarity or religious conviction. During the Cold War, great power rivalries often spilled over into Third World countries such as Korea and Angola. And armed conflicts are still being fed with the means of combat emanating from the multi-billion armaments industry, which is engaged in its own competitive struggle.

Paradoxically, the misery of subjugated and marginalised groups is one of the main causes of population growth among such groups.[17] That poor people have large families is a worldwide phenomenon.[18] We shall deal with the reasons in chapter 5. Population pressure again leads to an increased impact on nature: forests are chopped down, grazing is overstocked, agricultural lands are over-utilised, footpaths change into gullies, soil erosion takes away the topsoil, and water is polluted. The deterioration of the natural resource base again increases misery, thus leading to further population growth, further pressure against the system, greater security needs of the system, greater impact on nature, and so forth. Again it is a vicious circle, or rather a vicious network.[19] Throughout the book we shall refer to these four interlinked growth processes (diagram 2-4):

(a) in economic *centres* economic potency grows,[20]
(b) in economic *peripheries* the population grows,[21]

30 | **Part I - The structure of the global system**

(c) *between* centres and peripheries, as well as between competing centres, the security arrangements grow,
(d) underneath both centres and peripheries the deterioration of the *natural environment* grows.²²

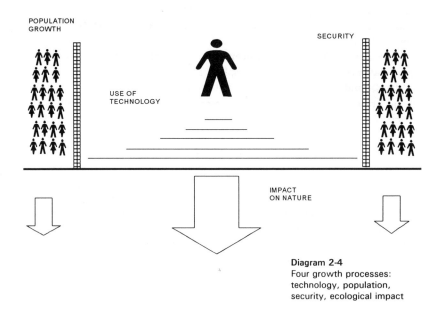

Diagram 2-4
Four growth processes: technology, population, security, ecological impact

> *Revision: Explain how the use and abuse of power by human beings lie at the roots of the economic problem as a whole.*
> *Application: South Africa's average wealth (Gross National Product per capita) corresponds with that of semi-industrialised countries such as Brazil and Mexico. Yet a small elite enjoys living standards comparable to those in industrial nations, while the incomes of large sections of the population correspond with those of some of the poorest countries in Africa. Can you explain this phenomenon by means of the tools provided in this chapter?*
> *Critique: (a) "What you say is nothing but Marxist ideology disguised as economic analysis. There is no alternative to rapid economic growth*

> *through the development of all available potentials." (b) "This model shows how the different aspects of our economic predicament, such as affluence and poverty, industrialisation and urbanisation, armed conflict, the nuclear threat, population growth and the ecological time bomb all hang together."* Which statement is, in your view, closer to the truth?

Section III: Towards comprehensive cost-benefit analyses

Economics has developed powerful tools of establishing the relation between the costs of an economic enterprise and its benefits. An undertaking is not "rational" if costs exceed benefits. Therefore the calculation of "marginal returns", that is, the additional benefits gained from additional allocations of resources, is fundamental for any successful economic concern.

By and large, conventional economics has not applied this insight to the economy as a whole.[23] In the first place, cost-benefit analyses are usually applied only to the *individual enterprise.* Every operation has "external costs". If the firm does not take them into consideration, they have to be paid for by somebody, whether now or in the future. To dump the toxic effluent of a chemical factory into a river, for instance, may be cost-efficient in terms of the balance sheet of the enterprise, but the costs of this "saving" is paid by farmers, fishermen, households, and non-human creatures downstream. It is now generally accepted, therefore, that external costs must be "internalised", that is, polluters must pay for the cleaning up of the mess they create. There should be no "quick-fix" repairs to the environment which simply cover up the harm done and cause unrecognised costs elsewhere in the system.

In the second place, external costs have usually been restricted to environmental costs, such as the pollution of a river mentioned above. But there are also *social* costs. A firm may become more cost-efficient when it replaces half of its work-force with machines. But what about the costs to those who have been retrenched? What about the costs to the society which has to cope with thousands of unemployed and unproductive people? Severance packages are a first step to internalise social costs, but these individualised compensations are, in many cases, nothing but palliatives. What about the indirect costs of industrialisation, such as chaotic urbanisation and social decay?

This raises the question of the distribution of costs and benefits. The beneficiaries are usually not identical with those who bear the costs. It is not sufficient to calculate costs and benefits; one has to go further and ask whose costs and whose benefits they are. Those of the producers? Those of the consumers? Those of third parties? Those of the society as a whole? Or those of future generations?

In the third place only *material* costs and benefits are taken into account. But what about costs in terms of the physical, psychological, cultural, communal and spiritual dimensions of life? Tobacco, alcohol, drug and casino industries may be very lucrative in material terms - but they are creating unfathomable havoc and misery in millions of human lives, families and communities. And what are the costs of utilising female attractions as a resource in the entertainment and advertising industries?

One could argue that economics, by definition, was only concerned about things that can be expressed in money terms. If this were true, one would have to concede that the economic picture of reality is truncated. To compensate for physical, psychological, social and spiritual damage can be very costly even in money terms, let alone in terms of human happiness, communal wellbeing and social peace. Decisions, if they are to be responsible, cannot be taken on purely material grounds.

Least of all should the financial dimension of life be allowed to crowd out all other considerations. In the modern economy, all things, including learning, intimacy and spirituality, are being subjected to the criteria of financial gain. The loser in this case is the richness of human life as a comprehensive whole. If marital commitment is replaced by commercialised sex, human individuals, families and communities have to pay the price.

A responsible society will, therefore, insist on inclusive horizons. Material costs and benefits must be seen in the context of the whole spectrum of needs. Individual costs and benefits must be seen in the context of costs and benefits to the community, to the society and to humanity as a whole. Present costs and benefits must be seen in the context of costs and benefits to the chain of generations extending, God willing, over centuries and millennia into the future. Human costs and benefits must be seen in the context of costs and benefits to the natural world of which humanity is a part, and on which humanity depends. This means that, if economics were to be restricted to the quantifiable and material dimensions of life, it could never be more than *one* of the tools of analysis in decision making.

The costs of cancerous growth

As we have seen, both the industrial economy in the centre and the population in the periphery have been growing faster and faster in recent times. Accelerating growth is dangerous in both cases.[24] *Industrial growth* leads to the depletion of scarce resources. Some resources, such as wood, is renewable. But this takes time, financial resources and collective will. At present more forests are cut down than planted.[25] Other resources, such as crude oil, cannot be renewed. Crude oil is an incredibly versatile resource which is used to make all sorts of materials, from medicine to clothing. But it is also the source of energy upon which the entire industrial system, including mechanised agriculture, is built.[26] It is irresponsible to squander such a valuable and strictly limited asset within two or three centuries.

Industrial growth also leads to pollution. Unless measures are taken to reduce toxic emissions, pollution levels rise with the growth of the industrial economy. These measures are expensive, and poor countries cannot afford them. What might happen to the eco-sphere when large Third World countries, notably China and India, seriously begin to industrialise is anybody's guess. Harmful effects include water pollution from industrial effluents, soil pollution from artificial fertilisers and insecticides, air pollution from gases and dust pumped into the atmosphere through chimneys and exhaust pipes, noise pollution, even temperature pollution.[27] Pollution means that the environment is poisoned. Fish perish in rivers; forests die because of sour rain; the ozone layer protecting our planet from ultraviolet radiation is eroded.

Population growth is dangerous for two reasons. In the first place, unchecked population growth is bound to outstrip the increase in agricultural production. The volume of agricultural production can still be increased, but the rate of growth of food production is slowing down.[28] If the growth of the population continues to accelerate, the two lines have to intersect sooner or later.[29] Then food surpluses turn into shortages and famine is the result. This has already happened in large parts of Africa.[30]

In the second place, population growth leads to increased pressure on the land. This has ecological consequences. Over-populated areas are vulnerable to soil erosion; grazing is depleted; water resources dry up; rivers are polluted; wild animals are eradicated; forests are chopped down. Flood waters are unchecked and cause erosion. Progressively the natural basis of human survival is destroyed. This impedes production and leads to further want.

The typical scene of rural misery includes denuded grazing areas crossed by numerous foot paths which turn into gullies; pitifully lean animals; meagre crops on eroded and exhausted fields; children with bloated tummies due to malnutrition; mothers spending their energy carrying water over great distances; animal droppings used for fuel and so on. It is not surprising that people are tempted to migrate to the cities in search for greener pastures. Once there, most of the poor end up in slums. In densely populated urban areas without public amenities the problem of pollution is even worse than in rural areas.[31] Where pressures on scarce resources increase and great discrepancies in standards of living emerge, social tensions grow and armed conflict can be expected. It leads to destroyed huts, burnt fields and endless streams of refugees which have to be kept alive by international agencies.[32]

In sum, economics will have to embark on comprehensive cost-benefit analyses which take all these factors into account. If it does not, we simply have to work on an alternative approach to economics.

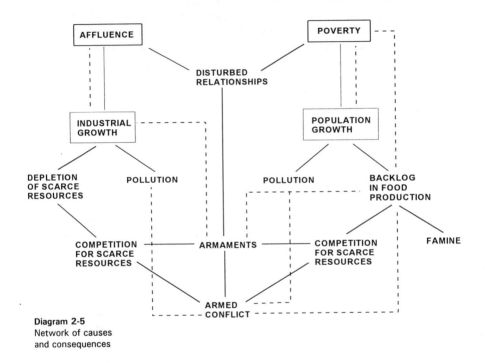

Diagram 2-5
Network of causes and consequences

> *Revision: Summarise the dangers of various growth processes in the world today.*
> *Application: Try to estimate the proportion of the Gross Domestic Product of your country which is spent on crime prevention and "national security". If half of that sum would become available by resolving conflict through a more balanced distribution of wealth among different population groups, how could it be spent to increase the prosperity of the nation as whole?*

> *Critique: (a) "Hardship, conflict and war have been the lot of humanity as long as it existed on this globe. All our moral indignation and all our ill-advised redemptive efforts will not change that." (b) "Jesus himself has said that the poor will always be with us. He also said that wars and famine will increase as we approach the end of the age. There is no way we can avoid what God has foreseen for this evil world." Compare these two statements and say how you would react to each of them.*

Section IV : Ethical considerations

There is no cast-iron law which says that there must be, and always will be, exploitation of fellow humans and destruction of nature. Perhaps the economic process does not need to be a "zero sum game", which means that one group (or nature) loses as much as another group (or nature) gains. Perhaps there is a possibility of "creating wealth" to the advantage of all partners, including the poor and nature. But, contrary to what most economists believe, this is not what happens at present.

In further chapters we shall reflect on the preconditions for such an economy. These preconditions are hard to fulfil, and humankind is currently not in the mood of fulfilling them. Here it suffices to indicate the direction:

(a) It should be possible to ensure that the utilisation of nature remains *below the threshold* of the capacity of nature to absorb the impact and to regenerate itself. Trees do not die out when their fruit are taken, or when surplus trees are cut. Cattle do not perish when cows are milked and oxen are slaughtered. This is how nature itself operates. Lions only kill what they need to satisfy their needs.

(b) It should be possible to ensure that human beings *cooperate* in the productive process to the advantage of all participants by increasing the level of production and by distributing the proceeds equitably.

(c) It should be possible to ensure that resources that *cannot be renewed* are used with circumspection and, ideally, only to the extent that substitutes are already in sight.

All this would have to imply that humans reduce both their growth in numbers and their levels of consumption. But why should they bother to do that in the first place? Why should they not take what they can, enjoy their lives, while it lasts, and forget about the consequences for their contemporaries, for other creatures and for future generations?

These questions lead us to the foundations of human existence: what is the *meaning* of reality in general and of human life in particular? Do we have a *right* to survive and prosper - and according to which criteria? Do we have the *authority* to do so at the expense of other humans and other creatures - and who has granted us that authority? What humans are supposed to be (human authenticity) and whether they live up to these standards (human integrity) raise issues which cannot be solved outside the realm of convictions, values and norms.[33] It is quite impossible, therefore, to exclude religion and ethics from economic analyses and reflections, if the latter is to be more than a mere legitimation of material self-interest.[34]

Let us summarise

This chapter served as a simple introduction to the economic problem. We began with a sketch of the *main phases of the economic process* which lie between the resource base and the rubbish dump, namely extraction, processing, distribution and consumption. In each case the application of effort and tools is critical. We again subdivided the category of human effort into physical labour as a source of energy and the mental factors of initiative, technical know-how and organisation. While goods and services flow from extraction to consumption, money as a means of exchange flows in the opposite direction. Economic hardships and imbalances result from failures on one or more of these levels. There can be a mismatch between resource base and the needs of the population; the intermediate stages can be inefficient; the distribution can be unsatisfactory, and the impact on the natural environment can destroy life chances.

In section II we placed these phases of the economic process into the context of *economic power structures*. Because of their superior intelligence and the shape of their bodies, humans can extend their power beyond immediate survival needs, either at the expense of nature, or at the expense of other human beings. The two directions can also be combined. Resistance of the disadvantaged necessitates growing investments in security arrangements. Increasing poverty is causally linked to population growth. Both have an adverse impact on nature. The links between these four growth processes - the growth of potential in economic centres, the growth of the population in economic peripheries, the growth of armaments and the growth of ecological deterioration - can be expressed in the centre-periphery model, which we are going to use extensively from now on.

In section III we challenged economics to conduct comprehensive *cost-benefit analyses*. In general such analyses do not take the environmental and the social costs of an enterprise into account. They also do not take non-financial costs into account. They are often oblivious to the fact that costs and benefits are distributed unequally: some reap the benefits, others bear the costs. They also do not care too much about the interests of future generations. To be responsible we have to develop a comprehensive systems-analytical approach which combines "hard" and "soft" data. There may be non-destructive and non-exploitative ways to generate wealth. But why should we bother? In section IV we maintained that, to serve humanity, economics cannot ignore the realm of convictions, values and norms.

Notes

[1] Economics tends to overlook this fact. See Daly & Cobb 1989:194-6.
[2] This fact points to an important characteristic of reality as we know it. In physics it is expressed in the law of entropy: the universe moves from infinite concentration of energy (at the time of the "big bang") to an infinite dissipation of energy. The implication is that wherever there is construction (= concentration of energy) there is a greater amount of destruction somewhere else in the system. We shall come back to that in chapter 13.
[3] Traditionally economists restrict capital to tools produced by humans; ecologists, in contrast, add "natural capital", that is, resources. Daly & Cobb 1989:72. We call the latter the "resource base" to avoid confusion.
[4] See for instance Caporaso 1987 for detail.
[5] This is measured by the marginal productivity of that input relative to the input of other factors of production.
[6] For more detail see, for instance, Fontaine 1992.
[7] For a critical discussion of the principle of comparative advantage see Johns 1985:149ff; Daly & Cobb 1989:209ff. We shall see in chapter 5 why the principle does not necessarily function as the text books predict it will.
[8] See chapter 12 for detail.
[9] Lev 25 - 27.
[10] For more detail see Basalla 1988; Pacey 1992.
[11] For a first introduction see Faber 1990:1ff.
[12] Cf Jan Pen in Krelle (ed) 1978:335ff.
[13] Heilbroner says that the change from domestic to centralised production 5000 years ago is "perhaps the greatest inflection point in social experience" (1988:85). He believes that exploitation is the "necessary condition for the achievement of civilization, at least so far as its material triumphs are concerned" (1988:87).
[14] Cf Brown Weiss 1992:385ff.
[15] Cf e.g. Daly & Cobb 1989:229.
[16] Even after the Cold War had ended, the United States invested $ 3 million per year in the "Star Wars" programme. *Mail & Guardian* Dec 9-14, 1994, p 13.
[17] Cf Jordaan 1991; Ehrlich & Ehrlich 1990; Duden in Sachs 1992:146ff; Bondestam & Bergstrom 1980.
[18] For example, in 1987 the total fertility rate of Blacks in South Africa was: 2,8 in urban areas, 4,2 in semi-rural areas, 5,7 in rural areas and 6,3 in "independent homelands". (SA Institute of Race Relations: Race Relations Survey 1988/89. Johannesburg: SAIRR, p 150.)
[19] Bedjaoui has a point when he says that the Third World "is prolific because underdeveloped, not underdeveloped because prolific" (1979:45). But causation is circular.

[20] Meadows et al 1992:33ff.
[21] Meadows et al 1992:23ff.
[22] For an introduction to different growth processes see Meadows et al 1992:3ff. "Present-day society is locked into four positive feedback loops which need to be broken: economic growth which feeds on itself, population growth which feeds on itself, technological change which feeds on itself, and a pattern of income inequality which seems to be self-sustaining and which tends to spur growth in the other three areas. Ecological humanism must create an economy in which economic and population growth is halted, technology is controlled, and gross inequalities of income are done away with." (Victor Furkiss 1974; quoted by Daly & Cobb 1989:21, who also add the "arms race that feeds on itself.").
[23] For a professional introduction see Turner, Pearce and Bateman 1994:93ff.
[24] Meadows, Meadows & Randers 1992:14ff.
[25] The Ivory Coast, for instance, has lost 3/4 of its rain forests within 20 years (Von Weizsäcker 1990:7).
[26] Daly & Cobb 1989:272ff.
[27] Burning oil and coal produces carbon dioxide which leads to the so-called greenhouse effect (Goodland 1992:171; Simpson 1990:52ff; Elliott 1994:27).
[28] On the law of diminishing marginal returns see Heilbroner & Thurow 1975:39; 139ff; Ruffin & Gregory 1983:452ff or any other text book on economics.
[29] Broadley & Cunningham 1992:20.
[30] The "green revolution" has severe ecological repercussions because it uses fossil energy and is achieved with artificial fertilisers and pesticides. These chemicals also distort trade relations because they are usually imported by poor countries.
[31] A recent report speaks of "Asia's choking cities", *Newsweek* May 9, 1994, pp 37ff.
[32] Broadley 1992:28ff. At the time of writing, Rwanda, which is one of the highest populated countries in the world, presented one of the worst cases.
[33] See for example Preston 1983; Wogaman 1986; Mieth 1980; Dussel 1988.
[34] "Futurologists" often close their books with a passionate appeal to activate the religious and moral resources of humankind. Examples are Meadows et al 1992; Mesarovic & Pestel 1974; Laszlo 1977.

Economic discrepancies

What is the task of this chapter?

Part I deals with economic discrepancies at a social-structural level. We began in chapter 2 with a brief overview of how economic imbalances come about, whether at a local, regional, national, or global level. The present chapter develops a model which expresses the nature of these economic discrepancies in various dimensions.

Section I begins with the *geographical* dimension of economic discrepancies. Everybody knows that in some parts of the world, such as North America, Central Europe and Japan, we find a degree of affluence unheard-of in previous history, while others, for instance Sub-Saharan Africa or India, are marked by mass poverty. This provides us with a visual model, namely the contrast between the "centre" and the "periphery", which we shall utilise throughout our further discussions.

In section II we consider the distribution of economic potency between certain *sectors of the population*. "Centre" now becomes a metaphor for population groups with high economic potency, while "periphery" becomes a metaphor for population groups with low economic potency. In section III we consider the relation between *income and need*, both in the centre and in the periphery. This enables us to formulate a definition of the concepts "affluence" and "poverty". The *vertical relationships* between two levels òf centres and their respective peripheries in the system will be discussed in section IV.

Section I: Geographical centres and peripheries

Even untrained people are able to observe that economic activities and opportunities are not distributed evenly over the surface of the earth, a region, or a country. There are areas in which the economy develops rapidly, and other areas in which it develops more slowly, if at all. Just compare Johannesburg, the South African metropolis, with a little settlement in the Kalahari desert. Concentration is the "most striking feature of the geography of economic activity".[1] To characterise this phenomenon we speak of the contrast between the *economic centre* (or core) and the *economic periphery*.[2]

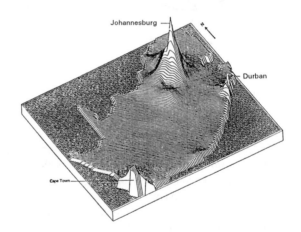

Diagram 3-1
GDP per square mile
in South Africa

Diagram 3-1 depicts a computer projection of Gross Domestic Product per square mile in South Africa, drawn up during an era of rapid economic growth in the 1970s.[3] We see a commanding centre around Johannesburg, two big sub-centres around Durban and Cape Town and a few other elevations. The rest is flat. A brief look at the atlas will reveal the same pattern in Swaziland, Kenya or France.

There are many reasons for the development of such centres in certain localities: some begin as administrative capitals; some develop on intersections between important trade routes; some are coastal outlets for international trade; some mushroom where mineral deposits were discovered; some owe their existence to rich soils and a thriving agricultural base.[4]

Once a centre is established, it may develop an inherent dynamic which leads to rapid economic growth while peripheral regions lag behind. Again there are many reasons for this: shorter distances make transport faster and cheaper; various industries located in the same neighbourhood complement each other; a greater population provides not only more labour and more sophistication, but also a market for a greater variety and a greater quantity of products; water, electricity and telecommunications can be supplied more economically and efficiently; recreational facilities, schools and health services are provided more abundantly, and so on. Obviously centre elites can also use their greater economic or political power to strengthen their position at the expense of less powerful groups and areas. All these factors contribute to a growing economic dynamic which sucks the potential of the environment into its vortex, and prevents the establishment of competing centres in the immediate environment.

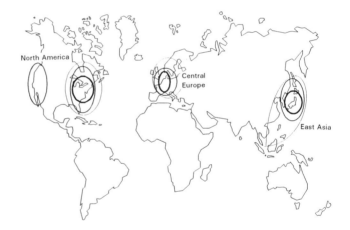

Diagram 3-2
The "Triad" -
the United States,
Central Europe
and Japan

The growth of centres at particular locations does not mean that they are necessarily there to stay. In the early phases of commercial capitalism the centre of gravity shifted from northern Italy (more precisely some city states such as Venice, Milan, Genoa and Florence) to the Low Countries (Antwerp and Amsterdam). With the onset of the Industrial Revolution and the growth of the British Empire, Southern England became the economic centre of the world. After World War II this centre declined in favour of Central Europe, especially Western Germany and the surrounding areas.[5] However, Europe as a whole lost its commanding position to the United States. At present it is not clear whether East Asia, possibly also a united Europe, will be able to challenge the dominance of the United States. Diagram 3-2 shows the location of the great three on the world map.

The discrepancy between centre and periphery is a universal phenomenon that can be observed in small and large areas. Here are a few examples:

(a) The *world* as a whole. As mentioned above, the United States is currently the dominant economic centre in global terms, while Western Europe and Japan are the most important subcentres. These three giants are sometimes referred to as the "Triad".[6] For some time the Soviet Union and its satellites constituted a rival system of smaller proportions, but the latter proved to be unable to compete with the dominant constellation.[7]

(b) A whole *continent*. The economic centre of Europe, for instance, stretches from Southern Scandinavia to Northern Italy and from Eastern France to Western Poland. All around this circle we find less developed areas: Northern Scandinavia, Scotland, Ireland, Portugal, Southern Italy, Greece, Turkey, Albania, Romania, and others.[8] Diagram 3-4 below gives a vivid impression of this fact. The industrial core of the European Union is what geographers call the "Manchester-Milan growth axis", which mainly moves along the Rhine valley.[9]

(c) A *subcontinent* such as Southern Africa. As diagram 3-1 shows, the Pretoria-Witwatersrand-Vereeniging complex (now called Gauteng) is the dominant centre, while the Cape Peninsular and Durban-Pinetown are important subcentres. Neighbouring countries are also part of the periphery.

(d) A *large country* such as Brazil. The Brazilian centre is located in the triangle between Sâo Paulo and the two adjacent cities of Rio de Janeiro and Belo Horizonte. Recife and Porto Alegre are important subcentres. Sâo Paulo alone accounts for almost 50% of the total industrial output of the country.[10] A similar concentration is found in the "manufacturing belt" in the North-Eastern corner of the United States, where in 1900 74% of manufacturing employment was located.[11]

(e) A *small country* such as Swaziland. Here the economic centre is situated in the area which combines Mbabane and Manzini.

(f) A *town, city or metropolitan region*. The "central business districts" (CBD) of Chicago or Johannesburg, for instance, are economically more developed than the suburbs. Recently commercial subcentres have sprung up in the latter. Great cities also tend to generate a number of satellites.

(g) A *farming district*. Here the centre may be nothing more than a village with a few shops, a garage, a post office and a police station. A number of such villages find their centre in a provincial town.

(h) Obviously the homestead is the centre of a *farm*. On large farms one may have watering points or outposts as subcentres.
(i) Finally, we have to point out the fact that *particular industries* tend to be concentrated in particular locations as well,[12] though they may spread as they mature. It has been shown that the restrictions imposed by national boundaries, tariff walls, or higher transport costs lead to a greater spread, while the open market leads to greater concentrations of such industries.[13]

The hierarchy of centres

It is clear that all these centres together form a *hierarchical system* in which each group of smaller centres forms the periphery of the next larger one. This overall system stretches from the great metropoles in the "Triad" over regional centres such as São Paulo, Singapore and Johannesburg to the smallest rural village. The organisation of this system typically follows a radial pattern: lines of transport and communication spread from larger centres over medium centres to smaller ones.

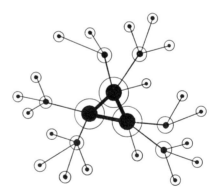

Diagram 3-3
The hierarchy of centres and peripheries

This network typically displays the following characteristics (see diagram 3-3).[14] Large centres compete with each other. But they also have a tendency to become integrated and form larger complexes. The merging of towns into great cities such as London or Berlin and the economic integration of countries in the European Union are good examples. Smaller centres, in contrast, tend to become satellites of larger ones. Each larger centre tends to develop a cluster of subcentres which are linked to the respective main centre.

Subcentres and their peripheries have less, if any, contact with each other. Between French and English speaking African countries, for instance, communication and trade links have been difficult. Lines of transport and communication between smaller centres often go via bigger centres. It is easier to travel from Johannesburg to London than from Ulundi to Mmabatho.[15] And when Argentinians and Nigerians meet they tend to do so in London, Paris or New York. The higher one moves up in the hierarchy of centres, the more efficient and dense the network of interaction becomes. The intensity of economic interaction decreases down the following list:

        within centres (where it is highest)
        between centres
        between centres and peripheries
        between peripheries.[16]

The interaction between larger and smaller centres tends to operate to the greater advantage of the more powerful and to the lesser advantage (or even the disadvantage) of the less powerful. Usually large centres grow at the expense of smaller ones.[17] If the scope of a major centre begins to cover an entire country, however, regional disparities begin to narrow down within that country. The Netherlands or West Germany, for instance, are located in the European centre and no longer show great regional inequalities.[18]

Developments also take place along the lines of *transport and communication* between centres, while the open spaces between them are left in the cold. Thus in South Africa you find lines of development going from the Witwatersrand northward towards Pietersburg, south-east towards Durban, south-west towards Bloemfontein, east toward Witbank, and west towards Rustenburg.[19] You will find the same phenomenon when you look at a map of the transport network in France.

The present global centre-periphery structure developed during colonial times. A colonial empire can be defined as a system in which one centre dominates a hierarchy of dependent sub-centres. The network of interaction is geared to the dominant centre. "All roads lead to Rome." That is why subcentres in the colonies, such as New York, Santiago, Lagos or Dar es Salaam, developed around harbours nearest to the dominant centre in the colonial motherland, that is London, Paris or Madrid. Basically, such colonial centres were transit points for administration and trade from which the interior of the colony was "opened up" in a radial fashion to become accessible to the colonial power. Some countries, such as Brazil, Nigeria, Tanzania and Malawi, have moved their capitals away from the coast to the interior at great cost. The economic effectiveness of such a move is not clear.

Dependent centres without an economic power base of their own are very vulnerable. When they are no longer useful for the dominant centre, their economic viability may deteriorate. For example, a thriving region based on sugar production developed in North-Eastern Brazil in early colonial times, with Salvador as its centre. Then the sugar market collapsed and the North-East has been the poorest region of Brazil ever since.[20] Instead, a new centre based on coffee has developed around Sâo Paulo. The coffee market also collapsed, but in the mean time secondary industries had established themselves there. Similarly the South African colonial centre of Cape Town was overtaken by Johannesburg, due to the discovery of gold on the Reef.[21] Port Elizabeth in the Eastern Cape is a deteriorating smaller centre.

There is a strong correlation between the economic potency and activity of human settlements and their geographical distance to the core. The nearer a group is situated to the core, the higher its productivity and income tends to be.[22] However, a secondary development can reverse this trend: the lure of urban standards of living attracts masses of rural people. The newcomers cannot easily be integrated in the formal economy and they end up in slums. Inner city areas deteriorate due to rising population pressure; the infrastructure is overpowered; the buildings become dilapidated; social decay leads to rising crime rates. As impoverished peripheral population groups move in, the affluent population moves out into more pleasant suburbs, often taking hotels, elite shops, offices, even some "clean" industries, with them.[23] Affluent groups may even acquire abandoned or neglected rural areas and develop them with their superior technology and capital into lucrative commercial undertakings, such as agro-business, tourism or forestry.

Typical characteristics of centres and peripheries

Before we move on, let us summarise some of the typical characteristics of centres and peripheries. Obviously there are many exceptions to the rules.

(a) In the centre, there is a *highly differentiated and integrated economy*, in which a large variety of specialised industries complement each other. In the periphery one often finds nothing but subsistence agriculture which means that families produce just enough for their own consumption. In peripheral countries the economy typically concentrates on one or two export products such as coffee or copper.

(b) In the centre, *capital* tends to accumulate rapidly. The result is that the *productivity of labour* also rises, unless other factors lead to inefficiency. In the periphery there is very little capital accumulation, and the productivity of labour is also very low.

(c) In the centre, the *infrastructure* (which includes things like telecommunications, transport facilities, financial institutions, water and electricity) is highly developed. In the periphery the infrastructure is poorly developed, if there is one worth mentioning at all.

(d) In the centre, we find a high level of *entrepreneurial initiative,* technological sophistication and organisational expertise. There are also large markets for *skilled labour.* Educational and specialised training facilities are well developed to cater for these needs. In the periphery there is little incentive to develop economic initiative. A poorly developed economy does not require high levels of sophistication and a lack of funds prevents the provision of training facilities.

(e) In the centre, one has to compete with others, so the pace of life tends to be hectic. Especially for those at the helm of economic enterprises *time* is a precious resource which is utilised intensively. In the periphery we find a relaxed atmosphere. On the other hand, high levels of productivity, union action and social policies have provided urban workers with more leisure to enjoy themselves after a hectic day at work, while peasants often have long working hours.

(f) Because more people have purchasing power in the centre, we find flourishing *consumer markets* for goods and services. In the periphery the population is often spread thinly over large areas. Families tend to be on their own. There is not much cash around and expectations are not aroused by the urban glitter. As a result, consumer markets are poorly developed. Where peripheral areas are heavily overpopulated in relation to their economic potency, people have no purchasing power and the effect on consumer markets is the same.

(g) In the centre, the average *standard of living* is fairly high but there are also great discrepancies in income between the affluent and the poor. In the periphery people tend to have relatively equal, but very low incomes. We shall have to qualify this observation considerably as we go along, because social policies can diminish income discrepancies within centre nations, while feudal structures, corrupt governing elites, or the emergence of offshoots of the centre in the periphery, can lead to vast income discrepancies in the latter. Within West Germany, for instance, such an equalisation had taken place. But unification brought about a serious imbalance between the former Western and Eastern parts of the country and billions are transferred to East to rectify the situation.[24]

(h) The centre is usually not only more developed in terms of industry and commerce, but also in terms of *agricultural* production. Part of the reason for this phenomenon is that in peripheries "cities exist to serve the farms", while in centres "farms exist to serve the cities". The closer one gets to the city the more intensive and productive agriculture becomes.[25] The state of Sâo Paulo, where the greatest industrial city in Brazil is located, is also the bread basket of the entire country. Obviously topography, climate and soil types also play a role, but the tendency is clearly visible. Diagram 3-4 depicts the intensity of agricultural production in Europe.[26] It is clear that it closely follows the intensity of industrial and commercial activity.

Diagram 3-4
Intensity of agricultural production in Europe.
The index of 100 is the average European yield per acre of eight main crops. [26]

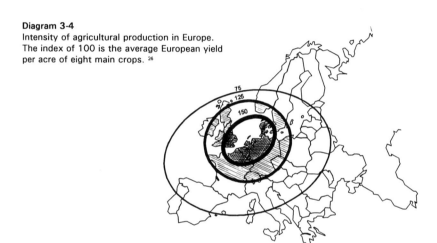

So far we have enumerated some economic characteristics of centres and peripheries. However, the difference between them can also be observed in *non-economic* dimensions of life:
(a) In the centre, a complex *social organisation* integrates great numbers of people. Relationships are rather impersonal. Most people do not fully understand the system in which they operate. In the periphery we usually find small scale communities where everybody knows everybody else. Relationships between them are well defined and based on tradition.
(b) In the centre, *families* are reduced to the basic core of parents and children, and even these tend to become isolated from each other. In advanced stages of dissolution, the "nuclear family" as such may break up and one-parent families emerge. Where individuals are on their own, they

enjoy a lot of personal freedom, but they can also be very lonely. They are also very vulnerable to impoverishment. Quite a number may get stranded.[27] In the periphery the extended family - which includes grannies, aunts, uncles and cousins - still survives in many places. Where the social network of the extended family is still intact, it protects its members from loneliness, destitution and despair.

(c) In the centre, *relations* between groups are restricted to certain functions and can easily change or cease, when they no longer serve their purpose. When dismissed or transferred, for instance, one loses all one's "friends" at one's place of work. Most relationships are casual and voluntary, and entail no lasting obligations. With little cohesion between them, people are like the grains of beach sand which can be swept into great masses by the tide of events and then disperse again. In the periphery relationships grow over long periods of time, especially in a tribal society, and are, therefore, far more holistic and stable.

(d) In the centre, people have to establish their *roles* and achieve their *status* in society. Where indigenous cultures are still intact in the periphery, roles are predefined and statuses are ascribed by tradition.[28]

(e) In the centre, due to specialisation and competition, *cultural activities develop* into a great variety of forms and some attain high levels of artistic quality. Major cities tend to share a cosmopolitan culture. In the periphery, cultural activities have a traditional and local flavour and usually do not go beyond the occasional dance or school concert.

(f) In the centre, people tend to be more *progressive* in political terms, while peripheral populations have the reputation of being conservative.

(g) In the centre, *religious beliefs* vary greatly and their impact is restricted to the private sphere. Values and norms, which guide social interaction and behaviour, take the form of secular and pragmatic conventions. Their only function is to make a common life in a pluralistic society possible.[29] In the periphery, people often share a relatively homogeneous set of convictions, which is considered to be binding and which pervades all spheres of life.

Our geographical observations have given us a first glimpse of the importance of socio-economic structures in the generation of affluence and poverty. Many people believe that these phenomena are caused solely by personal attitudes and actions. Thrift, diligence and expertise are supposed to lead to prosperity, while carelessness, lethargy and ineptitude are supposed to lead to poverty. Others again believe that affluence and poverty are chiefly due to oppression and exploitation. I do not deny that these are important factors. However, whether one is rich or poor also seems to depend on whether one dimension of the location of an individual or group in the overall economic power structure, whether within a country or in the world as a whole.

An example, which I was privileged to witness on a research trip, may serve to demonstrate this fact. During the last century a large group of Germans migrated to a rural area in Rio Grande do Sul, Brazil. Today, most of the descendants of these immigrants are poor peasants, while most Germans whose ancestors remained in their European heartland, now belong to the most affluent population groups of the world - and that in spite of the fact that their economy has been wrecked twice within a century during World Wars I and II. Moreover, Italian immigrants who came to Brazil later than their German counterparts, are economically ahead of the latter, while in Europe, Germans are generally more wealthy than Italians. It is quite unlikely that Brazilian Germans are inherently more lazy and disorganised than European Germans. The fact is that historical and structural factors have made it difficult for them to compete.

Revision. Describe the geographical "spider web" of interlinking centres on a world-wide scale.

Application. Take an atlas and compare the transport networks of various countries mentioned above, for instance, France, Kenya, Nigeria, Brazil, Swaziland, and so on. How do you account for the differences you observe?

Critique. (a) "Your description of centres and peripheries consists of vast generalisations which are pretty useless because, among other factors, they do not take account of social-cultural variations in different parts of the world, which surely make some people more efficient than others." (b) "Your analysis leaves one with the impression that affluence and poverty were due to some kind of uncanny geographical mechanism which operated universally; in fact they are the result of oppression and exploitation." What do these statements have in common? How do they differ?

Section II: The distribution of potency in the population

On terminology

The terms *centre* and *periphery* belong to geometry and refer to the two main dimensions of a circle. Because spatial terms are powerful visual symbols, they are also used in other contexts. We say, for instance, that a charming lady was at the centre of a party, that a politician occupied centre stage, or that an argument is peripheral to the issue being discussed.

So far we have used the term "centre" for a concentration of economic potency and activity in geographical space. The geographical phenomenon is, however, only the most conspicuous dimension of a more comprehensive social structural phenomenon. In modern societies economic potency and economic activity are normally concentrated in some groups of people more than in others. As a result, social prestige and political power also tend to gravitate towards those groups. The rest of the population has to be satisfied with a less glamorous position in life.

The relationships and the interaction between them are much more important than their particular geographical location. Therefore we now shift the emphasis from geographical space to *the structure of the society* itself. Because of the explanatory power of the centre-periphery model we apply the metaphor "centre" to the upper classes and the metaphor "periphery" to the lower classes. The transitional group between the two we call the "semi-periphery".[30] As we shall presently see, the income structure of these groups has the shape of a (dented) pyramid. Therefore, we could also speak of the "apex" and the "base" of a "social pyramid".

Potency discrepancies in the population

We begin with the relation between population and economic potency. While we concentrate on economic aspects, we should not forget that power is a comprehensive phenomenon. Political, economic, social, cultural, educational, psychological, medical, demographic, religious and other factors all combine to form an integrated network of power and privilege. Diagram 3-5 shows how the potency level rises from lower classes to the elites. Note that this is not a Lorenz-curve, as used in economics.[31] We simply relate income to population numbers: the vertical height (Y-axis) represents economic potency and the horizontal breadth (X-axis) depicts sections of the population according to their numerical strength ordered in the sequence of incremental income levels.

Diagram 3-5 also indicates that a minority of people have vast economic potency, while a great majority have very meagre economic potency. The ten percent of the population with the highest incomes (on the far right hand side of the diagram) have an income many times as high as the ten percent of the population with the lowest incomes (on the far left hand side). But the transition between these two groups is gradual. In other words, people who have neither very high nor very low incomes, form a gliding transition between the destitute and the affluent.[32]

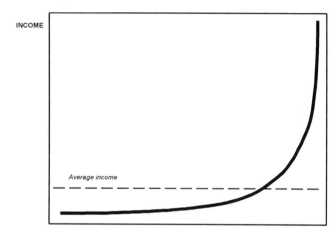

Diagram 3-5
Potency / income distribution

Actual income figures show, however, that in most less developed countries, as well as on a global scale, this transitional group is rather small. Therefore the existence of a transitional group cannot be used as an argument against the (rather crude) subdivision of the population into elite and lower classes. The diagram also shows that the further one moves towards the centre, the steeper the discrepancies in economic potency become, while the further one moves towards the periphery, the more equal the distribution of potency between people tends to become, though on a very low level.[33]

The shape of this curve can easily be substantiated statistically. To gauge economic potency various measures can be used. To indicate the distribution of potency *within a country* it is convenient to use *income per family* as a criterion of economic potency. Family income is only a rough indication of potency, because income can be derived from sources other than production, such as inheritance or pensions, but it suffices for our purposes. Diagram 3-6 depicts the situation in Costa Rica in the early 1970s.[34] In some cases, for instance in the United States during the Reagan years, the discrepancies are not only reflected in absolute income figures, but even in the growth rates of incomes, leading to exponential growth of the differences.[35]

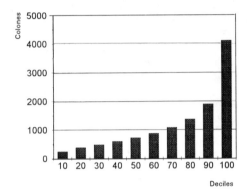

Diagram 3-6
Income distribution in Colones (local currency) in Costa Rica in deciles (tenths of the population)

At the *international level* it is convenient to compare the average production per person between countries. This figure is called the *Gross National Product per capita* (GNP pc). Diagram 3-7 depicts the global situation in 1993. Here all countries are lined up, as it were, in the order of their GNP pc.[36] Note that the vast majority of people, more than 4 billion of them, live in countries with very low average incomes. They form the global periphery. Rich countries, which form the global centre, have relatively few people. The highly industrialised countries which are united in the OECD, have less than 0,8 billion people. There are a few countries which form the transition between the rich and the poor, that is the semi-periphery, some at the lower end (for instance, South Africa and Brazil), some at the higher end (for instance, South Korea and Singapore).

The diagram gives an impression of the enormous discrepancies in income in the world today. But it does not yet tell the full story because, as the previous diagram has shown, there are also huge income discrepancies *within* countries. "Taking account of the income inequality within countries, the top 20 per cent of the richest people in the rich countries had incomes 150 times higher than the incomes of the poorest 20 per cent of people in poor countries."[37]

Chapter 3 - Economic discrepancies | 53

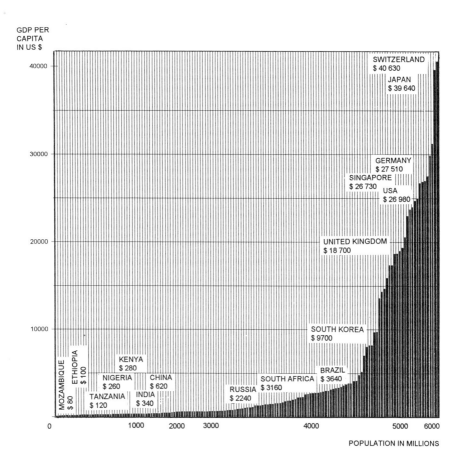

Diagram 3-7
Global distribution
of GNP per capita

The basic pattern of these curves remains the same all over the world. Within highly developed countries with social-democratic governments and well established trade unions, however, the curves flatten out considerably: the peak is lower, the bottom higher and the transitional phase more spread out. This simply means that such countries have achieved a greater balance in potency and income across the population. Unfortunately this cannot be done between countries.[38]

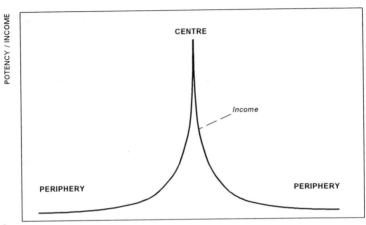

Diagram 3-8
The centre-periphery structure of potency or income distribution

To utilise the explanatory power of the centre-periphery model we now *shift the potency peak to the centre* of our graphical presentation (diagram 3-8). Note that this is no longer a conventional graph because there are positive values on either side of the peak and each of them represents one half of the population. Again we could depict a hierarchical network similar to the geographical one described in the previous section. Multinational enterprises typically display such networks.[39]

We now add another dimension to the relation between population and potency, namely *time*. The discrepancies between affluent and poor grow in time. This growth takes place in two directions: in the centre the population remains relatively static, while *economic potency grows;* in the periphery economic potency remains relatively static, while *the population grows.*[40]

The result is that the discrepancies in potency between centre and periphery also become greater. As the adage goes, "the rich get richer, and the poor get children!". Moreover, especially during the second half of our century both the growth of potency in the centre and the growth of the population in the periphery have been *accelerating*. "As it happens, both the absolute and the relative gap between rich and poor countries has widened in the last thirty years ... In 1960, the richest 20% of the world's population had incomes 30 times greater than the poorest 20% of the world's population. In 1990, this ratio was 60 times greater."[41] In terms of pure mathematics, these discrepancies are bound to grow in the medium term, even if the less developed countries had high growth rates and the industrial countries continued at their present rate of 3% per annum.[42] Similarly population numbers still grow exponentially in the poorest countries, although the general growth rate has slowed down in the Third World. Diagram 3-9 gives a three-dimensional impression of the two growth processes in time.

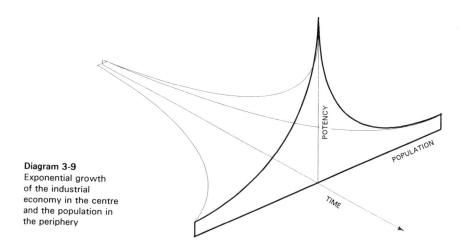

Diagram 3-9
Exponential growth
of the industrial
economy in the centre
and the population in
the periphery

When we speak of the accelerating growth of economic potential in centres we are referring to the complex phenomenon of evolving productive and organisational competence on a global scale. The most conspicuous manifestation of the phenomenon of "globalisation" is the rise of *multinational corporations*. The most powerful of these corporations have cast networks of social organisation over the globe which rivals that of the system of nation states, and the determining power of the internationally organised economy is beginning to overshadow the significance of international politics.

Irregularities

In real life there are many irregularities. The most regular and predictable factor is the exponential *growth of the population,* both in the periphery and in the centre, unless this growth is seriously affected by other factors such as famine, wars, epidemics or successful population control.[43] Thus the uncontrolled spread of AIDS, particularly in some African countries, may lead to great reductions of their populations within a few decades. Most demographers assume that there is a cycle in which a period of rapid population growth is followed by a period in which population numbers stabilise. This is called "demographic transition". Unfortunately there is no guarantee that this will happen in the poorest populations of the Third World.[44]

The *growth of potency in the centre* is more irregular because it depends on the business cycle. Boom is followed by recession. But the long term trend has been upward in the past. After World War II the Gross National Product (GNP) of the industrialised nations doubled every 12 - 15 years. "Industrial production has grow more than fifty fold over the past century, four-fifths of this growth since 1950."[45] We call this accelerating process "exponential growth".[46] Due to various factors, such as state overspending and high labour costs in affluent societies, growth rates have recently begun to slow down.[47] As we shall see in chapters 4 and 13, long term prospects for continued industrial growth are slim for ecological reasons.[48]

Before the recent "Asian disease", where a series of slumps followed each other, the "newly industrialising countries", especially the "young tigers" in South East Asia (Singapore, Taiwan, South Korea and Hong Kong), as well as certain regions in China, India, Indonesia and Latin America, had spectacular growth rates for quite some time.[49] Their circumstances have been rather unique and larger Third World countries may not be able to repeat their feat.[50]

The *relation between population numbers and potency* is irregular because of social barriers. In an open society, ambitious groups can climb up and better their position, while less successful groups can slide down the social ladder. We call this phenomenon vertical mobility. A feudal society, a patriarchal system, a caste system, a rigid class structure, racial discrimination, ethnic or religious group loyalties, and segmented labour markets all provide examples for barriers to vertical mobility.

Social policies may cause these patterns to change. In the West, the welfare policies of social-democratic parties and trade union action have contributed to a levelling out of discrepancies between the rich and the poor. In the East, Marxist policies have lead to greater equality. Some semi-peripheral countries, notably Taiwan, have built their economic success on a deliberate policy of social equity.[51]

Finally we have mentioned that centres and peripheries are organised in a *system of hierarchies.* This system also develops over time. As the centres become more integrated with each other, and develop more and more power, their spheres of influence also spread deeper into their respective peripheries.

Chapter 3 - Economic discrepancies | 57

Population groups and sectors of the economy which were once autonomous, are progressively integrated.[52] Obviously, some areas are less integrated into the system than others.

> *Revision: Describe how the centre-periphery structure of the economy evolves in time and mention a few irregularities in the pattern.*
> *Application: Can you explain the difference in economic potency between blacks in urban areas and blacks in rural areas in South Africa on the basis of the centre-periphery model?*
> *Critique: (a) "The recent history of the Afrikaner in South Africa shows that hard-working people can develop economically, even against severe odds. It is a moral, not a structural issue." (b) "The example of South African blacks demonstrates beyond any doubt that the poor are poor on account of greed and abuse of power on the side of the rich." What do these two statements have in common? With which would you sympathise and why?*

Section III: The relation between potency and need

Whether a group is wealthy or poor does not only depend on its relative income but on the discrepancy between its income and its needs. In rather oversimplified terms we can say that people are affluent when their income is higher than their needs, and poor when their needs are higher than their income. We can speak of an *affluence gap* in the former, and a *poverty gap* in the latter case (see diagram 3-10).

Classical economics does not operate with the concepts of potency and need, but with the concepts of supply and demand. It maintains that the concept *needs* refers to subjective emotions, which cannot be empirically established, nor expressed in monetary terms; it is also said that needs cannot be distinguished from wants and that wants are infinite.[53] When economists discuss market demand, they simply assume that the normal person has an income, that it is sufficiently large to make choices which are supposed to maximise one's utility or pleasure.[54] We do not subscribe to these contentions.

For the majority of humankind they are almost cynical. Is there no difference between the necessity to find the next meal for one's children and the craving for a third luxury car?

Need is, of course, a complex phenomenon which we shall analyse below.[55] But food and water are certainly objective needs; without them we do not only feel unhappy, but we die. It is indeed possible to establish what a minimum balanced diet for humans would cost in particular situations. And certainly needs and wants cannot be unlimited, otherwise humans would be divine and not subject to constraints of energy, time and space. In chapter 12 we shall argue that need and the capacity to fulfil that need should form the point of departure for economics.[56]

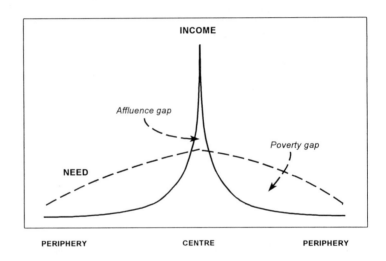

Diagram 3-10
The relation between need and income

In diagram 3-10 we plot a *hypothetical need curve* and superimpose it onto our income curve. This gives us an impression of the relation between income and need. We presuppose that, on average, needs are higher in the centre than in the periphery because the cost of living of centre populations tends to be higher than that of peripheral groups.[57] Think of rent, transport, compulsory contributions to insurance policies, food, clothing and so on. We also presuppose that the need curve flattens out gradually as we approach the centre, because the law of diminishing marginal utility comes into operation. This law says that the additional satisfaction or utility gained from additional units consumed declines as consumption approaches saturation point. The exact shape of this curve will differ widely from situation to situation and we cannot generalise. In countries which apply progressive taxation, for instance, the need curve may rise sharply as we move towards the centre. Progressive taxation means that the higher your income, the higher the percentage of your income demanded by the state.

The diagram shows that, as we approach the centre, income rises, yet needs rise faster than income and the poverty gap actually widens. This is the experience of many people who migrate from rural to urban areas in search of higher incomes. Although they may earn more money, they find themselves in greater financial difficulties than before. Only gradually the income curve catches up with the need curve until it eventually reaches the point where the poverty gap changes into an affluence gap. The diagram also indicates that the majority of the population experiences a poverty gap, while only a relatively small minority enjoys affluence. Of course, it is possible that the population of a whole country, such as Switzerland, can fall within the affluent minority in global terms.

Various types of material need

What are needs?[58] Theoretically we can distinguish between three types of material need: basic essentials, social expectations and personal wishes. Only *basic essentials* can be measured with some accuracy.[59] A formula is set up which includes all expenses that a normal family of a given size at a particular time and in a particular location has to incur to remain in good health. This includes a balanced ration, transport costs, rent, simple clothing, tax, school fees, and so on. No luxuries are allowed and it is assumed that the family budgets judiciously and purchases on the most favourable markets.[60]

Social expectations refer to the expenditure patterns a person occupying a certain status and playing a certain role in society is expected to follow. A managing director cannot arrive in a battered small car and meet his business partner at a street corner cafe; he would lose the contract. Social expectations are not geared to material sufficiency, but to social acceptability. Those who cannot meet social demands feel that they are frowned upon, not taken seriously, or despised. Actual levels where this happens vary considerably from culture to culture, from class to class and from group to group. Yet there is no reason why social expectations cannot be clarified conceptually and investigated empirically by means of questionnaires covering particular sections of the population in particular situations at given times. [61]

Usually we use the term "standard of living" when referring to social expectations. The standard of living is not what people actually spend, but what they are supposed to spend if they wanted to be "smart" or "in". This is largely determined by *reference groups* in society. Reference groups are usually local elites, but the "demonstration effect" can also cross national frontiers. African Americans, for instance, tend to form the reference group for South African Blacks.[62]

The third category includes all reasonable *personal wishes* of the members of a family or group. Personal wishes are not necessities like basic essentials, nor does the social environment expect one to fulfil them. They are gratifications of personal wants and desires. A desire can become a psychological obsession and this obsession can translate into a financial burden of the first order. Families can allow their children to starve for a television set or a car, because they are unwilling to go without these luxuries.

We speak of *reasonable* wishes. Obviously there are people with wants which are entirely unreasonable, such as the proverbial tycoon who derived pleasure from lighting his cigars with thousand dollar bills. It should not be impossible to measure total need statistically for certain groups of people in particular situations and at given times by determining the income level at which they begin to save and invest in considerable proportions.[63] It should also be possible to measure the actual utility of gadgets and opportunities in the possession of the affluent: which percentage of the contents of an expensive magazine is actually read; which percentage of available leisure time is spent on the yacht; how often is a toy played with?[64]

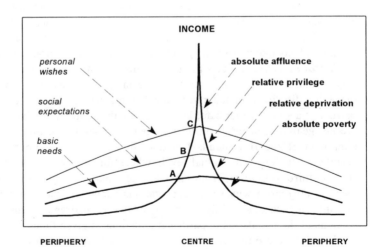

Diagram 3-11
Three types of need and four types of income-need relations

When we superimpose curves for the three types of need on the income curve, we obtain an interesting picture (diagram 3-11).[65] On the basis of this model we are able to arrive at four precise definitions:
(a) *Absolute poverty* is a situation in which income does not meet the level of basic essentials (below A on the diagram).
(b) *Relative deprivation* is a situation in which income meets basic essentials but not the level of social expectations (between A and B on the diagram).
(c) *Relative privilege* is a situation in which income meets basic essentials and social expectations but not all personal wishes (between B and C on the diagram).
(d) *Absolute affluence* is a situation in which income covers basic essentials, social expectations and all reasonable wishes and the surplus can be saved and invested or wasted (beyond C on the diagram).

In real life situations the three types of need are not so neatly stacked one on top of the other. People in absolute poverty also try to keep up with the Jones' and have intense personal wishes for which they may be prepared to go hungry.

Revision: Describe different types of need and show by means of a diagram how they can help to define affluence, privilege, deprivation and poverty.
Application: Try to locate your own position in the four categories depicted in the need diagram and think of other groups of people who might fit into the other categories.
Critique: (a) "Why make things so complicated! We know, after all, who the poor and who the rich are. Too much theorising is just a way of escaping from the practical demands for justice." (b) "You confuse the issues: personal desires and social prestige are not needs. Prestige is nothing but vanity and desires must be controlled. In contrast, needs must be fulfilled." (c) "You cannot distinguish between needs and wants. Is participation in sport a need or a want under stressful modern conditions? Nor can you distinguish between necessitities and luxuries. Is a refrigerator a necessity or a luxury?" Comment.

Section IV: Centres and their offshoots in the peripheries

Horizontal and vertical relationships

This section tries to shed light on the relation between interlinked centres and their respective peripheries. It has often been observed, for instance, that society in Third World countries is dominated by a wealthy modern elite, both in economic and political terms.[66]

There is an important distinction between vertical and horizontal relationships in society. *Horizontal* relationships are relationships between equals. The parties are on the same level concerning power, income and prestige. *Vertical* relationships are relationships in which one party has more power, income and prestige than the other. In the latter, one party is usually in a superior, the other in a subordinate position. If that is the case, the subordinate party is also dependent on the superior. This has far-reaching economic repercussions.

Looking at the potency or income curve in one of our diagrams above, we see that the further we move into the periphery, the more horizontal relationships become. People become increasingly more equal on an increasingly lower level of potency. In contrast, the nearer we come to the peak, the more vertical relationships become. This means that differences in power, income and prestige are far greater within the elite than within the underdog population groups.[67] More important, however, is the observation that the relation between the centre as a whole and the periphery as a whole is vertical by definition. Here we speak of the *class structure* of society.

The relation between dominant and dependent groups

The relation between dominant and subordinate groups is very complex. Johan Galtung, a Norwegian peace researcher, has designed a simple model which we shall adapt for our purposes.[68] Let us assume that there is a *centre region* such as Britain (C) and a *peripheral region* such as Kenya (P). In the centre region there is a group which represents the centre of that centre region, say the managerial elite in charge of multinational corporations based in London and Manchester. We call them the *centre in the centre* (cC). There is also a group which represents the periphery in that area, say the work force employed by those corporations in Britain and its supporting services sector. We call them the *periphery in the centre* (pC).

In the peripheral region there is also a group at the helm, say the local managers of the offshoots of British multinational corporations based in Nairobi. One can also think of the elites of the governing party. These people we call the *centre in the periphery* (cP). Then there is the peripheral population in that region, say the Kenyan workers and peasants. We call that the *periphery in the periphery* (pP). Diagram 3-12 explains the position:

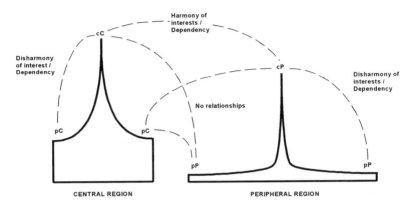

Diagram 3-12
Galtung's model

The relation between these different groups is characterised by two variables. The first variable is whether *harmony of interests* exists between the two groups. The second variable is whether a specific group, say group A, is *dependent* on group B, or whether group B is dependent on group A. To a certain extent every entity in a network of relationships is dependent on every other entity,[69] and this interdependence is used by some economists as an argument against "dependency theory".[70] But there is no question that Taiwan, for instance, is more dependent on the US than the US on Taiwan.

Let us go through the different relationships and see what the situation is:
(a) *Between cC and cP* there is harmony of interests because the offshoots of the corporations in Nairobi participate in the prosperity of their mother companies in London. Third World elites usually enjoy a standard of living similar to that of their counterparts in industrial countries. There is also dependency of cP on cC because the offshoots of multinational corporations are not autonomous and their industrial performance depends on capital, technology and expertise from abroad.
(b) *Between cC and pC* there is disharmony of interests. British workers want higher wages, while management and shareholders want greater profits; pC is also dependent on cC. However, British workers have built up collective power through trade union action. They have also formed political

parties which, in the past, have pressed for welfare legislation. As a result, pC has obtained a considerable share of the prosperity.

(c) *Between cP and pP* there is disharmony of interests. The urban elite and the rural peasantry vie for the meagre resources of the country. pP also depends on cP because all decision-making power in the country is concentrated there. In contrast to pC, however, pP was not able to build up collective power. Third World peasants are not organised, they cannot go on strike and, if they have a vote at all, it is not very effective because democratic institutions are not sufficiently developed. So pP has no share in the prosperity of cP.

(d) There is virtually no relation *between pC and pP*. All interaction between these two groups goes via cC and cP. This indirect interaction is marked by disharmony of interest. Why should the British workers want to pay more for their Kenyan coffee? There are, of course, exceptions to this rule but they may not necessarily be without self-interest. A higher wage level in the periphery, for instance, prevents large corporations from translocating their operations to low wage countries at the expense of jobs in their own countries.

(e) What has been said of the relation between pC and pP is also true of the relation between *cC and pP* and between *pC and cP*. All interactions between P and C go via cC and cP. A quotation sums it all up: "foreign aid is the transfer of money from poor people in the rich countries to rich people in the poor countries."[71]

The upshot of Galtung's model is that while both pC and cP get their share of the prosperity controlled by cC, *pP is left out in the cold*. This model is revealing, but it is still too crude for our purposes. We can make it more precise by adding two further considerations. First, both in the centre and in the periphery there is a dependent but relatively powerful, and therefore relatively privileged, work force. This group forms the bulk of the population in the centre. In the periphery it is usually only a small minority. Therefore it is sometimes called a *worker aristocracy*. The rest of the population consists of impoverished peasants. This means that there is a miniature replica of the centre system in the periphery, which forms a modern enclave within a traditional context.[72] The relation between the worker aristocracy and the peasants obviously resembles that between pC and pP discussed above under (d) above, unless industrial workers support their extended families in rural areas.

Second, there is some evidence that in the vicinity of a centre system a section of the peripheral population is forced out of its means of livelihood without being integrated into the centre system. This process is called

Chapter 3 - Economic discrepancies | 65

marginalisation. We shall discuss the reasons for this phenomenon in chapter 12. In the centre region this may be a small minority of unemployed and asocial people, but in the periphery region it can become very substantial. Just think of the great slums which surround every major city in the Third World. The relationship of this marginalised group to all other groups in society is again one of dependency and disharmony of interests. These people are often worse off than those who have at least a little piece of land to fall back on.

Revision: Describe the (revised) Galtung model and say which of the relationships are characterised by dependency and which by disharmony of interest.
Application: Try to name the population groups in your country which would fit into the four categories: cC, pC, cP and pP.
Critique: (a) "You seem to imply that all the leaders of the Third World are stooges and beneficiaries of Western capitalism. This betrays a lack of respect for the peoples of the Third World." (b) "There is no reason why the poor in the Third World cannot do what the poor in the West have done, namely, to pull up their socks and become competitive." With which statement do you sympathise? How would you respond to both statements?

Let us summarise

In this chapter we sketched the structural aspects of economic imbalances. In section I we dealt with *geographical concentrations* of economic activity and wealth. We call them centres. Surrounding these centres there are areas which do not share their prosperity. We call them peripheries. The centres are characterised by a vibrant urban, industrial and commercial civilisation. The peripheries are often rural areas where little development is taking place. The global system of centres forms a *hierarchical order,* which stretches from the great metropoles in the industrial nations to the smallest rural villages of the Third World. Smaller centres are dependent on the next bigger ones. The more dependent a centre, the more vulnerable it becomes.

In section II we applied the centre-periphery model to social and demographic realities. Centres and peripheries were now defined in terms of the relation between *population groups* and their respective economic potency. We saw that the centre population remains constant while their economic potency grows; the peripheral economy remains relatively stagnant while its population grows. For quite some time both growth processes have also accelerated. As a result, the income gap constantly widened.

In section III we introduced a second dimension into the picture, namely *need*. Where need exceeds income we have a poverty gap. Where income exceeds need, we have an affluence gap. By distinguishing between different types of need we came to understand the difference between absolute poverty, absolute affluence, relative deprivation and relative privilege.

In section IV we took note of the fact that the periphery is dominated by local offshoots of the centre system. We used Galtung's model to describe the *vertical relationships* between a dominant and a dependent centre and their respective peripheries. We saw that the periphery in the periphery region is least able to defend itself against the power play between the other three entities.

Notes

[1] Krugman 1993:5.
[2] The phenomenon has been discussed by economic geographers since a seminal work by Heinrich von Thünen in 1826. Gunnar Myrdal (1957) has pioneered research on the geographical disparities between rich and poor nations. The development sociologist R Behrendt has used it extensively. The neo-Marxist and the dependency schools (Dos Santos, Baran, Frank, Amin, Emmanuel), inspired by R Prebisch (1950 and *American Economic Review,* May, 1959), have contributed important aspects. The issue has also been taken further by economic geographers such as Smith (1977 and 1979) and Anzuck (1982).
[3] Compiled by J Browett; source: Smith 1977:20.
[4] Apart from Smith, I owe many of these insights to Smith 1979 and Anzuck 1982.
[5] Britain accounted for 34% of world trade in 1900, 18% in 1959 and 8% in 1986 (Liston & Reeves 1988:3).
[6] Purcell 1989:29.
[7] For an analysis of the collapse of the Soviet economy see Nürnberger 1998, chapters 4 and 5.
[8] Poverty in Catalonia, Ireland and Greece is twice as high as in the Benelux countries (Bosch et al in *Journal of Population Economics* vol 6/1993:247).
[9] Broadley 1992:69.
[10] Broadley 1992:126ff.
[11] Krugman 1993:11ff.
[12] Taking the US only, computer firms are concentrated in Silicon Valley, aircraft industries in Washington, carpets around Dalton, Georgia, financial services in New York, rubber in Akron, textile in Piedmont, entertainment in Los Angeles (Krugman 1993:53-67).
[13] Krugman 1993:76ff.
[14] Galtung 1971.
[15] Administrative capitals of former South African "homelands".
[16] Anzuck 1982.
[17] Gunnar Myrdal (1957) was one of the first to show that, in contrast to what static equilibrium theory taught, the interaction between unequal partners leads to disequilibrium.
[18] That is the reason for the phenomenon observed by Williamson that regional inequalities

Chapter 3 - Economic discrepancies | 67

first widen, then narrow down, as a country develops (Thirlwall 1994:132). In the case of West Germany, however, a peripheral region emerged along the border with East Germany. After unification the latter has become the German periphery. It is highly likely that this is a temporary phenomenon because Germany is located right in the centre of an enlarged European Union.

[19] For a map see Nürnberger 1988:26.
[20] The North-East has 27% of the population but 50% of the country's poor (World Bank 1990:28).
[21] See the maps in Nürnberger 1988:24, based on Browett and Fair 1974.
[22] Incomes in the Sâo Paulo region in Brazil and the coastal region in Nigeria are both 40% higher than the respective national averages (World Bank 1991:41).
[23] In South Africa group areas legislation has prevented this development for a considerable time (for detail see Nürnberger 1988:24ff). After the abolition of apartheid the surge into the cities has commenced in full force. Durban is believed to be growing at the rate of more than a thousand inhabitants per day. Sâo Paulo grows at the rate of 1300 per day. Most of the newcomers settle in "favelas" or slums (Broadley 1992:126ff).
[24] The normal income differential between core and periphery in the European Union is reflected by an average GNP pc ratio of 122:64 (Krugman 1993:94).
[25] Krugman 1993:13, quoting H H McCarthy.
[26] Diagram based on van Valkenburg and Held:1952 (Source: Chisholm 1979:89).
[27] In Brazil, for instance, there are more female-headed families among the poor (Fields 1980:155f).
[28] For the economic effects of these differences see chapter 8.
[29] See the epilogue to Nürnberger 1998, as well as chapters 6 and 8 of the present book for a discussion of the significance of this fact.
[30] The concept was apparently coined by Wallerstein 1974:349.
[31] Thirlwall 1994:13f; Samuelson 1989:644ff.
[32] The terms used to describe the two main sections of the population differ widely according to various ideological predilections. The affluent and powerful are often called the "ruling classes", the less privileged are called the "lower classes". Other terms are "elites" and "underdogs". Obviously there is a gradual transition between the rich and the poor. Western sociology therefore distinguishes between six classes: lower lower, upper lower, lower middle, upper middle, lower upper, and upper upper. This obscures the fact that, in global terms, the transitional (middle) class is quite small while the other two stand out. Marxists, in contrast, only speak of only two classes: the "bourgeoisie" (or the "middle class", or the "oppressors"), on the one hand, and of the "proletariat", (or the "worker class", or the "oppressed", or the "masses") on the other. This stark juxtaposition again neglects the importance of the transitional class. Actual income curves show that there is some truth in both contentions: while there is indeed a vast gap between the elite and the underdog, as the Marxists claim, it is also true that there is a sliding transition between the two, as Western sociology claims (see the diagrams below).
[33] For an empirical study of the precise shape of this curve in South Africa see Nürnberger 1988 39-62.
[34] Details taken from Fields 1980:188. Deciles are tenths of the population; colones refer to the currency of the country. For the racially differentiated income curve in South Africa, see Nürnberger 1988:49.
[35] Krugman 1984:130ff. His graph on page 135 shows that the income of the first and second quintile actually dropped, that of the third and fourth rose, the former by about 6%, the latter by about 10%. However, the income of the top 1% rose by roughly 105%, This groups siphoned off 70% of the entire growth realised from 1977 - 1989!
[36] The statistics are taken from World Bank 1997:214f (World Development Report 1997).
[37] Thirlwall 1994:19.
[38] For a full discussion of the social-democratic system and its achievements see chapter 5 of Nürnberger 1998; for a reflection on economic power distribution see the same book, chapter 8, section III, pp 197ff.

39 For detail see Gilroy 1993.
40 Obviously these are generalisations. In Niger, Nicaragua, Madagascar and Haiti there was a steep decline in economic potency, namely a negative growth rate of between -2.4% and -4.4%, while in China, Pakistan, Indonesia and Chad there was considerable economic growth in 1989-91, namely between 3% and 8% (World Bank 1993:238). The same is true for population growth. In Germany, Denmark and Belgium the growth rate is negligible (about 0.1%), and is expected to become negative in some cases, while in the US, Switzerland and the Netherlands it still grows at rates between 0.6% and 0.9% (World Bank 1993:289).
41 Thirlwall 1994:19.
42 See the revealing calculations of Thirlwall 1994:19f. To reach present First World standards, a country with a GNP pc of $1200 (which is one of the privileged ones) and a growth rate of 2% would take 95 years. If the industrial countries grow at 3% and such a country grew at 4% it would take 3 centuries to catch up.
43 While the growth rate of the population for Asia is 4.2%, and for Pakistan as high as 6.8%, it is only 2.8% in China (Broadley 1992:10).
44 Brown *et al* believe that the theory of "demographic transition" is not working in some of the poorest countries; they "get trapped in the second stage" and population growth rates "begin to overwhelm life support systems" (1987:20).
45 World Commission on Environment and Development (Brundtland Commission) 1990:4.
46 For the concept and the mathematics of exponential growth see Meadows, Meadows & Randers 1992:14ff.
47 Heilbroner 1988:66f.
48 For detail see Meadows, Meadows & Randers 1992.
49 Keesing 1988.
50 See, for instance, Vogel in Reynolds 1988:1ff. However, a World Bank study of 29 sub-Saharan countries, called *Adjustment in Africa*, points out that economic conditions in 1990 resembled those prevailing nearly 30 years ago in some of today's most robust Asian economies. *Newsweek* March 28, 1994 p 24. For a full discussion see chapter 6 of Nürnberger 1998.
51 See chapter 6 of Nürnberger 1996 and Fields, G: Assessing progress towards greater equality of income distribution; and Hicks, N: Is there are trade-off between growth and basic needs? In: Seligson 1984:292ff and 338ff.
52 E.g. Wallerstein 1974. See Johns 1985:276, 280ff; Caporaso 1987:2, etc.
53 " ...economists feel that the assumption that man seeks to satisfy ever-expanding wants is a true picture of the way individuals behave" (McKenzie 1985:17).
54 "The rational person will have maximised his utility after fully allocating his income and when his enjoyment of the last unit of each good consumed is equal". (McKenzie & Tullock 1985:33).
55 See Schäfer 1993. Cf Wisner 1988 for modern leftist approaches.
56 There are radical critics who debunk the very concept of need as a modern invention: necessity is turned into an evil, desire is turned into need; humans are no longer defined by what they are, what they can take, or what they can create, but by what they lack (Illich in Sachs 1992:88ff). The point is taken, yet need is not just a misplaced concept but a simple description of the human condition.
57 See the chart in World Bank 1990:27 which shows that the poverty line rises with average consumption levels.
58 For the philosophical discussions on this concept see Schäfer, G (ed) 1993.
59 Note that this definition of "poverty" is based on need, not income. There are some researchers who define poverty, for instance, as 50% below aggregate income or income of the lowest 10% (G K Duncan *et al* in *Journal of Population Economics*, vol 6/1993:217). This is not satisfactory, because in a very rich society even the lowest income groups do not need to be poor. It is also not satisfactory to base income needs on subjective views of respondents (K V Bosch *et al* in *Journal of Population Economics*, vol 6/1993:238), because this confuses absolute poverty with relative deprivation, which depends on the demonstration effect of reference groups. See below.

[60] In general, researchers agree that this "minimum subsistence level" or "poverty datum line" is unrealistically low because there are unforeseen expenses such as funerals, there is ignorance concerning family budgeting and the best buys in town, and there are understandable indulgences such as a bottle of beer or going to the cinema now and then. So they add a certain percentage to make up the "minimum effective level". For global figures see World Bank 1990:28ff. For a discussion of ways to measure poverty see World Bank 1990:25ff, and Ravallion 1991. For the early debate on the concepts see Ellison, P A, Pillay P N & Maasdorp, G G: *The povery datum line debate in South Africa: an Appraisal*. Durban: Univ of Natal, Dept of Economics, Occ Paper No 4. For a more recent discussion see Wilson & Ramphele 1989:13ff.

[61] The average income of groups lying between the absolute poverty datum line and the absolute affluence datum line could be taken as a rule of thumb. For various measures of poverty and relative deprivation see also B Delhousse et al: "Comparing measures of poverty ande relative deprivation". *Journal of Population Economics* 1993:83-102.

[62] See Nürnberger 1995:30ff.

[63] For a preliminary example see Nürnberger 1988:93ff.

[64] Classical economics explicitly denies that there are upper limits to what people desire. It is taken for granted that only scarcity keeps expenditure in check. But this assumption is questionable. Rostow, author of the classic liberal development theory, concedes that "Americans, at least, have behaved in the past decade as if diminishing relative marginal utility sets in, after a point, for durable consumers' goods ..."; so they choose to have larger families and turn to the arts (Rostow 1990:11.) Of course, there are individuals who squander financial resources senselessly, especially if they have won a jackpot and have never learnt to deal with vast financial resources. But they are not typical for the rich and it is statistics, not individuals, which matter. Even millionaires normally do not spend all their income, but only a fraction of it; the rest is invested. The classical explanation of investment, namely that people defer consumption now for the sake of greater consumption later, is obviously not convincing beyond certain levels of income. The reason for limited consumption in cases where financial resources are not the limiting factor, are not difficult to find. In the first place, human life is subject to the limits of time, space and energy (Mishan 1993:194). Wishes, which could theoretically all be fulfilled, have to compete for their chance within this circumscribed area. A tycoon cannot enjoy his yacht, jet around the world and play golf all at the same time, especially if he also presides over a demanding business empire. In the second place, the law of diminishing marginal utility says that the higher the quantity consumed the lower the additional satisfaction gained. The tenth luxury car is a bother rather than a boon. Sooner or later saturation must be reached. Because of the limits of time, space and energy, I believe that this is does not only apply for individual items, as is commonly assumed (Daly & Cobb 1989:85), but for total wants as well. What might be true, however, is that the drive to amass wealth is due to an infantile and basically pathological craving for prestige and power. Money can indeed become the object of unlimited craving (Heilbroner 1985:33ff,56). But this is a type of need which is not necessarily connected with material resources and which can be ignored for our present purposes.

[65] For empirical data in the case of South Africa see Nürnberger 1988:80-95.

[66] See Klitgaard 1988 and 1991; Andreski 1968; Nürnberger 1991:304-318 (more literature there).

[67] Differences in income among white South Africans, for instance, are much greater than differences among blacks, and even greater than differences between whites and blacks. McGrath 1983 328; Nürnberger 1988:47ff.

[68] Galtung 1971.

[69] Cf e.g. London & Williams 1990.

[70] For a description of this school within the context of social theories see Faber 1990:19ff; Johns 1985:263ff, 280ff; Caporaso 1987:29. For a Marxist critique of dependency theory see Weaver & Berger in Wilber 1983.

[71] Herman Daly in *Population and Environment* 15/1993:66f

Ecological deterioration

What is the task of this chapter?

So far we have dealt with economic imbalances. We now add the ecological dimension to the picture. By and large liberal (and Marxist) economists overlooked this dimension, or argued that the danger was exaggerated.[1] Environmentalists, in contrast, believe that a change of emphasis from economics to ecology will become inevitable as the situation deteriorates.[2] The natural world has a right of its own, but its destruction also deprives future generations of their means of livelihood.

In section I we give an overview of the problem. This leads us to an integrated model which combines economic growth, population growth and the growth of the impact of both on the environment. In section II we give an impression of the current magnitude of the impact of population growth and industrial growth on the environment and its likely effect on future generations.

Section I: The ecological problem in a nutshell

A rough idea of the impact of the global economy on the natural environment could be obtained if one multiplied population numbers with levels of affluence and again with levels of technological development.[3] In the centre, population numbers are low but the impact per person is high; in the periphery, the impact per person is low, but numbers are high. Obviously the actual weight of the impact greatly depends on the type of commodity or service produced and the technology used. Depending on circumstances, the effect may be equally destructive here and there.

Some ecological devastations, such as soil erosion, have locally confined effects, others, such as air pollution, have global repercussions. Nuclear radiation has particularly devastating transnational effects, as the Chernobyl disaster has shown. Global warming and the erosion of the ozone layer are other examples. But even local effects tend to spill over national boundaries. Witness, for example, the new phenomenon of "ecological refugees".[4]

When we add the ecological dimension to our previous model, it takes account of four factors: economic growth, population growth, growth of discrepancies in potency, and deepening ecological impact (diagram 4-1).

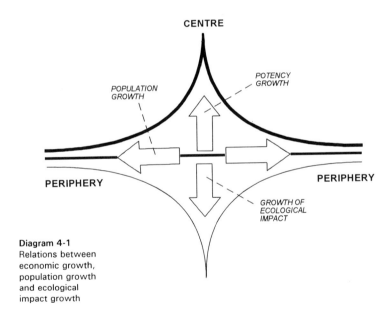

Diagram 4-1
Relations between economic growth, population growth and ecological impact growth

Note also that the model is meant to depict current *tendencies;* it does not presume to predict the future. It also represents a gross oversimplification. It does not, for instance, reflect the growth of conflict potential, the arms race and the social disintegration which are part of the entire syndrome. It also assumes that exponential growth will continue for ever, which is not possible in a limited world. Futurologists use much more sophisticated models.[5]

How does the problem arise?

Humans cannot live without taking from nature. As long as small bands of nomadic hunters and gatherers moved around in vast virgin areas, nature could absorb the impact fairly easily. But the emergence of agriculture and larger concentrations of people in fixed settlements had more serious effects on nature, even in antiquity. Witness the barren hills of Greece, where once flourishing forests were chopped down to build Phoenician fleets.[6]

Since the advent of the industrial era this impact has begun to assume frightening proportions. *Industrial growth* leads to accelerating depletion of non-renewable resources, over-exploitation of renewable resources and pollution of nature in general.[7] *Population growth* increases the pressure on the land, overgrazing, erosion, deforestation, slum settlements, and so on. When the periphery begins to develop in the direction of industrialisation and urbanisation its ecological impact increases.[8]

Accelerating growth cannot continue indefinitely in a limited world.[9] Sooner or later a peak must be reached; the only question is how close we are to this peak.[10] The problem of *pollution* receives most attention at present. In some areas, such as the destruction of the ozone layer, we may have reached the point of no return. But the depletion of precious *non-renewable* resources, such as crude oil, is just as serious. Even if we allow for the discovery of further deposits, the development of new materials, and the trend to use less bulk, the curve of rising demand for these resources will soon intersect with the curve of declining availability (diagram 4-2).

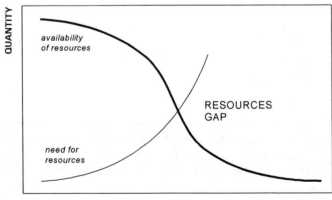

Diagram 4-2
Resource needs and availability over time

Chapter 4 - Ecological deterioration | 73

The looming scarcity of food is a case of over-exploitation of *renewable resources*. While the population continues to grow - at least in the poorest sections and at least for the time being - the growth of the capacity to produce food declines.[11] This is partly due to pollution and erosion, partly to the law of declining marginal productivity. Eventually, the rising population curve will intersect with the declining curve of agricultural production (see diagram 4-3).[12] Beyond that point the nutritional needs of the world population can no longer be guaranteed, regardless of how generously future generations might want to share their food.

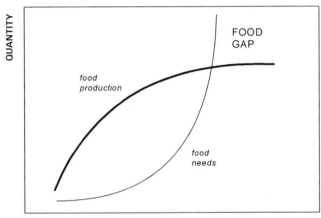

Diagram 4-3
Food needs and
food production
over time

The upshot of these considerations is that the present generation may very well go into history as the most affluent, and also the most peaceful, that the world has ever seen and will ever see. Science and technology have increased our life chances and our standards of living dramatically and may continue to do so until our "natural capital" is exhausted.[13] Then the great "morning after" may dawn and the entire social pyramid may begin to sink. If that happens, the rich may become much poorer on average, and the poor may go below the water line and drown. Needless to say, an increasingly violent scramble for dwindling resources may also ensue, unless humankind undergoes a fundamental learning process.

74 | Part I - The structure of the global system

> *Revision: Explain why the gap between the availability and the need for resources is likely to grow in the future.*
> *Application: Should virgin lands, for instance natural parks, in your country be opened up for human settlement?*
> *Critique: (a) "Human ingenuity will solve problems as they come; why become hysterical?" (b) "Resources are unlimited. When some resources become scarce, their prices will go up and industries will switch to alternatives. The 'limits to growth' and the 'depletion of resources' are unhelpful myths." Comment.*

Section II: The current dimensions of the problem

Population growth

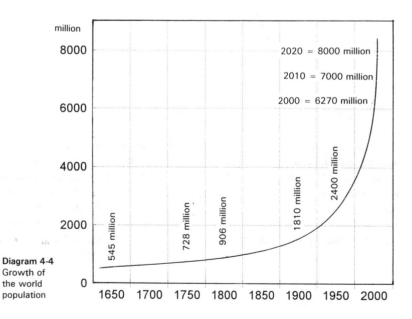

Diagram 4-4
Growth of the world population

According to one school of thought, the first humans emerged in Africa and gradually spread over the other continents. Here is one estimate of how humanity may have multiplied over time:[14]
- One million years ago there might have been about 125 000 people;
- 300 000 years ago they might have numbered 1 million;
- 25 000 years ago they might have numbered 3.34 million;
- 6000 years ago, the time when some humans began to settle in villages and towns, they might have numbered 5.32 million.
- 2000 years ago, at the birth of Christ, there may have been 133 million. Since then population numbers have risen dramatically - By the year 2000 humankind is expected to have reached 6.3 billion. What will happen after that is anybody's guess.[15] (diagram 4-4).

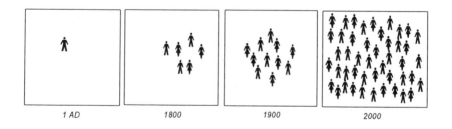

Diagram 4-5
Population density per square kilometre

The problem can also be expressed in terms of population density (diagram 4-5). When Jesus was born, the density was about one person per square kilometre; 200 years ago it was 6.2; 100 years ago 11; 50 years ago 16; by 2000 it will be 46.[16] It is a sobering thought that, when the biblical injunction to "be fruitful, multiply and fill the earth" was issued, which may have happened somewhere in the middle of the first millennium BCE,[17] the earth was still fairly empty. Nature was also hostile and dangerous, and needed to be "subdued" if humans were to survive. All this has changed dramatically since then.[18]

These figures give an impression of the growth of the population. More than that, they give an impression of the acceleration of that growth. If every couple has four children, and none of them die, the family will number 64 after 5 generations and a massive 2048 after 10 generations. It is not just arithmetic growth that is taking place here (thus 1+1+1+1 = 4), but geometrical or "exponential" growth (thus 1+2+4+8 = 15). Here is another indication of acceleration: humankind as a whole took about 50 centuries to increase from 5 million to 50 million (from about 8 000 BCE to roughly 3000 BCE);[19] the population of England and Wales will take 3 centuries (from 1700 to 2000) to increase from 5 to 50 million;[20] the population of South Africa will have grown from 5 to 50 million in the single century from 1900 to just after 2000.[21]

Acceleration is most frightening when not only the absolute numbers but even the growth rates increase. We speak of *super-exponential growth* in such a case.[22] In the year 1650 the world population grew at a rate of 0.3% per year, which amounts to a doubling time of about 250 years. By 1900 the growth rate was 0.5% per year, with a doubling time of 140 years. By the year 1970 the growth rate had reached 2.1%, with a doubling time of 34 years. In Kenya and Ivory Coast the growth rate was as high as 3.8%, with a doubling time of less than 19 years.[23] Fortunately the overall growth rates have fallen since then from 2.1 to 1.7%.[24] But there are still 19 countries with growth rates of over 3%, 13 of which are in Africa.[25]

Exponential growth cannot continue forever; a downturn is inevitable.[26] The only question is how it will happen. Humankind will have to reduce its numbers voluntarily, or face mass extinction through famine and pollution. Among some of the poorest of the poor, famine already strikes at regular intervals. It is estimated that 200 million people have died of starvation or hunger related diseases since Malthus. One billion people are believed to be chronically undernourished today.[27]

As mentioned before, there is a "positive feedback loop" between poverty and population growth.[28] In Japan and Germany the crude birth rate is 10 per 1000, in Ethiopia and Mali it is 50.[29] In Europe there are, on average, 1.9 children per family, which means that the population does not grow at all. In Africa, in contrast, there are 6.5,[30] which means that the population trebles within each generation, unless life expectancy declines.

But the link between poverty and population growth is *not inevitable*. Growth rates have recently fallen in many less developed countries. The most drastic measures have been taken in China where there are now 2.8 children per family, while just next door in Pakistan there are still 6.8.[31] It is difficult to condone the draconian Chinese population policies. The point is, rather, that it is not impossible to curb excessive population growth.

Food production

Famine related deaths occur in spite of the dramatic growth in food production over the last few decades, achieved by the so-called "green revolution". This is partly the result of unequal distribution of resources which we have analysed in the previous chapter. The reason for hunger at present is that people have no money to buy existing food. Soon there may be no food to buy or to give away any more, because the countries which have large surpluses today, and need to get rid of it, may have to keep those surpluses for their own use.[32] The phenomenal food surpluses of the West are not sustainable, because agricultural methods destroy the soil, deplete water resources, and depend on fossil fuels which will not last for ever. As fossil fuels become scarce and expensive, some of them may divert agricultural resources to energy generation: sugar cane, for instance, can be used to produce fuel.

In the meantime food needs grow rapidly. In fact, we have just about reached the stage where total food production balances out with total needs. Estimates indicate that "recent world harvests, if equitably distributed and with no grain diverted to feeding livestock, could supply a vegetarian diet to about 6 billion people." If the world diet had a 15% animal component they could feed 4 billion. Less than 3 billion could be fed with a Western diet with a high meat component.[33] Since there are already 5.3 billion people on earth today, we have just reached the limits of the possibility to supply everybody with a balanced diet.[34]

It is virtually certain, therefore, that the earth will not be able to supply the 10 - 12 billion people expected to be alive just a few decades from now with a balanced diet, let alone a Western type diet which includes a high percentage of meat. Obviously these figures conceal vast regional discrepancies in food production which cannot easily be bridged. While there are massive agricultural surpluses in Europe, for instance, there are at least 19 countries in the world which cannot meet their food needs, even under the best possible circumstances.[35]

Can food production not be enhanced further? Unfortunately not much. Agricultural production is subject to the law of declining marginal productivity: the first bag of fertiliser pushes up the harvest considerably, the second less so, the fifth may even poison the soil and cause a decline in output.[36]

Again, technology may alleviate, but not solve the problem. It is true that grain production has risen from 600 to 1800 million metric tons per annum between 1950 and 1985. And it is also true that there is still considerable land which can be brought under cultivation. Better agricultural methods can still push up yields in many developing countries. But most of the possibilities of expanding food production have been utilised by now.[37] "Yields in the highly industrialised countries are nearing their practical limits."[38]

In fact, our resource base for food production is not expanding but contracting fairly rapidly. *Fish resources,* for instance, are depleted through dragnet fishing.[39] The resources of the oceans are already exploited to 90% of their capacity.[40] Bluefin tuna, for example, declined 94% between 1970 and 1990. Technology has not helped to increase the resources but to "find and catch every last fish."[41]

Much of our *agricultural land* is marginal in terms of rainfall and fertility and should not be used for cultivation. Due to population growth, agricultural land is lost to residential areas, transportation networks, industrial sites, recreation, military installations, and so on. Overcultivation and overgrazing lead to erosion and exhaustion of fertility. When virgin forests are chopped down, for instance in the Amazon basin, the land may give impressive yields for one or two seasons and then become barren. Deserts grow by 6 million hectares annually. Topsoil is lost at the rate of 25 billion tons per year, roughly equivalent to all of Australia's wheat lands.[42] For every ton of agricultural produce, South Africa loses 20 tons of soil on average.[43]

Land is also being lost to pollution and erosion. Fertile soil is degraded by fertilisers and insecticides. Emissions of chemicals from industrial smoke stacks and exhaust pipes settle on the earth. In the Krakow region of Poland, for instance, 60% of the food produced is deemed unfit for human consumption because of heavy metal pollution derived from brown coal firing.[44] Increasing ultra-violet radiation, due to the destruction of the ozone layer and a warming of the atmosphere may affect plant and animal growth. "There has been an overwhelming excess of potentially cultivable land throughout all human history, but within about thirty-five years (the last population doubling) there has arisen a sudden shortage."[45]

The scarcity of *water* is another problem. Irrigation farming, the industrial use of water, rising domestic consumption levels and urban population growth all combine to make fresh water a dwindling resource. Ground water is being depleted through boreholes.[46] Oceans and fresh water reserves are polluted by oil tankers, industries and waste. If the ice in the polar regions begins to melt, large fertile areas along the coasts will be swamped.

Reliance on technological advances may be deceptive. The massive surpluses produced by the United States and Western Europe today are the result of state subsidies to the agricultural sector. The shortfall is covered by taxes taken from the powerful industrial and commercial sectors of the economy. In other words, it is an artificially boosted, "parasitical" sort of production, which could not pay its own way on the open market. The industrialised countries are rich enough to be able to afford such a luxury, but not the poorer countries.[47]

Moreover, the impressive agricultural output of the industrialised countries is largely based on energy and chemicals derived from fossil deposits - which will run out precisely at the time when food shortages begin to escalate! "Sometime during the next 40 years the cost of oil will necessarily rise to the point where the present agricultural system will collapse."[48] Before that, however, costs of modern agricultural inputs will rise beyond the reach of most poor communities.

All these developments do not make it likely that agriculture will be able to supply the food needed for a growing population much longer. While the growth of the population *accelerates,* the growth of agricultural production *declines.* Sooner or later the two growth curves will intersect and overall shortages will be the result, whether there is equitable and efficient distribution or not (see diagram 4-3).

Industrial growth - depletion of resources[49]

There are three major problems with industrial growth: the overexploitation of renewable resources; the depletion of non-renewable resources, and the pollution of the environment. Perhaps we should add systemic dangers as a fourth category. Increasing organisational complexity may spiral out of human control and develop major dysfunctions, or lead to catastrophic breakdowns. The extent to which a globalised economy, or nuclear weapon systems for that matter, depend on the reliability of computer controlled communication systems is frightening.

Deforestation is a good example of overexploitation of renewable resources. There are virtually no primary forests left in Europe. The United States has lost 85% of its primary forests. In the tropics, half of the original forest cover is gone.[50] Ivory Coast has lost 3/4 of its forests in only 20 years.[51]

Many materials, such as metals, are limited and cannot be renewed. As ore grades decline, energy use, waste and pollution increase. "Even if there were no further growth, present rates of material use would be unsustainable in the long term."[52] Fortunately metals can be recycled - and that has to be done everywhere and consistently. But even that is not possible without considerable financial and ecological costs. Other non-renewable resources, such as crude oil refined into petrol, cannot be recycled. Petrol disintegrates when utilised. Experts also point out that gene material, which could become important in the future, is being lost on a vast scale. It is estimated that, by the year 2000, 50 000 species will be extinguished *annually*.[53]

The least we would need, if the whole of humankind was to industrialise fully, is clean, safe and affordable energy. Fossil fuels (oil, gas and coal) are particularly critical in this regard. Their rapid depletion represents a total loss for future generations. It is rarely recognised that the mighty industrial machine, which emerged only during the last two centuries and to which we primarily refer when we speak of "growth", is built almost entirely on these precious deposits:[54]

(a) As *sources of energy* they form the backbone of many kinds of industrial activities, including commercial agriculture, including even the activity of preventing pollution and cleaning up the environment. Solar based sources of energy seem to have their limitations, and nuclear energy is prohibitively dangerous for the future of life on earth.

(b) As *raw materials*, oil and coal form the basis for the production of consumer goods from plastic containers to synthetic fibres, to medicines. They have become equally indispensable for the production of capital goods. Oil, gas and coal must be valued among the most precious resources with which humankind has been endowed.

Obviously the tap will not suddenly be turned off. But scarcity leads to higher prices - that is a simple economic law - and industrial production will become more expensive.[55] While technological advances will undoubtedly increase the cost-effectiveness of production, they cannot spirit away all costs, because even the most energy-efficient plant still uses energy. Unless there are dramatic break-throughs in the development of cheap and ample sources of energy, material standards of living will decline in the future. Costs will first escalate beyond the reach of poorer societies; eventually richer societies will also have to cut back their consumption quite drastically.[56]

Rising costs also imply that industrialised and industrialising societies will not feel the pinch equally. Because of the vast differences in economic potency and military power, the needs of the rich will probably continue to be satisfied and it is the poor who will suffer, as they always do. Where both poverty and income discrepancies increase, conflict potential may also increase. In some countries, notably in Africa, the era of mass famine and civil strife seems to have already begun.

If this is the case, it seems to be irresponsible to burn up fossil fuels indiscriminately within a few short decades and deprive all future generations of this valuable asset. They are virtually irreplaceable. One could perhaps commit large agricultural resources (such as sugar cane) to the production of fuel, but this would happen at the expense of food. It would also cost energy.

Humankind has utilised the continuous flow of energy from the sun on the earth's surface for 100 millennia of its history; it is now plundering the priceless treasures stored underground in a rush for gain.

At current rates of consumption known reserves of oil and gas would be exhausted within half a century, coal reserves within four centuries.[57] This does not take into account the fact that, when oil and gas run out, coal may have to supply the balance. This would drastically reduce the life span also of this source of energy. And if Third World countries industrialise in real earnest, as most economists suggest they should, coal reserves will be depleted even sooner. Unfortunately coal is also the worst air polluter of the three kinds of fossil fuel. Moreover, while rates of consumption rise rapidly,[58] exploration and extraction become more expensive and yield less and less results.

Objections

Objections against these conclusions are not convincing. It is plain logic that in a scenario of uninhibited exponential growth, escalating needs will overtake the declining availability of resources, while the ecological impact will reach dangerous levels. The fact that the predictions of "prophets of doom" have not materialised in the past is no source of comfort.

Let us look at some of the counter-arguments. Some experts have seriously claimed that the deposits of oil, gas and coal are "virtually unlimited". The truth is that the oil crisis of the early 1970s has led to a flurry of explorations which resulted in the discovery of new deposits and the utilisation of marginal sources. This has led to a formidable glut on the oil market.[59] But a glut simply indicates that too much of the precious resource has been pumped out of the ground for the market to absorb at present.[60] If a millionaire is foolish enough to spend all his money today, nothing will prevent him from being a pauper tomorrow.

It has been argued that the known reserves of crude oil have gone up from 65 billion tons to 124 billion tons in the two decades since the oil crisis,[61] and that there may still be double the amount of known reserves lying undiscovered in the earth's crust.[62] This is quite possible, but it does not mean that available resources have increased, or that they can all be burnt up without loss for future generations and the biosphere, or that exploration and extraction costs will not soar, let alone that the deposits in the earth's crust are inexhaustible.[63] Whether oil resources will be depleted within half a century or within two centuries is really not the point; what is a century in terms of human history! For all we know there will be people around two or three centuries from now, and they may curse our generation one day for having ruined their prospects. What about "intergenerational equity"?[64]

Modern technologies, it is argued, use less energy and resources than they used to. But that does not mean that the "link between growth as measured by GNP and energy consumption has been broken".[65] It is true that greater efficiency will make a difference. It has been calculated, for instance, that current Japanese and European levels of efficiency could cut the energy bill of the United States by half and even this level of efficiency can be increased drastically.[66] But the US population is in no mood to subject itself to such a discipline. And if it were, this would still not mean that the energy bill would not rise exponentially. If present trends continue, the savings through better technologies will be more than offset by the number of would-be entries into the club of industrialised nations.

The other argument is that science and technology will advance rapidly enough to find alternatives to all resources which are depleted. This optimism usually goes hand in hand with faith in the market mechanism: as the supply of one resource runs dry, its price will go up, the comparative price of alternatives will go down and so it will become economical to explore and exploit these alternatives. As cotton has been replaced by synthetic fibres, copper wire will be replaced by glass fibres and oil by nuclear fusion as a source of energy.

This argument is not without a bit of inadvertent cynicism. If an alternative becomes cheaper in comparative terms this does not necessarily mean that it becomes cheaper in absolute terms. On the contrary, with dwindling supply the price of both could rise steeply. Who will be able to afford the alternatives? There is no doubt that the economic elite will find ways of remaining comfortable, but what about the rest of humankind? By the time the bulk of the Third World is ready to industrialise on a large scale, cheap forms of energy may have been depleted and new sources may not be within their reach.

But will economies of scale and technological advance not lower the prices also in absolute terms? Yes, they will. The development of technologies to derive energy from the sun, wind, stored water, ocean tides and biomass has progressed appreciably.[67] In 1970, solar energy needed a capital investment of $ 150 per watt; in 1990 the cost had dropped to $ 4.50 per watt. Another fourfold decrease would make it competitive with coal-fired power stations. Solar hot water systems and wind-powered electricity have already become competitive in particularly suitable locations. Hydro-electric power is clean.

Chapter 4 - Ecological deterioration | 83

The development of alternative sources of energy should, therefore, receive more support than it does at present. The powerful interests of the oil companies and derived industries seem to be the major obstacle. Yet the potential of such alternatives to offer comprehensive and lasting solutions should not be overestimated. Alternatives do not come without a price - both financially and ecologically.[68] Dams, solar panels and windmills also have ecological repercussions. Moreover, the construction of storage dams, solar panels and windmills requires suitable locations, huge capital outlays, a high level of technological efficiency and again - a lot of energy. There are regions of the world where these requirements simply cannot be met. "Biomass energy is only as sustainable as the agriculture or forestry that produce the biomass."[69] The flows of solar and wind energy depend on weather conditions. They have a fixed peak and cannot be relied upon to support indefinite industrial growth.

Nuclear fusion may one day provide an answer, but we cannot predict whether and when this method will become technologically feasible, at what cost, and with which repercussions for the environment. Nuclear fission has already brought us to the brink of global disaster and we cannot afford to play with fire.[70] The problems of radioactive waste disposal, nuclear accidents and the possible proliferation of nuclear weapons have not been solved and this development is extremely hazardous.[71] As fossil fuels run out, industrialisation will turn to nuclear energy. Asia already seems to follow this path with a determination which is deeply disturbing.[72]

It is gratifying to know that the future holds possibilities which we cannot fully appreciate today. Hope is the prerogative of the human spirit and it should not be dampened. But hope is not the same as foolhardiness. The point is: as long as we cannot be sure of viable alternatives, it would be prudent not to squander fossil fuels, but wait for substitutes to arrive before exhausting existing reserves. To gamble with the future of posterity is irresponsible.

As mentioned above, another resource that will become scarce is fresh water. In many countries this has already become the first priority problem. It has been said that the wars of the future will be fought over water supplies. There are 214 multinational river systems in the world which are potential conflict generators.[73] The conflict between Israel and its Arab neighbours is largely about water. Industrialisation and rising living standards go hand in hand with escalating water consumption. If hundreds of millions of hitherto poor people begin to use flush toilets and bath tubs, unimaginable quantities of water will be needed.[74] With growing industrialisation and the use of artificial manure and pest controls in agriculture, water resources are also increasingly being polluted. This brings us to the next consideration.

Industrial growth - pollution[75]

Many experts argue that waste and pollution present a much more serious problem than the depletion of resources. The spaces where waste and pollution are deposited are called "sinks". The problem is that these sinks are neither unlimited nor watertight. In the last two decades considerable progress has been made with some toxic wastes, such as lead in petrol, DDT, dildrien and nuclear pollution in industrialised countries. Sulphur dioxide emissions have been cut by 40%.[76] Yet four problems remain:

(a) Cleaning up often simply means that toxic materials are transferred from one sink to another, say from water to air. High smokestacks simply spread pollution from the neighbourhood to wider areas.
(b) Avoiding or cleaning up pollution is expensive and often unaffordable for poorer societies.[77]
(c) Many gains are neutralised by higher growth rates. The United States, for instance, has halved municipal effluent pollution at the cost of $ 100 billion over 20 years, yet the volume of effluent has doubled during the same time. Similarly, the increase in emissions through growing numbers of cars exceeds the reduction of car emissions achieved by greater efficiency.[78]
(d) The ecological concern has not become part of the economic concern. The mandate of the institutions created to combat pollution is restricted to fighting the symptoms of the problem, while the roots are not tackled: "... much of their work has of necessity been after-the-fact repair of damage: reforestation, reclaiming desert lands, rebuilding urban environments, restoring natural habitats, and rehabilitating wild lands."[79]

Pollution too follows the centre-periphery pattern. In the industrialised countries the true extent of the problem is largely concealed from the public eye. They have the education and the financial resources to avoid, or clean up, some of the worst forms of pollution. They can also transfer polluting industries to Third World countries with less stringent environmental restrictions.[80] Poor communities, in contrast, struggle to satisfy their daily needs. Poor states have balance of payments problems. They may have to service their debts with greater exports of primary products such as wood. Under such conditions pollution and resource depletion are secondary concerns. At present some of the greatest ecological devastations happen in the former Second and Third Worlds.[81]

Even the most ardent growth optimists concede that there are, at present, four major ecological concerns:[82]

Chapter 4 - Ecological deterioration | 85

1. *The greenhouse effect,* or the heating up of the earth's atmosphere.[83] It is caused by increasing levels of such gases as carbon dioxide, carbon monoxide, methane, nitrous oxide and chlorofluorocarbons in the atmosphere. The increase is estimated to range from 0.2 to 0.5% per year. The problem is caused by modern industrial and agricultural activities, especially the burning of fossil fuels. These gases prevent the heat generated by the sun on the earth from escaping into space in the form of infra-red radiation.

The result may be that as the atmosphere becomes warmer, rainfall will decline, severe water shortages will occur, many of the most fertile agricultural regions will be parched, the ice in the polar regions will melt down, the sea level will rise and submerge many settled and cultivated regions. There are many additional feedback loops, such as the emission of methane gas from melting tundra soils which will increase the greenhouse effect.

The implications of the full industrialisation of humankind are staggering. In a modern economy such as Germany, 70% of all carbon dioxide emissions are derived from traffic.[84] The motor industry is one of the most important economic sectors in industrialised countries. Even where the highways are already choked, as in Germany, economists shudder when car sales go down. At present only 8% of the world's population owns a car.[85] Imagine what would happen to the atmosphere should the rest of humankind achieve First World levels of motorisation - with First World motor sales still expanding![86]

Because there are so many unknown factors, many scientists and economists dismiss these predictions as alarmist. They suggest that we might wait and see what happens before considering a drastic reduction in industrial activity, because the latter would have severe repercussions for the economic wellbeing of humankind. The problem is that by the time the damage becomes more visible, it may be too late to control it. Moreover, there is no guarantee that people will be willing to act once the facts are known. Experiences with maritime resources, for instances, have shown that people are generally more interested in their short term profits than their long term responsibilities.

2. *The destruction of the ozone layer.*[87] Ozone is a gas which forms a layer between 10 and 50 kilometres above the earth and which filters out much of the ultraviolet rays with which the sun is bombarding the planet. In excessive quantities, this radiation is harmful to organic life, causing, among other things, skin cancer and the weakening of the immune system. Decreases in the quantity of ozone in this layer endanger the entire food chain and threaten all life on earth.

Scientists have discovered that this is precisely what is happening. Chlorofluorocarbons (CFCs) are gases which act as catalysts in breaking down ozone molecules. A catalyst facilitates a reaction without being changed. So a single chlorine atom can destroy up to 100 000 ozone molecules in a continuous chain reaction. CFCs are used in the manufacture of foam products, spray cans and refrigeration. Seemingly enough quantities of them have escaped into the higher atmosphere to cause a massive gap in the ozone layer above the antarctic and severe erosion over the arctic during winter. At other times of the year the gap disappears. This means that the process dilutes ozone in the rest of the stratosphere.

The scientists who have discovered the calamity, the activists who exerted pressure, and the politicians who finally agreed to limit the production of CFCs deserve unreserved praise. It is one of the rare ecological success stories.[88] Yet this human ingenuity is overrun by powerful economic interests. First the wait-and-see attitude prevailed and valuable time was lost. Even now that the evidence seems to have become clear, international treaties have not placed an immediate and total ban on the production of chlorofluorocarbons, as they should have done, but only called for a gradual reduction.[89] The profit of great enterprises is still considered to be more important than the future of humankind. Moreover, it is an isolated symptom which is being treated. What about treating the disease?[90]

3. *Acid rain.* Rain is made acid by the emission of sulphur dioxide and nitrogen oxides from burning fossil fuels in factories, cars, heaters and so on. In some heavily polluted areas rain has been found to be more acid than lemon juice. In the short run acidity does not seem to affect humans directly, but the indirect effects may accumulate. In the meantime it certainly affects forests, which are another precious asset of humanity. Again the problem has generated scientific bickering and wait-and-see attitudes, though forests in Europe already show severe symptoms of distress. Obviously the doubts and denials are motivated by economic interests, as in the other cases mentioned.

4. *Waste disposal.* "The world is running out of places to dispose of its waste - ordinary refuse and household garbage, toxic industrial waste, and nuclear waste."[91] This is a problem which feeds on itself: extremely wasteful and waste generating consumption patterns in industrial countries are served by an industrial "throughput" which is kept going by energy, whether generated with fossil fuels or nuclear plants. Poor societies obviously produce much less waste per person, but as their numbers increase, the problem surfaces there as well. Most poor societies also have no sophisticated garbage removal systems in place. Just think of the garbage heaps accumulating in slums! Moreover, constant attempts are made to dump the toxic waste of industrialised nations into Third World countries, or to shift whole industries which produce toxic waste to these countries.[92] Governments and corrupt officials

often secretly condone the practice to gain lucrative compensations on behalf of their unsuspecting subjects.

Fortunately a lot can be done about household waste through processing and recycling. Japan only places 20% of its waste into land fills, while the United States dumps 90%; Japan recycles 95% of its beer bottles, the US 7%.[93] But what about cutting down on waste in the first place? One of the great sources of garbage is excessive packaging, from which not the consumers, but the producers benefit. In Germany a law has been passed which forces manufacturers to take back their packaging materials - with dramatic effects on their overall volume! Measures like this upset the economy, but why should society suffer to prop up the profits of an elite?

The problem of toxic industrial waste, such as heavy metals, is more difficult to solve. Effluents have traditionally been channelled into rivers, killing the fish, creating filth and stench, reaching the oceans and creating more havoc there. Toxic waste has been placed into land fills, from where it may reach the ground water. If burnt, it may cause problems in the atmosphere, unless all the gases and ashes are collected, which just shifts the problem elsewhere, besides being very expensive. Much research is devoted to finding acceptable means of toxic waste disposal. One of these is biodegradation, a process in which microbes eat up organic waste. But in the end the problem must be tackled where it originates.

An emerging problem is space debris caused by defunct rockets and satellites in outer space.[94] But, as mentioned above, the most worrying problem is nuclear waste. Wherever it is deposited, nuclear waste continues to emit radioactive radiation for thousands of years. The United States alone is expected to have produced 50 000 tons of nuclear waste within half a century, that is since nuclear power was first developed in the 1950s. This does not include the radioactive material contained in huge stockpiles of nuclear weapons. With industrial growth spreading this figure will multiply. "It is one of the ironies of our age that in the name of national security, we may be poisoning ourselves."[95]

In addition there is a constant danger that obsolete or inefficient nuclear power plants may leak radioactivity into the atmosphere, or blow up altogether. The accident of the plant at Chernobyl in the Ukraine may just have been the beginning of an escalating series of catastrophes as nuclear power plants grow older and disintegrate. This accident alone has rendered a fourth of Belorus permanently uninhabitable; it will continue to cause the deformation of countless fetuses, impair the immune system of children, and lead to hundreds of thousands of cancer deaths over the next 25 000 years.

It also needs to be emphasised that, in spite of the end of the Cold War, the immense danger of nuclear weapons has not been overcome as yet. Most of the stock piles still exist. India and Pakistan have just joined the nuclear club. A lucrative black market for nuclear material and technology from the former Soviet Union seems to develop. More and more nations, perhaps even clandestine political movements, may soon get into the possession of such weapons. The detonation of only a few of these nuclear arsenals would spell disaster for humankind and much of the rest of nature.

Pollution control has become popular in the industrialised world. There are powerful "green" lobbies. For business environmental conservation is becoming a profitable public relations exercise. Governmental policies and intergovernmental agreements are relatively advanced, especially in the Triad. There are a number of gratifying success stories. While all the measures alleviate the problem, and must be supported as such, they do not go to the roots. Economic growth continues to be the paramount priority on the agenda of even the richest countries. The wasteful use of resources, especially energy, minerals, land and water, is not covered in the treaties. They do not address the simple polluter and consumer.[96] Pollution control also involves high administrative costs and demands high levels of efficiency. Poorer countries in the Third World cannot afford the type of controls enforced in industrialised countries. Because of more lenient environmental laws, polluting industries and toxic waste can easily be transferred to these areas. If we are to overcome all these problems, a more fundamental rethinking is necessary.

Ecology and armed conflict

Humankind is fast depleting its resources. We have mentioned energy and water.[97] If global warming leads to the flooding of coastal regions all over the world, as experts expect, vast agricultural and urban areas will be destroyed.[98] Desertification (for instance in the Sagreb region), deforestation (for instance in Bangladesh, the Amazon, or Ivory Coast), erosion (for instance in South African black rural areas or in North China), and depletion of local fish resources all lead to the large scale migration of "ecological refugees" in search of grazing, agricultural land or urban sources of income.

Competition for scarce resources leads to rising conflict potential. The economic roots of a conflict are often overshadowed by more visible political motives, but ultimately people fight over life chances and power. Where conflict over resources builds up, it often rekindles ethnic hostilities with all their intensity and bitterness. Liberia, Bosnia, the Sudan, Sri Lanka and some former Soviet Republics are examples.

Chapter 4 - Ecological deterioration | 89

Ambitious and threatened elites of poor nations commit vast portions of their meagre resources to armaments. In highly industrialised nations war has become too dangerous to be a political option, but in poor countries arms are extensively used to settle disputes, whether in the form of international tensions, civil wars, revolutionary violence, or social unrest. The result is that further resources are wasted, further environmental destruction takes place, less food can be produced, and more people turn into refugees. And again it is the poor who suffer first and most.

The tighter the situation, the more survival instincts assert themselves, the greater the potential for violence. At the same time the destructive power of weaponry has escalated dramatically. If the "lethality index" of a sword or a spear - the most common weapons of our ancestors - is taken as 20, the index of a modern assault rifle is 4200, that of a light machine gun 21000, that of a fighter bomber with cluster bombs between 150 and 200 million, that of a guided missile with nuclear warheads 200 billion.[99] Modern weaponry has become freely available even to poor societies - often in the form of "development aid". As the lure of firing power increases, people lose their sense of solidarity, their regard for the dignity of others, let alone for the interests of future generations.

As each national elite tries to protect its interests against rivals, restive populations, organised crime, and unwelcome immigrants, resources, capital, labour power and research are channelled into the production or acquisition of sophisticated weapon systems rather than to goods which could be utilised for human consumption.[100] During the first decade after independence, the average annual growth rate of military expenditure in Africa has been in the region of 20%.[101] Compare this with the dismal performance of economic growth on this continent!

Armed conflict makes development illusory and destroys whatever development has already taken place. Potentially rich countries, such as Angola or the Southern Sudan, have been devastated beyond recognition and will take decades just to regain their pre-independence levels. Probably the most afflicted are those who have been driven from their homes by war and violence. According to the United Nations Commissioner for Refugees there are currently 18 million cross-border refugees and 24 million displaced persons within their own countries.[102]

In a sense the great powers, with their vast arsenals, are the greatest culprits. During the Reagan Era accelerated military build-up was a deliberate strategy to force the Soviet Union to its knees in economic terms. The tensions and rivalries between the great industrial and commercial powers have often spilled over into Third World conflicts. The Near East, Central America, Korea, Vietnam, Afghanistan and Angola are recent examples.

Moreover, to protect their interests and sell their military hardware, industrial nations have contributed vastly to the military buildup in the Third World and continue to do so. The arms trade provides irresponsible leaders with the means to subdue their subjects and protect their interests. But the lucrativeness of the trade constitutes a disincentive for peace and democracy in the corridors of power. Corrupt and dictatorial regimes, such as Mobutu's regime in the former Zaire, have been propped up for decades by the "democratic" West.

Because it is highly technology- and capital intensive, the arms industry does not create jobs in significant numbers. Because it devours great capital investments it has to find markets for its products. It has also become an important earner of foreign exchange. So all market opportunities are ruthlessly exploited. Profit remains the prime consideration, whether these opportunities are presented by dictators, warring factions or organised crime syndicates. Nobody seems to be concerned about the fact that fields and livestock are destroyed, that masses of people turn into beggars and refugees, and that the First World then has to feed starving people through aid agencies.[103]

In a vicious feedback loop the biosphere again suffers from conflict.[104] The Americans destroyed large areas of forest in Vietnam during the war. The Iraqis incinerated the oil wells in Kuwait when they withdrew. Large regions of the Third World, for instance in Mozambique, Afghanistan, Iraq, Somalia and Cambodia, have become no-go areas because land mines kill and maim people decades after the respective wars have ended. Even in times of peace there are ecological consequences. The military-industrial complex has become a major part of the industrial economy in many industrial countries. As such, it contributes substantially to the depletion of resources and the pollution of the natural environment. The impact of military exercises can be added to the account.

Obviously the use of modern weapons leads to increased pollution. The most terrible situation would materialise if the vast arsenals of nuclear, chemical and biological weapons would ever be used in a full scale war.[105] It is a foregone conclusion that, in such an eventuality, humanity and most other forms of life would be either eradicated or condemned to a painful process of extinction over time.[106] But we do not even have to think of such an extreme eventuality. Chernobyl was a minor catastrophe if compared to the ongoing consequences of nuclear tests conducted on the earth's surface in the former Soviet Union and China.[107] The Russians have dumped nuclear waste taken from nuclear submarines in the Japanese sea as late as 1993. While Western nuclear powers have been slightly more careful, the long term effects of their tests should not be underestimated.

So what must be done?

Humans are not supposed to be the victims of their fate; they can mould their future. They are not supposed to be the slaves of their desires; they can balance costs with benefits. They are not supposed to live at the expense of other humans; they can distribute sacrifices and rewards equitably. They are not supposed to seek satisfaction in material waste and destroy their own habitat; they can keep within the limits of the resources provided by nature and find fulfilment in non-material treasures. The capacity to think, live and act rationally and responsibly belongs to the dignity of the human being. The facts we have discussed present humankind with three basic challenges:

(a) *Population growth* in the periphery must be arrested. Population quantities must change into quality of life for all.
(b) *Industrial growth* in the centre must be arrested. Economic activity must switch from material quantity to cultural quality.
(c) *Participation* in productive activity must be balanced out between centre and periphery.

To the extent that these three tasks are effectively tackled, the other two growth processes, *ecological impact* and *conflict potential,* will subside as a result. In part V we shall attempt to spell out the meaning of these tasks in greater detail.

Revision: Summarise the main ecological concerns of our times.

Application: When you see the muddy water in a river, smell the exhaust fumes in the streets, or find empty beer cans on a lawn, do you associate these experiences with the global processes sketched above?

Critique: (a) "There is enough food in the world for all, it only needs to be distributed fairly." (b) "The poor, who consume next to nothing, are not causing the problem of pollution, but the rich; so why mention population growth at all?" Comment.

Part I - The structure of the global system

Let us summarise

The economy is a subsystem of the eco-system as a whole. Recent developments in this subsystem have seriously begun to disrupt the eco-system of which it forms a part. In section I we gave an *overview* of this problem. We saw that the relative ecological impact in any area is the product of population numbers multiplied by levels of economic throughput. This enabled us to draw up a model which showed the linkages between economic growth, population growth and growth of ecological deterioration. We also drew attention to the fact that current developments are not sustainable in the long term.

To gain an impression of the magnitude of the problem, we gave a few details of the two processes which cause most concern: the growth of the population and the growth of the industrial economy (section II). Both have assumed phenomenal proportions in the last century or two, and the rate is still accelerating. It seems unlikely that the growth of *food production* can keep pace with population growth in the long run. In fact, a food crisis already exists, though it is concealed by agricultural overproduction in the West.

The growth of the industrial economy leads to an increase of the rate at which *non-renewable resources* are depleted, *renewable resources* are over-exploited and the *natural environment* is polluted. The most dangerous developments at present are the greenhouse effect, the destruction of the ozone layer, acid rain, and the accumulation of toxic waste. Competition for dwindling resources as well as increasing economic discrepancies heighten the level of conflict potential in the national, regional and global system. With technological advance, weaponry becomes ever more sophisticated and destructive, adding to depletion, pollution and misery, even when it is not used.

Notes

[1] The idealist vision of liberal economics is expressed by Rostow: "But on one point Marx was right - and we share his view: the end of all this is not compound interest for ever; it is the adventure of seeing what man can and will do when the pressure of scarcity is substantially lifted from him." The "age of mass-consumption", shared by "billions of human beings", will "become universal" within "a century or so" (Rostow 1990:166.)

[2] "It is my thesis that the beautiful days of naive economic consensus are numbered. The growth of consumption has reached its limits. What the richest 10 per cent of the global population consume in terms of energy, area, water, air and other natural goods - directly and indirectly - cannot be extended to the remaining 90 % without the ecological collapse of the Earth. And yet this `standard' is the declared goal of development. Against this ecological collapse the 'invisible hand' is of no avail." Von Weizsäcker 1992:7. See also Ehrlich & Ehrlich 1996.

[3] "Impact = population x affluence x technolgy" Meadows et al (1992:100).

[4] See: *United Nations Environmental Programme 1985: Environmental refugees*. New York: UN.

[5] See, for instance, Meadows et al 1992:104ff.

Chapter 4 - Ecological deterioration | 93

[6] Von Weizsäcker 1990:83.
[7] Macneill 1989:10. For the difference between renewable and non-renewable resources see Meadows, Meadows & Randers 1992:47ff and 66ff.
[8] On living conditions and ecology in Asian "megacities" see *Newsweek*, May 9, 1994, pp 36-44.
[9] For detail see, inter alia, Meadows et al 1992.
[10] See Meadows et al 1992:11 for the growth curve of material standards of living: they all peak before the middle of the next century and decline more or less dramatically thereafter. None of them depicts continuous growth.
[11] For detail see Ehrlich & Ehrlich 1996:65ff.
[12] For an example of such a curve see Broadley 1992:20.
[13] For the concept of "natural capital" (introduced, as far as I know, by Myrdal) see Daly & Cobb 1989:71; Ehrlich et al 1993:1f; MacNeill 1989:27.
[14] Bähr 1992:40.
[15] Broadley and Cunningham expect 8 billion by the year 2022, assuming that the growth rate will go down (1992:8).
[16] Bähr 1992:40.
[17] Gn 1:28 belongs to the socalled Priestly source, which scholars believe to have been written during or after the exile of the Jews in Babylon.
[18] For a powerful statement of the issues see Korten's reflections on the "end of innocence" (1990:213ff).
[19] Bähr 1992:40.
[20] Broadley & Cunningham 1992:19.
[21] Nürnberger 1988:54 - sources are quoted there.
[22] Meadows et al 1992:23.
[23] World Bank 1993:288f.
[24] Meadows et al, ibid.
[25] World Bank 1993, ibid.
[26] For a dismantling of popular myths concerning population and food see Ehrlich & Ehrlich 1996:65-89.
[27] Ehrlich et al 1993:2.
[28] Meadows et al 1992:40.
[29] World Bank 1993:290f.
[30] Broadley and Cunningham 1992:10.
[31] Ibid.
[32] Brown & Kane 1995; Ehrlich et al 1993:3.
[33] Ehrlich et al 1993:4. Cf Meadows et al 1992:48.
[34] Ehrlich et al 1993:27.
[35] Meadows et al 1992:50, quoting the FAO.
[36] See for instance Ruffin & Gregory 1983:451ff, Heilbroner & Thurow 1975:39f, or any other standard text book on economics.
[37] Brown and Kane 1995.
[38] Meadows et al 1992:50.
[39] For an analysis of sustainable fisheries management see Turner, Pearce and Bateman 1994:205ff.
[40] Ehrlich et al 1993:5. For more detail see the Special Report "Fished Out" in *Newsweek*, April 25,1994, pp30ff.
[41] Meadows et al 1992:185f.
[42] MacNeill 1989:5; cf Ehrlich et al 1993:1.
[43] Huntley et al 1989:38.
[44] Elliott 1994:29.
[45] Meadows et al 1992:51.
[46] See Reisner 1993:5.
[47] Recently agricultural subsidies have been cut back in the US and in Germany peasants are paid for not cultivating their land.
[48] Daly & Cobb 1989:273.
[49] For a dismantling of popular myths concerning non-living resources see Ehrlich &

Ehrlich 1996:91-105.
[50] Meadows et al 1992:57ff.
[51] Von Weizsäcker 1990:7; cf Simpson 1990:132.
[52] Meadows et al 1992:85.
[53] Von Weizsäcker 1990:130.
[54] Daly & Cobb 1989:1963.
[55] For the "law of increasing cost" see Heilbroner & Thurow 1975:40.
[56] There are phenomena which suggest that we may have passed the threshold already. In the United States, for instance, the younger generation is no longer as affluent as the post World War II generation and some see a "generation war" in the offing. I owe this insight to a TV programme called *Adam Smith* screened in South Africa in June 1994. Part of the problem is that the American society as a whole has lived beyond its means through continuous and massive budget deficits for too long and somebody will have to foot the bill. That could still be a temporary problem, which is not directly linked to the depletion of resources. But a solution to the problem will certainly demand a tightening of belts.
[57] Daly & Cobb 1989:11. In 1989 known reserves of oil could last for 41 years at current levels of consumption; gas reserves for 60 years.
[58] Between 1860 and 1985 global energy consumption grew sixty-fold. Between 1970 and 1990 it more or less doubled (Meadows et al 1992:66, 68).
[59] Sunter 1987:37.
[60] The Club of Rome was not wrong, as Sunter assumes (1987:37). In the new edition, Meadows et al (1992:183) explain how the glut had come about: it was a typical overshoot oscillation. Shortages in the 1970s led to an overshoot of production capacity in 1982. By 1985 oil prices plummeted. Sunter quotes the same facts, but his conclusion is completely different: "The book ... ought to have been titled *The Growth of Limits* to indicate that we always underestimate the ingenuity of mankind in finding ways around limits" (1987:37). That is the standard technocratic optimism, which simply cannot hold its own against overshoot theory.
[61] Simpson 1990:95f.
[62] Meadows et al 1992:71.
[63] Meadows et al 1992:68.
[64] Brown Weiss 1992:385ff.
[65] MacNeill 1989:10.
[66] Meadows et al 1992:75f.
[67] For the following see Meadows 1992:76ff.
[68] "Renewable energy sources are not environmentally harmless and they are not unlimited" (Meadows et al 1992:76).
[69] Meadows et al 1992:77.
[70] "Ten years ago, nuclear power was till seen by many as the logical replacement for fossil fuels. But the explosion at Chernobyl in 1986 sounded the death knell for the industry." (Brown et al 1993:xvi). This judgment was premature.
[71] "The use of nuclear power as an energy source is sheer folly. It surpasses the ecological impact of large-scale energy production from coal, which is already devastating, by several orders of magnitude, threatening not only to poison our natural environment for thousands of years but even to extinguish the entire human species" (Capra 1983:254). It would take 5 times the time since the Neanderthaler lived 100 000 years ago for nuclear waste to lose its radio-activity (Capra 1983:263).
[72] *Newsweek*, June 13, 1994:20ff.
[73] Bächler et al in *Der Überblick* vol 30/1994, p 8.
[74] Water consumption in South Africa's Gauteng region (Pretoria-Witwatersrand-Vereeniging complex) rose from 90 million cubic metres in 1940 to 800 million cubic metres in 1980 and estimates in that year were that it would reach 7000 cubic metres by the year 2020 (Van der Riet 1980:85). Experts already dream of pipelines from Central Africa to the economic heartland of Southern Africa.
[75] For a dismantling of popular myths concerning pollution see Ehrlich & Ehrlich 1996:125ff, 153ff.

Chapter 4 - Ecological deterioration | 95

[76] Meadows et al 1992:87f.
[77] Meadows et al 1992:89; Von Weizsäcker 1990:112.
[78] Meadows et al 1992:88.
[79] World Commission on Environment and Development (Brundtland Commission) 1990:10.
[80] Remember the Bhopal catastrophe in India in 1984.
[81] The ecological problems caused by obsolete and unclean technologies in Eastern Europe are a nightmare; nuclear plants in the former Soviet Union are notoriously unsafe (witness the Chernobyl disaster); the Ivory Coast lost three quarters of its rain forest in 20 years and the Brazilian Amazonas rain forest is threatened by a severe assault (von Weizsäcker 1990:7); air pollution is highest in Third World cities like Mexico City and Kairo (von Weizsäcker 1990:112); many Third World megacities are losing the battle against urban waste, or soil erosion caused by slums on steep hills; desertification of steppes in the Zagreb has reached catastrophic proportions; deforestation of highlands leads to floods in the plains; water supplies are a headache in vast areas of the Third World, and so forth. While the industrial countries are mainly responsible for air pollution and toxic waste, "about 90% of the extinction of species, soil erosion, deforestation, desert destruction and desertification currently takes place in the developing countries" (Ibid). "Meanwhile, the industries most heavily reliant on environmental resources and most heavily polluting are growing most rapidly in the developing world, where there is both more urgency for growth and less capacity to minimize damaging side effects" *(World Commission on Environment and Development - Brundtland Commission* 1990:5).
[82] The following is based on Brennan 1991:88ff. Cf Meadows et al 1992:89ff; von Weizsäcker 1994:41ff.
[83] Meadows et al 1992:92ff.
[84] Von Weizsäcker 1990:84.
[85] Meadows et al 1992:78.
[86] Triad = US, Europe and Japan.
[87] Meadows, Meadows & Randers 1992:141ff.
[88] Meadows et al 1992:159.
[89] The calamity also exposes the phoniness of pseudo-scientific solutions to the problem of pollution: some scientists have suggested that one could clean the atmosphere by means of giant lasers atop mountain tops - instead of stopping pollution where it is generated!
[90] Cf von Weizsäcker 1994:21ff: "Why pollution control is just not enough".
[91] Brennan 1991:112.
[92] Meadows et al 1992:92.
[93] Brennan 1991:113.
[94] World Commission on Environment and Development (Brundtland Commission) 1990:19.
[95] Brennan 1991:118.
[96] Von Weizsäcker 1994:21f.
[97] Meadows et al 1992:54ff.
[98] Simpson 1990:52ff.
[99] Freeman & Yahoda 1978:346, 353.
[100] Apostolakis 1992:93.
[101] Freeman & Jahoda 1978:355.
[102] Information gleaned from a TV programme screened in South Africa in June 1994. For more detail see Broadley 1992:28ff.
[103] It has come to light, for instance, that South Africa has sold weapons worth $4,5 billion to Iraq during the war against Saddam Hussein. *Mail & Guardian,* July 8, 1994:5. Throughout the time of isolation Armscor has been one of the few growth industries in the country. It is unlikely that for all its commitment to peace the new government, which is in great need of cash, will reduce the capacity of the South African military-industrial complex.
[104] "The arms race ... pre-empts resources that might be used more productively to diminish the security threats created by environmental conflict and the resentments that

are fuelled by widespread poverty" (*World Commission on Environment and Development* [Brundtland Commission] 1990:7).
[105] Albrecht, P & Koshy, N (eds) 1983. *Before it's too late: The challenge of nuclear disarmament*. Geneva: WCC.
[106] Prinx 1984; Malcolmson in Carlton et al 1989:11ff.
[107] 509 nuclear tests with a combined power of 20 000 Hiroshima bombs have been detonated on the Soviet test site Semipalatinsk in Casachstan. As a result, every third child is born either dead or deformed in the neighbourhood. Chinese nuclear tests at Lop-Nor in the province of Xin-Jiang, which continue unabated, are reported to have led to over 200 000 deaths among the local population due to radiation. Cancer and leukemia are the main causes. Radio-activity does not respect national boundaries. With every test in China infant mortality in Alma Ata, the capital of the neighbouring former Soviet Republic of Casachstan rises up to 40%. (Abidin Bozdag in *Der Überblick* 30/1994 p 38).

Causes of economic imbalances

What is the task of this chapter?

In the two previous chapters we designed a model depicting the processes which lead to growing economic and ecological imbalances. To suggest possible solutions, we must understand how such imbalances come about. That is the task of this chapter.

We begin with a *model of causation*. This model can be adapted to all kinds of situations - different continents, countries, regions, sectors of the economy, groups within such sectors, even individual behaviour.[1] The mechanisms discussed occur in many variations, emphasise some aspects more than others, operate in various degrees of intensity and apply to some situations more than to others. Our aim is to provide versatile tools which can be applied to particular cases as needed.

Economic and ecological imbalances, such as affluence, poverty or environmental exhaustion, refer to the availability or the lack of material resources. This is the special domain of economics. In chapter 1 we have argued, however, that economics and ecology must be seen as aspects of a multidimensional whole. Therefore we treat the broader context of *non-economic* factors first (Section II), before we come to *economic* factors (Section III).

Section I: A model of causation

The model of causation presented here is a working hypothesis which further deliberations will substantiate. It has the following elements:

(a) There are structures and processes in the centre which lead to the rapid growth of economic potency and therefore to the advantage of the centre over the periphery. The cumulative effect of productively invested capital or the development of sophisticated technology are cases in point. We call these **structural C-factors**.

(b) There are structures and processes in the periphery which lead to the inhibition of the growth of economic potency and therefore to the structural disadvantage of the periphery. Where there is a lack of capital, for instance, capital cannot grow. We call these **structural P-factors**.

(c) There are structural mechanisms which cause the interaction between the centre and the periphery to work to the greater advantage of the centre and the lesser advantage, or even the disadvantage, of the periphery. The interaction between a more powerful and a less powerful partner we call "asymmetrical", "unbalanced" or "vertical" interaction.[2] The centre can, for instance, sell its manufactured goods to the periphery on credit. The debt has to be repaid with interest. If interest rates rise substantially, the periphery is in financial trouble. We call these **structural C/P-factors.**

These three kinds of factors reinforce each other in a series of feedback loops. They can be "vicious circles" or "golden spirals". Myrdal (1957) has introduced the concept of "circular and cumulative causation",[3] futurologists like Meadows *et al* speak of "feedback loops".[4] Once institutionalised, structural factors operate almost automatically. We can compare them with machines which have been constructed to perform certain functions.

Of course, a machine can also be constructed differently. It is human beings who set goals and who construct machines to achieve their goals. They also determine how machines should operate. They can slow down a machine; they can let it move in a different direction; they can decide to do without the machine. So there is another set of factors, namely human decisions and actions. We call them *volitional factors*.

Volitional factors depend on the perceptions, assumptions, beliefs, values, norms, goals and procedures which the groups concerned have internalised in the socialisation process or acquired through experience. These again depend on their basic convictions and interests. We refer to this whole realm of reality as "mental structures", or "collective consciousness", or "mindsets". Part II will deal with this dimension of human reality.

Volitional factors also operate on the three levels mentioned above:

Chapter 5 - Causes of economic imbalances | 99

(a) There are volitional factors in the centre which enhance the growth of economic potency, that is, **volitional C-factors,**
(b) there are volitional factors in the periphery which inhibit the growth of economic potency, that is **volitional P-factors,** and
(c) there are volitional factors in the interaction between the centre and the periphery which lead to the greater advantage of the centre and the lesser advantage (or disadvantage) of the periphery, that is, **volitional C/P-factors.** Diagram 5-1 shows how these factors are related to each other.

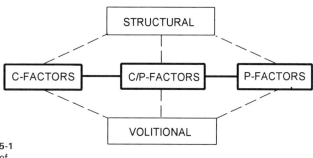

Diagram 5-1
A model of causation

Note that there is a dimension which is not included in the model, namely the dimension of *time*. Obviously all these factors operate in the context of continuing history. And because processes such as population growth, capital growth or technological development have a cumulative effect, they tend to accelerate. We speak of "exponential growth" in such cases.[5]

This model is of great importance for understanding the debates on economic development and international trade. The *liberal* school tends to emphasise the contrast between factors inherent in the centre and the periphery, that is C- and P-factors, and to neglect interaction factors[6]. The *radical* schools (whether the Marxist or the dependency version) tend to emphasise interaction factors, that is C/P-factors, and to neglect inherent factors.[7] In chapter 10 we shall discuss the significance of such ideological biases. Because ethics is concerned with justice, it has also tended to emphasise C/P-factors at the expense of C- and P-factors. Our model combines the valid concerns of both perspectives.

Likewise, the *materialist* approach emphasises structural mechanisms at the expense of volitional factors, while the *idealist* approach concentrates on volitional factors at the expense of structural mechanisms. The former says:

minds are formed by structures; so change structures, and the minds of people will follow suit. The latter says: change can only be effected by people; so change the minds of people and structures will follow suit. As we shall argue in chapter 6, factors on both levels interact in a complex and reciprocal relationship and again our model integrates the valid arguments of both approaches.[8] A similar distinction is made been values and technicalities. Theologians, among others, emphasise the former, and economists the latter.[9]

Revision: Make a sketch of the model of causation and see whether you can explain all the different factors.

Application: Do you know of a slum or a particularly poor suburb in your neighbourhood? Try to figure out which kinds of factors, as depicted in the diagram, could have led to the formation of these conditions.

Critique: (a) "Everybody knows that poverty is caused by oppression and exploitation. Why do you try to conceal the obvious behind complicated 'models'?" (b) "Humans will always be different; there are different tastes, different work habits, different expectations. Why don't you allow people the freedom to live the way they want to live?" What do these two views have in common? How would you respond?

Section II: Non-economic factors

The social structure

Development theory has increasingly recognised the importance of efficient institutions. To a large extent, development means institutional development. The most progressive laws, for instance in South Africa or Brazil, do not mean much if the law enforcement agencies or local authorities are not capable of their effective administration.

In chapter 3 we have mentioned the fact that in industrial countries we find a large scale and complex type of social organisation which integrates great numbers of unrelated and culturally divergent individuals and groups into a highly productive economy. In rural Africa, in contrast, we find small, simple and culturally homogeneous communities, which live from subsistence agriculture.

Social organisation in the periphery remains more or less constant over decades, if not centuries, while the organisation of the centre constantly unfolds. Organisation is not just an outward arrangement of objects, but happens in the *minds* of people. They have internalised certain patterns of behaviour and quite unconsciously act according to them. People in cities are used to complex relationships and constant flux, while rural populations follow relatively stable traditional patterns. Centre people are conditioned by the values needed by an industrial society, such as punctuality, efficiency, and precision, while peripheral people are much more relaxed in their approach to life. It does not matter to them whether a meeting begins an hour late, or whether a furrow is not straight. All this has a great impact on production.[10]

Organisation means *integration*. In the periphery we find small, autonomous social units which may be tightly organised within themselves, but which are relatively isolated from the centre and from each other. The higher we go up in the hierarchy towards the centre, the more complex and compact the networks of relationships become. In the periphery the social hierarchy peaks in the elders the of clans and tribes, though it extends beyond the living to the deceased; in the centre there is a system of interlocking elites linked by highly efficient *communication networks*. These elites operate in specialised fields such as business, politics, science, education, research, even religion. As we have seen in chapter 3, the whole system is organised radially from highly integrated centres to loosely attached peripheries, while between the peripheries themselves there is little interaction. This phenomenon can easily be manipulated by the centres to their own advantage, for instance through a divide-and-rule policy.[11]

The social structure of industrialised regions is remarkably similar the world over: individuals are basically on their own and free to establish relationships as they wish, to move about within the economy, and to develop initiatives according to their gifts and desires. We call this *social mobility* - whether horizontally between different pursuits, or vertically between status ranks. The results are high levels of *specialisation* and *competition*. In contrast, peripheral regions display a great variety of forms, such as the caste system in India, the feudal system in much of Latin America, or the tribal system in Africa and Polynesia. All these systems bind the individual into predefined roles, statuses and relationships and thus inhibit the development of individual initiative, specialisation and mobility. While they grant belonging and security, they are hamstrung by collective land tenure, taboos, dependence on divination, inherited authorities and religious observances. We shall come back to these factors in chapter 8.

Economic potency is directly linked to the level of development, integration and efficiency of the network of social institutions. The current emphasis is on "lean and clean" institutions. Unwieldy bureaucracies and pervasive corruption destroy the efficiency of the institutional framework of a society. Social evolution in the centre has created an economically, politically and professionally strong middle class from which the ruling elite is derived, to which it returns after being in office, on which it depends, to which it is accountable. Instances of corruption cause social scandals, loss of legitimacy and legal action. State bureaucracies are balanced out by a powerful private sector which sets standards of efficiency and profitability. There is a strong trend towards privatisation of state enterprises to improve efficiency.

Because civil society is relatively undeveloped and economically weak in peripheral countries, post-colonial elites tend to consolidate their power and their privileges in the state apparatus. The colonial and post-colonial impact has severely disrupted traditional institutions, left a vast vacuum of power and created a small privileged elite who has appropriated that power. A vast abyss of influence and expertise opened up between the lofty heights of leadership and the bulk of the population. The absence of effective democratic controls led to the development of institutionalised corruption. Lucrative channels of illicit income are protected rather than exposed. Lack of accountability leads to bureaucratic inefficiency, financial mismanagement, and misallocation of scarce resources for the self-aggrandisement of rulers and their clienteles.[12] The following quotation may give an impression of what happens in what is called a "prebendal system":

> In a prebendal system, public office is treated as a personal (revocable) grant from the ruler. The beneficiary is expected to exploit the "rent" potential of the office ... and must reciprocate by giving the ruler absolute loyalty and a share of the "rent". In turn, he subcontracts the granting of subordinate prebends, subject to appropriate clearance ... The sanctions on a non-conforming elite member are normally loss of the prebend; sanctions on non-conforming citizens are starker. Overall enforcement is often provided by the army or by paramilitary "security" forces.[13]

Of great importance is the degree of *economic freedom* in a society. "Free enterprise" has been regarded as the basic prerequisite for a vibrant economy in affluent societies. Even where a welfare state transformed much of the proceeds of the private sector into social benefits, entrepreneurial initiative remained the corner stone of the economy. Third World governments, in contrast, have tended to try and run the economy to a greater or lesser extent. Where private entrepreneurs simply do not exist in sufficient numbers, the initiative of the state seemed to be indispensable. By and large the results have been negative. The disaster of former socialist states in Eastern Europe finally made it clear that economic power must be given back to the people. As we shall argue later, the role of the state is only to make certain that competition does not lead to undue concentration of economic potential and large scale marginalisation.

Culture and education

The social system is intimately connected with the culture of the population. Cultural factors have, in the past, often been overemphasised as causes of development or the lack of it. In the last few decades societies with widely differing cultural heritages have managed to enter the industrial age. Yet it would be wrong to ignore culture altogether. The technological civilisation of the centre is subject to accelerating change. Progress is the watchword, not tradition, and the future is more important than the past. In the periphery, the culture is much more stable: the heritage of the fathers sets the pace, not the innovative spirit of an avant garde; the wisdom of the elderly carries more weight than the impatience of the youth; the proven is sought, not the novel. Again we shall discuss these phenomena in chapter 8.

Culture is transmitted through *education*. But educational principles are also derived from relevant cultural values. In the centre, children are challenged to perform, to compete, to excel. Their curiosity and activity are evoked and encouraged. In the periphery, children are socialised into relatively stable inherited patterns. Because social conditioning happens during the first years in life, differences in collective mentality are programmed into each new generation and the cultural discrepancies are growing steadily.

Due to the migrant labour system in South Africa, the economically productive population moved to the centres, while the old, the sick and the young remained behind. The children were looked after by worn-out grannies with bygone mindsets. The youth was not socialised into the modern world of active adults. Being relatively isolated, they developed an inflated perception of their performance. Never having faced the competitive spirit of the urban environment, they tended to believe in luck rather than hard work.

Obviously there are also external factors at work in the field of education. The entire social atmosphere is different. While pre-school children in the periphery learn to hunt with slings, centre kids play computer games. Children in the centre enjoy all the educational facilities they need to develop their gifts. The meagre educational resources of Third World countries are concentrated in prestige universities accessible only to the elite. In peripheral regions there is a chronic shortage of financial resources, personnel and training. Class rooms, furniture, books and stationary are all in short supply. Many children come to school without having had breakfast.

Often the teachers themselves have not learnt to spell or to add correctly. Wrong educational conventions are spread and perpetuated through inadequately trained teachers.[14] Mathematics and science are not taught very proficiently, if they are taught at all. As a result, the weakness in these subjects perpetuates itself from generation to generation. Class rooms are overcrowded. If a modern curriculum is offered, it does not fit into the life world of the children. They learn abstract knowledge by heart, cannot utilise it in their lives, and promptly forget it again.[15] Parents are unable to support the efforts of the teachers at home because their educational standards are lower than those of the children. Only a fraction of the young generation attends school, only a fraction of that fraction continues beyond the first few years.

This again has *psychological* repercussions. The goal of education in the centre is to make children inquisitive, critical and self-reliant in their judgments and decisions. Freedom, initiative and responsibility are the watch words. In many traditional cultures an authoritarian education leads to a dependent personality structure. Children are supposed to obey and keep quiet, not to ask questions. In later life, people with this kind of education do not feel competent to assume responsibility but depend on their superiors for leadership. The emphasis on tradition saps their self-confidence. They have never learnt to develop and utilise their own possibilities.

Illiteracy leads to narrow horizons. One is insecure outside the restricted circle of one's familiar world. One is suspicious towards the unknown. The spread of the transistor radio has opened up these horizons somewhat, but it cannot substitute for profound educational formation. When youths break out of the authoritarian structure, they are not sufficiently equipped to fend for themselves and easily lose their bearings. They have not learnt to take over responsibility for their lives and their environments, nor to adapt to rapidly changing circumstances in a creative way. Anomie and delinquency may be the result.

Language too may become a formidable barrier. Traditional Third World languages have not normally developed the conceptual tools to deal with mathematics, the natural sciences, technology, commerce, and communications networks. If English is used in the public realm, those who do not speak it as their mother tongue are at a disadvantage. And how will English as the language of the internet and the flood of American information found there effect future developments? The principles of equal dignity, self-respect and cultural identity seem to demand mother tongue preservation and education. However, English speakers may simply outperform non-English speakers in

Chapter 5 - Causes of economic imbalances | 105

the modern world and mother tongue preservation may appear to be economic romanticism.

Stability and adequacy of sustenance

The centre population is privileged enough to enjoy ample *social securities* such as unemployment benefits and pension schemes. Financially, people do not have to depend on the extended family. In poor societies there is no money and no framework to cater for such institutions; so the extended family has to provide economic security for its members.

This leads to greater social cohesion, which is essential in a marginalised and vulnerable community. Belonging and protection are of great psychological and sociological significance. Traditional Africa is often envied by Westerners for the community spirit prevalent in a tribal context. But economically speaking, such a system has adverse effects. The large network of social obligations prevents people from enjoying the fruit of their labours. Every financial gain immediately melts away into the many claims within the extended family. As a consequence, people lose their motivation to produce. Men and women, who go to the cities for work, often abscond to escape the dragnet at home.[16]

Biological factors also play a role. In the centre, nutritional standards, recreational facilities and health services reach high standards. In the periphery the population is often dismally poor. Malnutrition can lead to permanent brain damage among children. Vitamin and protein deficiencies can lead to apathy and lethargy. Weakened bodies have to walk great distances, carry heavy burdens and perform hard labour on the fields. In tropical countries a disagreeable climate takes its toll. Dangerous diseases, lack of hygiene and medical facilities lead, in some areas, to the permanent sickness of great parts of the population. In some African countries there are already thousands of AIDS victims and AIDS orphans to be looked after. Where so little energy is left, people plod through life listlessly and often find consolation in alcohol and drugs. Obviously this further saps their energies.

In traditional societies, *women* usually bear the brunt of the situation. They carry the daily water supply for the family on their heads for kilometres, till the land, repair the homestead, cook the meals, go through regular pregnancies, care for the toddlers, spend whole nights to fulfil ritual obligations. Their work is taken for granted and there is no escape from the endless drudgery. Who can blame them if they take things as easy as they can! There is no rush; one walks and works with an air of measured dignity, relishing the little pleasure of chatter wherever the opportunity presents itself.

Population growth

The elite in the centre keeps its numbers low. That is one of the secrets of its strength. Economic potency is not diluted through frequent divisions of the estate, nor by the demands of too many dependents. Parents want their children to reach at least their own level of wealth and proficiency, so they spend large sums of money on education. Because higher education is costly, rich people cannot afford too many children. People have the information, financial means and self-control to prevent pregnancies.

For many centre couples, children are a source of frustration because life among modern elites is hectic and filled to the brim with other ambitions, obligations and sources of satisfaction. Children have to compete with other demands on time and energy. For a professional woman, especially, it is a great sacrifice to become pregnant, leave her job and raise children. There are no grannies around to look after the kids. A mother's status in society may immediately drop from that of a successful social figure to that of a housewife. All these factors contribute to the trend towards smaller families in economic centres.

In the periphery the situation is quite different. In pre-technological cultures, progeny was an economic asset because it supplied labourers and warriors. The costs of having another child were negligible compared with the gain. Life expectancy was low. The survival of the clan was under constant threat because of infant mortality, famine, disease, frequent wars and a hostile natural environment. Births had to make up for all these losses. It was natural to attach religious significance and social status to a large progeny under such circumstances.

In modern times the situation has changed dramatically. In many instances, rural areas have reached their carrying capacity and population growth leads to increasing poverty. The result is migration to the cities. Under urban-industrial circumstances, large families have become an economic handicap.[17] To feed a large family both parents have to seek employment. The children are insufficiently cared for. They may even be taken out of school prematurely to help earn a living; some never make it to the school at all. Their education may also be disturbed by unrest. Their chances of finding lucrative employment on this basis are slim.

What do they do with their time and their youthful energies? They hang around in dreary streets and receive a street corner education. They begin to smoke, use drugs, get involved in gang warfare and become sexually active at an early age. In marked contrast with the situation in undisturbed traditional cultures, girls become pregnant shortly after puberty and regularly after that. Immature young men abandon their sweethearts once they are with child and

Chapter 5 - Causes of economic imbalances | 107

feel no further obligations. Illegitimate children become the rule rather than the exception. Single parent households headed by grandmothers often perpetuate themselves over generations and constitute the poorest of the poor.

In rural areas the situation is not much different. There may already be too many people employed on the land, a situation called "disguised unemployment". Additional workers are not an asset but a drain on resources. In a modern state, civil protection has been taken over by the army and the police force. Due to modern medicine, the mortality rate has decreased. Most of the arguments for large families, therefore, are no longer applicable.

Yet because traditional cultures are inherently conservative, the old value system lingers on long after it has become counterproductive. One does not dare to manipulate sanctified natural processes. One believes that children are a gift and that pregnancies should not be avoided. Children still signify social status. Men consider a large progeny to be a sign of virility. The status of women too depends on the size of their families. Female emancipation is largely unknown or perceived as urban decadence. Sex education, family planning devices and discipline may all be lacking. It is taboo to talk about sexuality. These factors combine to bring about rampant population growth.

It must be conceded that there are still substantive reasons for large families among the poor today. The most important factor is the *absence of old age security*. Among the urban poor, two or three children may be too poor to support their elderly parents. Parents also cannot be sure how many of their children will abscond. Secularisation, breakdown of parental authority and dissolution of family loyalty have left contemporary youths with few communal scruples. In order not to be stranded in old age, therefore, parents tend to maximise their progeny. The new phenomenon of AIDS eradicating the productive age groups of a whole population leads to immense demographic distortions.

In a generally cheerless life, children are also one of the few remaining sources of joy. Similarly, when you find yourself on the lowest stratum of a competitive society you only have your progeny as a source of pride. They are also the only persons which you are entitled to lord it over to boost your self-image. One can also make them take over household chores at an early age. Compared with these benefits, the financial sacrifices seem to be slight. Neglected children tend to find food somewhere in the neighbourhood. Education is not necessarily a priority.

However, all these benefits may seriously backfire. Abandoned mothers are forced to seek employment and often dump their children in the laps of their grannies. The latter may not necessarily be competent to deal with modern adolescents. Street children turn into roaming gangs who find methods of obtaining food and money. Where there is large scale unemployment, the younger generation may suddenly pitch up at home and consume the meagre pensions of the elderly, assuming that there are pension schemes in operation.

In the society as a whole, overpopulation, environmental destruction and the decline in food production have made population growth one of the most serious economic predicaments of the Third World. The *relation between breadwinners and dependents* becomes more and more precarious. Paradoxically this is a concern shared by centre nations. But while in the centre the aged begin to outnumber the productive population, in the periphery it is the youth. More schools, medical facilities, law enforcement agencies and employment opportunities are needed in the periphery than in the centre. Yet the periphery lacks the means to provide them. Where there is economic growth at all, it is neutralised by growing numbers. Where the economically most active population is eradicated by AIDS, as in some African regions, the problem is hugely compounded.

In rural areas overpopulation leads to ecological destruction. Those who migrate to the cities often end up in slums, where they are exposed to polluted air and water, unhygienic living conditions, noise, filth, lack of living space, delinquency and crime. The general condition of the population spirals downwards, and this again saps its energy and determination to address the problems.

Revision: Briefly describe some of the "non-economic factors" which contribute to economic discrepancies between centre and periphery.

Application: Check whether you can trace some of these factors in your own social environment.

Critique: (a) "Far from being a set of cultural and economic disadvantages, African traditions could teach the derailed Western mind once again what it means to be human." (b) "Your description proves that the Third World will depend on the benevolent leadership of the First World for a long time to come." (c) "Some ethnic groups simply do not have the genetic prerequisites to cope with a modern economy, otherwise they would have followed the West long ago." How do these statements differ from each other? How would you react to each one of them?

Chapter 5 - Causes of economic imbalances | 109

Section III: Economic causes - factors of production

So far we have dealt with the broad environment in which the economy operates. Coming to the more specifically economic picture, we begin with factors located *within* the economic structure of centres and peripheries respectively. They are largely connected with what economists call "the factors of production". In contrast, section IV will deal with factors of a structural nature which are located in the *interaction* between centres and peripheries.

Factors of production

Production is a complicated process in which a number of *factors of production* are combined in such a way that they yield the desired results. They are depicted in diagram 5-2. We have redesigned the outdated scheme found in economics text books to obtain a more comprehensive picture.[18] At the top and at the bottom of the diagram we place the basic prerequisites of production, namely time, space and infrastructure. On the left side we have mental and social factors and on the right hand side material factors of production. To get an impression of what is at stake let us go through this list.

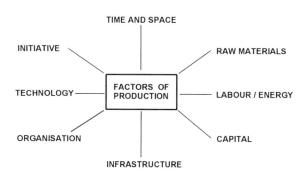

Diagram 5-2
Factors of production

Time, space and infrastructure

Time and space are needed for anything to happen at all. In a shrinking world, space is increasingly becoming a problem. Geographically, a rural periphery has usually more *space* at its disposal than the centre because, in comparison with other factors of production, there is a lot of land. Therefore production is "extensive", which means that labour and capital are thinly spread over the surface of the land. A good example is cattle ranching. There are exceptions, of course, for instance when rural peripheries are heavily overpopulated. In contrast, geographical space is precious in the centre and needs to be utilised to capacity. Thus production is "intensive", which means that labour and capital are concentrated. Think of all the high rise buildings which go up in the "central business districts" of large cities because land is so expensive.[19]

The same is true for *time*. Although life expectancies have risen, lack of time has become a major preoccupation among the leaders of a modern economy. The periphery has "more time" at its disposal than the centre - not in the sense that it has more hours each day, but in the sense that less events and actions are packed into those 24 hours. Where there is disguised unemployment, life is relaxed. It may even be boring. Where there is overt unemployment, people often do not know how to spend their time. In the centre of the centre time is a precious asset which is planned and fully utilised, whether for work or for pleasure. In the periphery economic processes are slow and intermittent. In the centre, they are fast and continuous. The result is that much more is produced per unit of time.

But there is also another side to the coin. When the peripheral population ends up in the centre, it cannot afford the high price of space. Rich people have much more space to work, to live and to relax. Just compare the houses, gardens and golf courses in affluent suburbs with the crowded conditions in slums; or compare a modern factory with a workshop located in an old garage. Except where they are driven by competition and ambition, centre employees also have more hours of leisure. In many parts of the world peasants have to work from sunrise to sunset while their urban counterparts knock off at five. People who have had time to relax again may be able to work more intensely and efficiently.

At the bottom of diagram 5-2 we find the term *infrastructure*. It refers to goods and services which create the essential environment for production. They include transport, communication, energy, water, financial institutions, and so on. Some scholars add "social infrastructure" here: schools, recreational facilities and hospitals. Usually social infrastructure is supplied by agencies other than the productive enterprises themselves, for instance by the state. All these goods and services are more developed in urban-industrial centres than in rural peripheries. A builder in the centre, who runs out of rafters, for instance, picks up the telephone and within hours the material is on his premises. His colleague in a remote rural area without proper roads and postal services may have to wait for weeks.

In modern times *telecommunications, information and knowledge* have replaced "capital, land and labor as the primary defining feature of the production process in advanced industrial capitalism."[20] Top industrial, commercial and financial enterprises communicate with each other through cellular telephones, fax machines and electronic mail - all linked up in global satellite networks. There is a continuous flow of information on global markets, share prices and financial policies, allowing instant reactions to changing situations. Due to increased mobility, aspects of production can be farmed out to the most profitable locations and translocated at short notice.[21]

Due to lack of funds and expertise, peripheral production in the Third World is not integrated in this global network. But even in highly industrialised countries, power differentials make themselves felt. "As knowledge-intensive industries are transformed into technologically based oligopolies, new barriers to entry are likely to be posed for latecomers, and the ability of small- and medium-sized enterprises to retain their independence diminishes." They are drawn into a "global network of interfirm linkages" by collaboration agreements, which "structures their future development options."[22]

Consumers too have no commensurate clout.[23] George Gilder maintains that "rather than pushing control to Big Brother at the top as the pundits predicted (Orwell for example) the [electronic] technology by its very nature pulled power back to the people."[24] Nothing could be further from the truth. In fact, access to computer technology is restricted to an elite, there is a new wave of manipulators around, and "the people" are more manipulated in the information swamp than ever before.

This brings us to factors of production in the more narrow sense of the word. On the left hand side there are *mental* factors, on the right hand side there are *material* factors. Because mental factors refer to collective consciousness, they also operate as social factors. We discuss them one by one.

Mental and social factors of production

Among the mental factors *initiative* is the most important.[25] It is usually discussed under the heading of "entrepreneurship". Where people develop initiative, most other constraints can be overcome. Because the development of potency is a dynamic process, a constant drive to develop new methods and reach greater objectives is a great asset. Initiative depends on self-confidence, will-power and independence. But underlying these are convictions, interests, values, norms and goals. We shall come to these in Part II.

The situation in the centre is more conducive to the development of initiative than the periphery. In the periphery life is quiet; in the centre it is hectic. In the periphery social relations are more important than progress; in the centre competition forces people to outperform others. In the periphery initiative is viewed with suspicion; in the centre it is rewarded with success and prestige. People with zeal do not find fulfilment in the periphery. So they move to the centres, where they develop their gifts. Those who remain in the periphery, are no longer challenged to excel. Economic systems run by the state tend to suppress initiative. This is one of the reasons for the failure of socialism in Eastern Europe and for economies in the Third World where the state has been the main actor.[26]

When it comes to *technological expertise and innovation*, the picture repeats itself.[27] To be effective, technological advance must be "embodied" in capital, that is, it must be translated into equipment.[28] Obviously the level of technology is a major factor in the development of economic potency. It makes a difference whether you count with your fingers or use a computer. In the centre, the systematic development of technology has long been institutionalised: competition compels firms to improve their methods and train their workers. Technological advance is cumulative; therefore technology develops faster and faster. A new computer may be outdated within a year. In the periphery, people tend to cling to traditional methods. They also lack the means to develop more sophisticated gadgets and techniques. As a result, production in the periphery is lagging behind more and more.

The next mental factor is *management and organisation.* In the periphery, one finds relatively simple patterns of organisation: to run a shop or a small farm is not too difficult. In the centre, production processes become larger and more complex; hence they require sophisticated management skills. The same is true for social organisation. In a rural village everybody knows everybody else - and has done so from childhood. Just compare this with the intricate system of impersonal relations which are typical for industrial and commercial centres!

Material factors of production

Among the material factors, *raw materials and energy resources* seem to be most basic, because human beings cannot create something out of nothing.[29] Fertile land and a good climate led to the emergence of centres in ancient Mesopotamia and Egypt. In modern times, centres developed in the vicinity of mineral resources such as iron ore and coal. More recently the geographical location of resources has become less important. Oil wells have often been far removed from their places of utilisation; modern production uses less and less bulky material, and fast transport routes have been developed to secure those raw materials which are still needed. Switzerland and Japan are examples of countries which have attained high levels of economic development although they have not been endowed with large deposits of mineral resources. In contrast, Third World countries endowed with rich natural resources are usually not able to utilise them for their own production, but have to sell them to industrial nations. The latter process the materials and sell them back to the Third World as manufactured goods at a price which reflects the "value added".

Labour as a factor of production has two functions.[30] In the first place it is a supplier of *physical energy*. In modern times this function becomes less and less important, because energy can be generated more cheaply and more abundantly by machines utilising coal, oil and nuclear reactions. Economic centres have developed these energy-generating technologies to high standards. Therefore they need less and less unskilled labour.

Secondly, labourers are carriers of *expertise*. Expertise is linked to technology which we have discussed under the previous heading. In modern times it is the quality, not the quantity of work that counts. Machines need to be designed, operated, maintained and repaired. To perform these tasks workers learn to do specialised jobs. They must be motivated to be efficient and reliable. Skills take time and effort to acquire; so they are scarce and attract higher salaries. Labour becomes an expensive factor of production, with the result that more is invested in the training and remuneration of workers, but fewer workers are employed. More and more unskilled labour becomes redundant. In this way whole sections of the population may become marginalised, and because peripheral workers are usually unskilled workers, they are the first to suffer.

Capital is investment in tools which help humans to produce, such as machines, warehouses, vehicles, fertilisers and so on.[31] The formation of capital is subject to technological progress. The effectiveness of capital depends on the amount of capital invested and its productivity, or the "capital-output ratio".[32] Obviously there is a close link between capital, technology and expertise, therefore economists often speak of "human capital" when they refer to education and training. Ecologists, on the other hand, powerfully argue that capital should include not only "humanly produced means of production", but also *natural resources,* thus any "stock that yields a flow of goods and services."[33] This means that a country which chops down its forests and calls this activity "economic growth", actually consumes its capital stock.

The higher the level of productively invested capital, the higher the *productivity of labour*. A hundred labourers with sickles cut less wheat than one labourer driving a combine harvester. The former we call labour-intensive production, the latter capital-intensive production. Productively invested capital snowballs. The reason is that capital helps to produce a surplus beyond the immediate needs of the producer. Labourers with sickles need all their income to feed their families, while owners of mechanical harvesters produce more than they need for their upkeep. After paying off their machines, they can use the proceeds to buy more such machines. The growth of the capital-labour ratio is called "capital deepening".[34]

As a factor of production, capital is plentiful in the centre and it is growing faster and faster. Its technological quality also improves rapidly. There is little capital in the periphery, and what does not exist cannot grow. The sophisticated machinery of the centre often cannot be utilised under peripheral conditions and the development of "intermediate" or "appropriate technology" has become an important concern.[35] Moreover, the centre is able to produce its own capital goods, such as machines, while the periphery has to import them from the centre, which often causes balance of payments and debt problems.

To summarise, we find a greater *quantity,* greater *quality* and greater *variety* of factors of production in the centre than in the periphery and the difference between them grows faster and faster. The available factors of production are also more efficiently utilised in the centre than in the periphery.

Distribution

On the one hand, distribution depends on the easy flow of goods and services from producers to consumers and the flow of money in the opposite direction. The catalysts of this flow are *commercial undertakings* which form the link between producers and consumers, and institutions which provide *services,* such as channels of communication, transport facilities, credit facilities, and so on. In the centre both of these catalysts are highly developed, while they are often absent, rudimentary or inefficient in the periphery.

More fundamental to distribution, however, is the fact that production depends on market demand. What cannot be sold, also cannot be produced, otherwise producers go bankrupt. Market demand depends not only on *need,* which is plentiful in the periphery, but also on *purchasing power,* which is not so plentiful. Purchasing power again depends on productivity and the distribution of earnings. We shall deal with this aspect in chapter 12.

There is very little purchasing power in the periphery because, compared to that of the centre, its contribution towards production is meagre. Where productivity, income, purchasing power and, therefore, market demand are lacking, you cannot sell much. As a result, centre producers find their chief markets not in the periphery, where the need is, but in the centre, where the purchasing power is. This is a "vicious circle" which keeps the peripheral economy at low levels. In contrast, the centre is awash with productivity, thus income, which can be tapped to boost production. Instead of vicious circles we have "golden spirals" which force the economy up and up.

Some of the factors mentioned under production also have consequences for distribution. In the centre, there are "bottlenecks" (= constraints) caused by the lack of skilled labour, expertise and management, which push up the incomes of professional elites. Semi-skilled workers in centre regions are organised in trade unions; thus they are able to negotiate from a position of strength. Lower classes at least have the franchise in industrialised nations. They can form socialist parties which lobby for greater equality of income.

In peripheral regions, there is an oversupply of unskilled labour. Much of it takes the form of "disguised unemployment" in agriculture, which means that the work done by five people could be done by two, if they used all their time and energy.[36] Where there is no demand for labour, people have to be satisfied with low wages, or lose their jobs. The opportunities to organise themselves in trade unions and negotiate on a collective basis are limited, and democratic institutions are either absent or do not operate effectively.

Revision: Make a sketch depicting the main factors of production and say briefly why the centre is better off in most regards than the periphery.

Application:20You go through a busy street and notice that some people in rags are begging for work while others with brief cases rush to their destinations under great stress. Can you use some of the insights gained in this section to explain this phenomenon?

Critique: (a) "In your view the misery of the periphery popu~lation is all its own fault. In fact, the centre takes more than its share of the common product and leaves the periphery destitute." (b) "You seem to say that it does not matter whether one works hard or not. If one happens to belong to the centre population one will automatically be wealthy. In fact, prosperity cannot be generated without serious effort." Which of these two statements do you consider to be nearer to the truth?

Section IV: Economic causes - asymmetrical interaction

Power in the economy

So far we have dealt with factors which are located within centres and peripheries respectively. Now we come to factors which belong to the interaction between centres and peripheries. Such factors include colonial oppression and exploitation; the suction of peripheral factors of production into the vortex of the centre; the penetration of the periphery by the centre economy; unbalanced trade relationships, the debt crisis and the abuse of power.

Where production cannot keep pace with growing need, people are bound to end up in poverty. For a long time, therefore, development theory has concentrated on the necessity of upgrading factors of production. But the interaction between centre and periphery is just as critical. This dimension has been emphasised by Marxists and dependency theorists during the last few decades, but its importance has often been ignored, even denied, by main line economists.

The interaction between centre and periphery is "vertical", or "asymmetrical". The word asymmetrical means "unbalanced". One would expect that interaction between unequal partners would *balance out* their potentials: when you connect two water containers, the water will flow from one to the other until it reaches the same level in both. The theory of equilibrium between supply and demand on a free market and the theory of comparative advantage, which says that all partners must benefit from free trade, sustains the perception that market interaction indeed creates such a balance.[37]

Unfortunately, precisely the opposite happens in reality.[38] Potencies are flowing *from the periphery to the centre*. Let us use another insight from physics to make the point, namely the so-called "parallelogram of forces". If a ship is loaded more heavily on one side, it will tilt and even the part of the load on the other side may slide and make the imbalance worse. If a heavy, fast moving lorry has a head-on collision with a light, slow moving passenger car, the car will be carried a long way in the direction of the lorry, not the other way round.

Imperialism

But why should discrepancies tend to become worse when economies interact with each other? The standard explanation of Marxist theory is that the rich abuse their power to exploit the poor. For reasons to be discussed in chapter 10, this argument is more popular among disadvantaged population groups than among elites. And indeed, conquest, plunder, oppression and exploitation have taken place on a massive scale. In classical empires the imperial centre may plunder the conquered regions and exact tributes, taxes, labour and raw materials from them. The backbone of this system is the military and organisational superiority of the centre over the periphery. In such cases one can speak of overt oppression and exploitation.

Such practices have caused great distress to oppressed populations, but they do not adequately account for the discrepancies which exist between centres and peripheries in modern times. Imperialist exploitation causes havoc to a peripheral economy, but it does not lead to sustained growth of wealth in centre regions. Extorted wealth is usually consumed or wasted rather than channelled into a continuous flow of production.

Portugal exploited its colonies, but it is still relatively poor today. Similarly, Spain ransacked Latin American civilisations but remained a backward country until it jumped onto the band-wagon of modern European developments. Sweden and Switzerland, in contrast, never had colonies to exploit, yet they are among the richest countries in the world today. Japan reached its present wealth only after its colonial empire had been smashed in a devastating war. Germany had colonies only for a short time, they were economically not very lucrative and its economy was devastated twice after its colonial spell had ended. Yet the country possesses Europe's most powerful economy. In contrast, Britain's economy has been ailing after World War II, in spite of the fact that it was at the head of a global empire and was victorious in the war. Its current prosperity developed only after the Empire had been shed.

These examples show that colonial exploitation *does not lead to long term wealth*. But the transfer of potencies from the periphery to the centre can also take place through more subtle mechanisms and without the use of military might. Let us look at some of them.

Suction of factors of production into the centre

In chapter 3 we have mentioned some of the reasons for economic concentrations to materialise. We have seen that centre growth is a self-sustaining mechanism. It is fed by greater market density and factor variety, shorter transport routes, more efficient infrastructure and "technological spillover". Technological spillover is the cumulative effect of people picking up new ideas in an environment of dense information and vibrant dialogue and developing them further. Robert Hall believed that "a city and a boom are essentially the same thing - one in space, one in time."[39]

Once established, the dynamics of centre economies feed on themselves. "The circular relationship in which the location of demand determines the location of production, and vice versa, can be a deeply conservative force, tending to lock into place any established centre-periphery pattern." However, when certain thresholds are passed new centres can spring up and old centres can collapse. Expectations such as a gold rush may cause people to move, for instance, from high income locations to low income locations, constituting a market to which production then responds by translocation.[40]

Because of its technological and commercial superiority the centre is able to draw the factors of production from the periphery into its vortex. This happens, for instance, with the *mental factors of production*, such as initiative, sophistication and skills. People with ambition, intelligence and education are attracted to the centre because they want to develop their gifts to the full and cannot do so in the periphery. Moreover, top people demand top facilities which only the centre can provide: better schools, better hospitals, smarter shops, a greater variety of recreation. Once these people have arrived in the centre, they are trained for specialised functions in the industrial and commercial economy of the centre. They cannot use such skills in the periphery. And even if they could, the remuneration paid by the periphery could not compete with the income offered in the centre. So they are permanently lost to the periphery. We call this phenomenon "brain-drain".

The same is true for *labour in general*. Because the productivity of labour is higher in the centre, employers can afford to pay higher wages and trade unions make sure that they do. So incomes are higher in the centre - and that in spite of the fact that the hours of work in the city are much shorter than on the farms. For that reason alone people tend to move to the cities. Able-bodied men and women may also become migrant labourers and leave their children and their fields to the elderly. Younger people, in particular, are attracted by the "glittering lights" of the cities because they find life on the farms boring. In this way the periphery loses its most important asset, its active labour force.

When rural people move to the cities, however, they are unable to compete with those who have been conditioned to live in the cities. They lack training, experience, contacts and the pace required in their new environment. If they are employed at all, they may begin as casual and unskilled labourers. More often than not they may have to make do in the "informal sector", where they render simple services to the more privileged. They end up in slums, cannot afford to send their children to school, nor to feed themselves properly.

Established city dwellers will not be eager to help them get onto their feet because they are potential rivals on the labour market. So these people do not readily become part of the centre population but remain a part of the periphery, only they now live within the geographical area of the affluent centre. For the city they present, to a large extent, not an asset but a social problem of proportions it cannot handle. To the degree that great Third World cities try to upgrade their slum areas, the avalanche of migrants from rural areas increases, neutralising any gains made. The influx decreases only to the extent that urban conditions deteriorate.

These observations need to be qualified. Some evidence suggests that people who move to the cities from the rural areas manage to find jobs more readily and climb up the social ladder more easily than unskilled and unemployed urbanites. The reason for this phenomenon may be that those who leave the rural areas and move to the cities are the "cream" of the peripheral population, who successfully compete with centre dropouts. They are more highly motivated and have lower expectations than established low rank city dwellers. They are not only more willing to work for low wages, but they also have the necessary drives and potential gifts to outcompete those who have already drifted to the bottom of the urban social system. So it may well be that the lower ranks of the *urban* population end up in the slums.

On the other hand, the financial power of the centre spills over into the periphery. As the poor move into run-down areas of the cities, the more affluent move out into the suburbs because they are repelled by the misery and the crime of the inner city. The city grows at the expense of the countryside. Moreover, abandoned rural areas are bought up by capital owners from the centre and cultivated on a commercial basis. These undertakings are run by trained managers who apply research results, utilise state assistance, exploit their greater access to credit facilities, and explore new markets.

The remaining peasants are unable to compete with these modern enterprises. When the latter make tempting offers they sell their land and move to the cities as well. This means, in effect, that *land as a factor of production* is transferred from the periphery to the centre. Cultivation with machines on commercial farms also deprives landless rural workers of their means of livelihood and they too swell the ranks of the slum dwellers.

Suction can also be observed with regard to *raw materials*. Historically speaking, centres have often sprung up on the sites of great mineral deposits, such as iron ore, coal, copper or gold. Once a dynamic economy has established itself in such a location, however, the sites of newly discovered deposits find it hard to compete with the existing site. Although rich gold reserves were found in South Africa's Free State Province many decades ago, the site has not become another Johannesburg. New sites do not have the financial institutions, the infrastructure, the technology, the skills, and the developed organisational framework which are at the disposal of established centres. So it is these centres which extend their tentacles to exploit the newly found deposits to their own benefit.

Raw materials are also derived from *agriculture*. It is obvious that the greatest demand for agricultural products exists in the areas of high population concentrations. It is here that storage, processing and marketing becomes economically viable. The closer one moves towards an urban-industrial complex the less the transport costs, the easier the delivery, the more lucrative the farming operation. So efficient and intensive agricultural production also tends to move towards the centre.

In the same way the *capital* of the periphery is sucked up by the centre. Returns on capital investment in the periphery are usually meagre if compared with those of the centre because of their higher productivity. The result is that the little capital, which the periphery is able to generate, is attracted to the centre. Conversely, larger firms in the centre may buy up rural outlets, or establish their offshoots in the villages. Local businesses may be unable to compete against such firms.

On the international scene, further factors come into play. Because of widespread poverty, the breakdown of the social fabric and extreme income discrepancies within the population, many Third World countries tend to be politically unstable. Revolutionary leaders love to claim that foreign enterprises are responsible for the misery. For embattled rulers the temptation to nationalise successful concerns is great because, in the short run, such a move is both lucrative and popular. However, the result is a flight of capital from poor to rich and stable nations.

Even *infrastructure* is subject to suction. We have seen in chapter 3, for instance, that the transport and communications network is organised in a radial fashion. It is strongest between centres and least developed between peripheries. The reason for this phenomenon is that interaction with another centre is more lucrative than with a periphery. But this pattern has its consequences: transport and communications remove obstacles to trade and facilitate interaction. Gradually, two centres of approximately equal strength are integrated with each other, while a weaker centre is attached to the stronger. The periphery is reduced to the hinterland of the centre, and eventually it may be absorbed by the latter.

In all these cases, therefore, the centre has greater power to attract, integrate and bind factors of production than the periphery. If an investment of such factors takes place in the periphery at all, it is usually an investment by the centre. In other words, it is an offshoot of the centre in the periphery. Such enclaves are not very stable. If the returns are not satisfactory, the investment is withdrawn and the little centre in the periphery collapses. Let us look at this phenomenon in greater detail.

Penetration of the periphery by the centre

The corollary of suction is penetration. As the peripheral system disintegrates, the centre takes over both the production and the distribution of goods and services for the periphery. That means that the organisational network of the centre penetrates the periphery. The centre establishes local outlets for its products in the periphery. They are connected with their mother firms by means of transport links and lines of communication. Repair workshops and storehouses for stocks and spares are built. As their markets grow, they may establish assembly plants and factories, which take over the production of less sophisticated parts of the product (local content). They enter into contracts with local businesses. They press for the development of the necessary infrastructure by the host environment, or supply it themselves. For specialised tasks mother firms send their own expatriate executives and skilled labour. The presence of expatriates again calls for increased diplomatic representation, quality accommodation, financial institutions, schools, recreational facilities and so on.

Many Third World leaders lure overseas enterprises into the country by granting them special concessions. If circumstances for investment are favourable and large amounts of relatively stable capital are invested in the peripheral country, it may generate a dynamic which builds up local infrastructure, entrepreneurial skills, managerial competence, technological expertise and marketing networks. As a result, local capital may grow and an economic miracle may occur. In the case of the "Asian Tigers" such developments have laid the foundations for self-sustained economic growth.

However, this will not happen automatically. In fact, the recipe may seriously backfire. The centre will only invest in the periphery if the returns there are higher than at home and offset the greater risks of the offshore location. Therefore concessions must be substantial, and they can only be granted at the expense of local competitors and the local population at large.

To satisfy the demands of the investors for improvements in infrastructure, the state often takes up loans which may not generate corresponding incomes but ensnare the peripheral economy in a debt trap. In return for favours, officials may demand substantial bribes. Corrupt politicians and officials add such bribes to the money they take from the state and build up private accounts in Switzerland to safeguard their future. Appropriated state funds are, of course, derived from the tax paid by the local population.

To make the point, let us consider a *worst case scenario* in which the periphery has no control over these developments. Leaders are technologically incompetent, and their people are puzzled by what the foreigners are doing. Enterprises bribe local politicians to condone whatever practices best suit their interests. Capital goods are imported from the mother country; all relevant research and development is done in the mother country, and crucial decisions are made in the mother country. The periphery cannot sustain the resultant economic dynamic without the organisational network of the centre. Whatever has been built up collapses when the foreigners depart. Their departure usually signifies that the operation is no longer viable, or too risky in terms of global market conditions.

This structural dependency can take on bizarre forms. In modern times, capital has become internationally mobile, labour has not, at least not to the same extent. Centre firms divide the productive process into technology intensive and labour intensive phases. For the labour intensive phases, for instance the assembly of radios, they find a place in the Third World where there are masses of unemployed and unskilled people who are willing to sell their labour at rock-bottom wages. Under such conditions the firms can afford to employ only the "cream" of the local labour force, namely young women between 15 and 25 years of age. Once they are exhausted, they are dismissed. The centre utilises the labour from the periphery but avoids its influx and the social obligations going with it.[41]

This method has been made possible by the development of cheap and fast transport facilities, the division of the production process into small units, and the greatly reduced bulk of modern products, especially in electronics. The effect of the practice is that the centre obtains a factor of production, namely labour, much cheaper than at home. The centre in the centre gains. The periphery in the centre loses out, because the practice creates unemployment in the mother country. The centre in the periphery, that is the government and the small modern sector, which both need foreign exchange, may benefit. That is what motivates them to welcome the establishment of such enclaves. But for the periphery as a whole the benefits are negligible. The best workers are drawn from their traditional contexts into the centre enclave. The wages they earn barely keep them alive. No capital formation takes place, and no indigenous industries are developed.

The production enclave may result in a fiasco for the periphery. Such an enclave inevitably leads to the establishment of a small *centre in the periphery*. Much of the cost of its development - houses, roads, schools, administrative facilities, shops, transport and so on - is shouldered by the peripheral economy. The rural environment begins to be oriented towards the

Chapter 5 - Causes of economic imbalances | 123

little centre. It becomes more and more dependent on the market which the centre provides. But then, if the firm finds a more convenient place, or if the local workers begin to make demands, or if the state raises its taxes, the firm packs up its facilities and moves elsewhere. The little centre collapses and leaves a whole environment in a state of economic decay. The same happens, for instance, when military bases of industrial nations are shut down.

It may also happen that a semi-peripheral country, such as South Africa, applies *progressive labour laws,* rendering its labour more expensive relative to those of comparable countries, such as India. Then a firm producing soap, for instance, will translocate its operation from South Africa to India, thus undercutting the gains in welfare reached in South Africa.

Another new factor is that, due to an increasingly competitive market, enterprises have to respond to rapid changes in demand and to small scale opportunities. This is made possible by *flexible automation*: "reprogrammable machinery can be used to produce a great variety of products and for profitable small batch production" so that "economies of scope" replace "economies of scale".[42] Such automated industries are no longer farmed out but located as close to the great markets as possible for reasons of control and rapid response. Thus the comparative advantage of the periphery based on its cheap labour largely evaporates.

Asymmetrical market interaction

Commercial superiority of the centre leads to asymmetrical market interaction with the periphery.[43] In rudimentary forms this had already surfaced in precolonial times. Early European traders became rich by exchanging cattle for rifles, or land for copper wire. These examples show that commercial superiority is often based on technological superiority. In modern times this has become crucial: unprocessed raw materials from the Third World are traded for manufactured goods from the industrialised nations. Manufactured goods represent a growth in value over against raw materials. The periphery cannot produce these goods on its own. If it wants to purchase manufactured goods, it has to pay for this added value by selling more of its raw materials.

The centre may not want these raw materials, at least not at the old price. In the last few decades raw materials have become more readily available. Many industrial nations, particularly the United States and the former Soviet Union, have huge reserves of raw materials of their own which they can utilise when Third World countries ask for a higher price. This recently happened with crude oil. There is also a glut on world markets for such resources as iron and copper ore.

At the same time modern technology tends to use less bulk. Compare an old fashioned adding machine with a modern pocket calculator! The centre also finds more and more alternative materials. It can, for instance, replace copper wire with glass fibres in telephone cables. Greater supply and lower demand brings about lower prices for raw materials offered by the periphery.

Meanwhile manufactured goods become more and more sophisticated and the value added rises. Production costs in the centre also rise because of higher salaries and capital outlays. The gains of higher wages and dividends remain in the centre, while in the periphery the relative price of products sold by the centre rises. Thus the periphery sells cheaply and buys expensively and this constitutes a transfer of resources from the periphery to the centre.[44]

Moreover, the centre is often able to stipulate the terms of the deal. It may grant credit to the periphery on condition that the latter buys its own products, employs its own experts, and uses its own transport systems. Thus credit is often given with strings attached.[45] Once the periphery buys certain brands, it must continue with these brands because it needs spare parts, upgraded versions which fit into the system, and advanced training to operate the machines. The periphery is no longer free to hunt for the best bargain on the open market.

During the Cold War political conditions for such "development aid" played a great role as well. Even today the motivation of aid is to win friends, not to help needy countries. "More than half of US bilateral assistance in 1991 was earmarked for five strategically important countries: Israel, Egypt, Turkey, the Philippines and El Salvador. With five million people and a per capita income of $1000, El Salvador received more US assistance than Bangladesh, with 116 million people and a per capita income of only $210."[46]

Many Third World countries have little more than agricultural products to offer on the world market. However, their agricultural production is relatively undeveloped and its quantity and quality cannot readily compete with the scientific-technological farming methods found in industrialised countries. Free trade would be to the advantage of the most efficient producers and the Third World would lose out.

But that is not the worst problem. Industrialised countries have the financial means to subsidise and protect their farmers at colossal rates.[47] There are no commensurate means of boosting agricultural production in the Third World. On the contrary, in many cases agriculture is the main source of state income. The artificial distortion of the market in rich countries then leads to the build-up of huge agricultural surpluses which need to be disposed of. If they are dumped in Third World countries, even in the form of food aid, the products of local farmers are crowded out and the competitiveness of peripheral agriculture is undermined even on its local markets.[48]

Modern information technology and the enormous growth of *financial capital* in recent times have made the situation of Third World economies particularly vulnerable. Financial capital is extremely volatile. News about market developments spread through the global system with split-second speed and huge financial resources are shifted from place to place with the click of a computer mouse. Although financial capital is "virtual capital" in the sense that it does not consist in factories or infrastructure, its impact has become overpowering. A country like Mexico, and more recently some of the famous "Asian tigers", have become the victims of abrupt capital withdrawals. Speculators exploit even the smallest currency fluctuations between countries to make gains. And the financial capital market itself is a "bubble" which can burst at any time.[49]

These occurrences have highlighted the degree to which governments, labour and citizens in general have become exposed to the vicissitudes of the financial market. At the same time it has become clear that a globalised economy is vulnerable to organisational breakdown. A highly complex and internationally integrated system is fast growing beyond the capacity of any government or organisation to control, direct or prevent it from imploding. The only hedge against such eventualities lies in the relative autonomy of particular sectors and regions.

Comparative advantage

Free trade does enhance efficiency, but to the benefit of the centre and at the expense of both the periphery and, because of increased throughput, the natural environment. But does the law of comparative advantage not state that free trade operates inevitably to the benefit of both partners, even if one partner is able to outperform the other in all respects?[50] No, not inevitably so. The law of comparable advantage is a theoretical construct, indeed one of the fetishes of conventional economics, whose validity must be tested against actual cases.[51] It seems to work perfectly well where the partners are roughly at the same level of development. But the greater the discrepancies between the two, the more problematic their relationship becomes.

(a) Even where free trade operates to the advantage of both partners, it may still operate to the greater advantage of the more powerful economy. This led some economists to argue for "aggressive support by a nation's government of the international competitive position of home firms."[52] They call it "strategic trade policy". Ironically, this policy has been applied without hesitation by centre nations, for instance in the case of the agricultural policy of the European Union. Its aggressive application was also one of the secrets of the success of the "Asian Tigers".

(b) There are discrepancies of such a magnitude that they render comparative advantage meaningless. Countries can be "too small to accommodate industrial districts or too unequally endowed with capital and labor to make up for their deficiencies through trade."[53] A country may even be so poor that it has nothing at all to offer to a highly developed country in return for much needed imports. Say, for example, that Tanzania wanted to import trucks and telephones, but could offer only sisal and coffee in return. What would happen if the market for coffee was swamped by other producers and sisal was replaced by nylon fibres? But this extreme case only highlights a general trend. It is a well known fact that the bulk of world trade flows between Triad economies (US, EU and Japan) precisely because poor countries do not have enough to offer.[54]

(c) There is also no guarantee that the law of comparative advantage will actually be allowed to operate in a free market. The theory presupposes that the partners figure out which products should be traded against each other. But in a free market there are no such agreements. American firms will naturally try to export not only trucks but also nylon ropes to Tanzania. We shall see in chapter 12 that aggressive market penetration by the centres may eradicate peripheral production altogether. Even Tanzanians prefer cheaper and more durable nylon ropes above their own sisal ropes, or imported plastic containers above locally produced earthenware! If the peripheral economy takes up loans to pay for these products, it ends up not only in a dependency trap, but also in a debt trap.

A particularly interesting case of the operation of the free market is the globalisation of financial capital. As mentioned above, financial capital can be moved with the click of a computer mouse from Kuala Lumpur or Johannesburg to New York or London. There are speculators, organised in "hedge funds", who deliberately target currencies they consider to be weak, make them crash, and earn billions of dollars in the process.[55] The socio-economic impact of such upheavals has been enormous and it is clear where the "comparative advantage" lies in this particular case of free trade.

The legitimation of this praxis by liberal economists is particularly revealing: "But blaming these banks and funds for doing what they do is the equivalent of blaming a leopard for eating sheep."[56] Indeed, in a globalised free trade it seems to be in order that some humans act like leopards devouring other humans like sheep! Of course, the theory of free trade implies that the sheep are also free to eat the leopards if they wanted to - which is a remarkable kind of logic!

The debt burden

An increasingly dangerous form of asymmetrical interaction between centres and peripheries is the foreign debt of Third World countries.[57] When these countries begin to engage in industrial production, they have to import machinery, build up a modern infrastructure, develop markets, and attract both experts and capital. All this costs the peripheral economy dearly. The small modern enclave in the periphery wants to enjoy industrial products imported from the centres. As a result peripheral countries tend to import more than they export. Foreign debts have to be repaid with "hard currency", that is foreign money which is not subject to local inflation. So the centre does not suffer, but the periphery is saddled with a *balance of payments problem.*

To cover the shortfall governments tend to *raise loans* in rich nations or to approach institutions such as the International Monetary Fund and the World Bank.[58] Banks in industrial countries may grant cheap credits and governments may underwrite loans. Such services are then interpreted as "development aid". This is a misnomer because in many cases such "aid" consists of credits to purchase goods and services from the creditor countries and must be repaid with interest, whether they benefited the poor country or not.

If the money is used for consumer items, the loan does not increase the productive capacity of the poor country, but the productive capacity of the rich country. Larger projects, which are meant to boost the productive capacity of the peripheral country, do not always succeed in doing so, because they are artificial constructs, not grown economic organisms.

Due to the rise in *interest rates* such loans can become a devastating "debt trap" for the peripheral economy.[59] In some cases the better part of the annual GNP has to be transferred back to the "donor" country. In this way the centre siphons off whatever surplus the periphery may be able to achieve. Often more loans have to be raised to repay existing loans, the poor country becomes "addicted" to loans and the situation progressively worsens.[60]

Some governments have recently *threatened to default.* When the problem grew to proportions which might have undermined the global financial system, a few tentative steps were taken by centre nations to alleviate the burden. But so far they are mainly intended to safeguard centre interests. Debt repayment is ostensibly essential to establish the creditworthiness of such countries. However, "these adjustments have been made at the cost of compressed consumption and wages, lower investment and output, and frequent recourse to inflationary financing of government deficits."[61] Governments were required to cut down on health, social securities and education, which was not only a retrogressive step in terms of welfare but also undermined long term development prospects.

Centre nations also began to live beyond their means on a large scale, this time at the expense of a future generation. The US "managed in less than a decade to transform itself from being the world's leading creditor nation to being its leading debtor".[62] The most powerful economy in the world is sucking up vast international capital resources and squandering them in senseless overconsumption. "The total current account deficit (before official transfers) for all Low Income Countries in 1987, with their nearly three billion people, was some $ 17.6 billion. The current account deficit of the United States, with some 240 million people, was $142 billion for the same year."[63]

Causes of the debt crisis

International debt in its current magnitude is a fairly recent phenomenon. During the oil crisis the financial windfalls of oil-rich countries piled up in banks and had to be recycled into the global economy. Money was cheap, controls were lax, and Third World governments availed themselves of the opportunity. Due to rising US budget deficits under Reagan, interest rates soared and debt servicing became prohibitive.[64]

Third World trade has been liberalised under First World pressure, while *protectionism* is rising in developed countries, particularly against imports from the Third World.[65] When peripheral export to centre countries is reduced, there is no money for imports from these countries, whether capital or consumer goods, nor can the periphery repay its debts. Due to higher interest rates, debt repayments take ever higher proportions of Third World export earnings. Peripheral economies are increasingly geared towards export to service their debts rather than to satisfy local needs. When they succumb under the weight and threaten to default, the First World banking system, and thus the world financial system, becomes vulnerable.

In response First World lenders get tough on Third World debts.[66] The austerity measures imposed by the structural adjustment programmes (SAP) of the World Bank and the IMF are at least partially meant to "increase prospects that the international banks could recover their bad loans."[67] The deterioration of social services and general wellbeing in the debtor country leads to political, social and economic instability and plays into the hands of radical saviours from the right or left.[68] In response the First World sends in arms to prop up shaky governments - whether they are particularly democratic or not is not the issue - and the spiral of military spending continues.

After the end of the Cold War this vicious circle has partially subsided, but its effects will continue to create havoc in Third World countries for a

Chapter 5 - Causes of economic imbalances | 129

long time to come. Moreover, the weapons trade continues to flourish. It is clear how the irresponsibility of banks, Third World governments and some economic superpowers have all joined forces to upset the global economy.

Apart from that, the world financial system has led to enormous anomalies. Governments often held dollars as reserves and wealthy citizens hoarded them as hedge against reckless hyperinflation. "As a result more greenbacks were circulating offshore than were held by Americans in their purses, wallets and cash registers. Many of those dollars come back as investment, amounting to a huge interest-free loan to US citizens; the US government can keep running large balance-of-payments deficits as long as the global appetite for dollars persists."[69]

In contrast with popular perceptions among the rich, the funds flowing into the Third World from the industrial countries at present are *ridiculously insignificant*. The United Nations have set a target of the payment of 0.7% of the Gross National Product of developed countries for development aid. What is one percent in the total turnover of a centre nation! Yet, in 1987 only 5 countries had reached that target. The amount paid by the United States was 0.21%, by Japan 0.3%, by the United Kingdom 0.29%.[70] By all standards, these figures should be a source of shame and embarrassment for the citizens and governments of richer countries. Meanwhile, debt servicing has reached such magnitudes that more money flows from the poor to the rich countries than in the opposite direction. Already by 1986 the net outflow from poor countries amounted to US $30.7 billion.[71]

As mentioned above, Third World countries may gear their commercial production to the *export market* to meet their external debts with hard currency. Thus, instead of producing food for its local population, Brazil produces soya beans for export to the United States and Europe, where it is used to fatten cattle for American and German consumers. Other countries produce cocoa, coffee, sugar, timber or cloves for export. In this way the use of fertile soil is diverted from food and clothing, which would satisfy the basic needs of the peripheral population, to products which satisfy the luxury needs of the centre. Meanwhile foods becomes scarce in the periphery, their prices rise and again it is the poor who suffer.

In their desperation, governments of Third World countries often resort to the *tactic of printing more money* to cover their debts. This leads to rampant inflation. In some Third World countries the inflation rate has been 100%, in others 500% or 1000% per year. While fiscal indiscipline and mismanagement in the Third World have been major causes of the malaise and cannot be condoned, the genesis of the problem should not be overlooked.

The use and abuse of power

For the Western elite, meanwhile, there is no reason to be unduly alarmed by the plight of Third World populations. If it is assumed that all people are self-responsible, and if competition is the rule of the game, those who cannot cope only have themselves to blame. The poor deserve to end up at the bottom of the pile, if they do not pick up their socks and show what they are worth. In this frame of mind, a new kind of *feudal attitude* emerges, which believes that differences in performance simply reflect the natural class structure of society.

But will *the market* not balance out the economy?[72] In chapter 12 we shall see why it cannot. Increasing concentrations of money power in the centre lead to an increasingly skewed market. Apart from structural mechanisms, which we have analysed above, this power can be used consciously and unscrupulously. A powerful party can either influence the market in its own favour, or it can bypass the market altogether and achieve its ends through "extra-market operations." Both these types of activity undermine the liberal theory that the market mechanism balances out the economic system.

There are many ways in which *the market itself can be manipulated.* Most of these derive from the existence of monopolies (= a single supplier), oligopolies (= only a few suppliers) or monopsonies (= only a single buyer). These are far more typical for the modern economy than the utopian ideal of "perfect competition" with which every textbook on conventional economics begins to develop its theory. Anti-monopolistic legislation is found in all Western countries, but it has not prevented the formation of huge business conglomerations. Takeovers of small companies by bigger ones and the subsequent building of vast economic empires are the order of the day.

Most prominent among these empires are the *multinational corporations.*[73] Although there are millions of small enterprises and hundreds of thousands of medium-sized companies around, it is power which counts in the market place. The financial clout of the top multinationals surpasses even that of most peripheral states, while economic enterprises in the periphery are small and fragmented. They are no match in the competitive game.

One must concede that at times there are compelling structural reasons for the formation of multinational corporations. Some of the most sophisticated items of modern technology, such as micro-processors, can only be developed by a handful of these giant corporations. Even greater European economies have to cooperate to develop new fighter aircraft or send communications satellites into orbit. One must also concede that, if they so wish, great corporations can generate space for a number of smaller undertakings which supply them with certain intermediate goods and services. But these considerations do not remove the fact that multinational corporations *wield power* and that uncontrolled power is generally used in self-interest.

Where there is a concentration of bargaining power either on the supply side or on the demand side, the price and quantity of commodities, labour, raw materials, and so on, *can be dictated* by the more powerful party. Products can be withheld to create market shortages and thus attain higher prices; credit facilities can be granted or withheld; only a certain type of product can be marketed; products can be deliberately designed to have a short life span (planned obsolescence); new needs and wants can be created through aggressive advertising and salesmanship; secret agreements can be made between suppliers. All these practices presuppose financial power, sophistication, organisation, political influence, the right connections, and so on, which are concentrated in the centre rather than in the periphery. As a result, the periphery is the loser in the game.

Some extra-market activities are *blatantly immoral,* even illegal. Some of the more common practices are threats to life and property, bribery and corruption, false promises, libel, spreading of rumours and disinformation, provision of misleading data, insider trading and fraudulent claims in advertising. Great companies in Western countries have discovered the importance of image building for business success, and thus invest in welfare, sports events, research organisations, cultural activities, and so on. But company profit remains the motive. So corporations with a respectable reputation in Germany or the United States *can afford* to be ruthless in covert international transactions, or in the Latin American jungle, where there is no effective law enforcement agency and no good name to be spoilt. Often enough, centre elites have not shied away from the most vicious atrocities, including genocide in some cases, in pursuit of their economic and political interests.[74] This happened during colonial times, under apartheid, and during times of war. It still happens behind the shield of military dictatorships and drug lords.

There are other uses of power which appear to be morally justifiable, or at least *perfectly legal.* Large corporations may impose a set of rules to be followed by all their subsidiaries and employees. They may enter into trade agreements, stipulating that the supply of a product is subject to certain conditions. They may launch costly promotion drives. They may offer commodities below cost so as to oust a smaller competitor. They may dump a product on a market in great quantities at ridiculously low prices to gain market entry or market dominance. From here it is just a small step to the rise of mafias that paralyse some economies with their extortion of "protection money", infiltrate the army, police and judicial system, and exploit the weaknesses of a spoilt generation with the lure of drug abuse. The deepest problem, as Stackhouse sees it, is that the "spirituality" of the corporation simply does not entail public responsibility but adjusts itself to conditions depending on business opportunities. With that it denies the heritage of Western democratic values.[75]

The state is not always very helpful in this regard. In many cases it has created monopolies in its own name. In other cases it has subjected the marketing of products to price controls advantageous to interest groups in the economy. The agricultural sector is a case in point. Powerful interest groups have their lobbies in the corridors of power. They may gain tax concessions, preferential treatment in the supply of infrastructure, price fixing, or controls on labour, from the government in power.

The costs of running a state are high. The interests of those who provide most of the revenue are likely to receive highest priority in policy making. During election campaigns, parties commit themselves to further the interests of those who finance the campaign. Even if the periphery population has the franchise - which is not the case in many Third World countries - it may not be able to make effective use of their democratic rights because centre groups within the country are better organised and have more financial clout. On the *international* level the periphery has no democratic leverage at all because there is no overarching world government. Big business, in contrast, has an international network at its disposal with which to further its interests. Rich countries can blackmail poor countries into submission. Food aid is an example. Many impoverished countries depend on the agricultural surpluses generated in the West.

Again: the arms trade

The *arms trade* is another example of potential blackmail. Many Third World governments need arms to protect themselves, not primarily against their aggressive neighbours but against their rebellious subjects.[76] Their own economies are unable to produce the sophisticated arms needed to win a modern war. To withhold or supply arms is one of the great levers applied against Third World countries by the great arms producers, the US, Western Europe and the former Eastern Bloc.

Military expenditure has siphoned off incredible sums of badly needed money which could have been used for development purposes. "In recent years, Southern countries have spent approximately $200 billion per year for military purposes, four times the amount that they were receiving annually in economic aid. Furthermore, these expenditures have been growing at a rate four times greater than their non-military spending."[77] And that does not take into account the devastations of the numerous violent conflicts themselves which have laid waste one country after the other.

The same economic growth could, of course, be achieved with investment in human and social infrastructure: education, health services, sanitation, roads, appropriate technology, agricultural projects, dams, and so on.

Military spending harms economic growth and welfare even in the industrialised nations.[78] The United States is reported to have spent roughly $19 trillion on defence, $5,8 trillion of which on nuclear weapons. The latter figure has exceeded spending on welfare, state medical insurance, health and education. It was also reported that this sum would have been sufficient to provide every household with a new Rolls-Royce. The utter senselessness of this expenditure is revealed by the fact that, when the US defence secretary stated in 1964 that a nuclear force equivalent to 400 megatons would be sufficient to cause assured mutual destruction with the Soviet Union, the US stockpile already totalled 17 000 megatons.[79]

Also in the Third World there is overwhelming evidence of the existence of a "trade-off between Milex (military expenditure) and spending on health, education, social security and welfare in the region".[80] Between 1965 and 1980 Africa's military expenditure has shown the fastest growth rate in the world, tripling in real terms, while the global increase was "only" 50%.[81] The military always receives priority, as can be seen from the fact that the defence budget remains constant in times of prosperity and austerity.[82]

To cut down on military spending would be in the interest of the periphery. But the periphery within Third World countries has no say over the allocation of funds, nor has the future generation, nor can the natural environment make its cry of agony heard effectively. For better or for worse, the use of power is determined by the interests of those who wield it - unless they wake up to their global responsibilities.

To overcome mass poverty and marginalisation, *power relations* in the global economy must be balanced out. The centre dynamic is a great boon to humankind, but it must be redirected to serve humankind as a whole and not just an elite. To avoid environmental destruction it must also be redirected from quantity to quality. The potency of the periphery must be enhanced in such a way that the natural environment and the resource base are not destroyed. This is only possible through the combination of development and population control. The abuse of power in all its forms must be subjected to legal controls. The development of mental, cultural and social structures conducive to the attainment of these goals is, perhaps, the most intractable and the most urgent problem humankind will have to face in the coming decades.

> *Revision:* Briefly describe the phenomena of colonial plunder, suction, penetration, asymmetrical market interaction, the debt crisis and the arms trade. Which of these do you consider to have the most serious consequences for the periphery?
> *Application:* Can you discern some of the phenomena described as "suction" in the relation between rural and urban areas in South Africa?
> *Critique:* (a) "Just concede for once that the predicament of the Third World is due to the incompetence of its leaders! There are newly industrialising countries in South East Asia for whom the 'production enclaves' you described above have become opportunities to build up economies which grow at an unprecedented rate and which begin to compete with Western interests. Why drag every positive action of the West towards the South into the mud?" (b) "It is just not good enough to shift the blame for the plight of the poor countries upon their leaders who desperately try to find a way out of a situation which has been thoroughly disorganised by the ravages of colonial oppression and neo-colonial exploitation." (c) "The debt crisis is due to the avarice and megalomania of Third World leaders. Nobody forces anybody to take up loans. But once money has been borrowed it must be paid back." Which statement is nearer to the truth? How would you respond?

Let us summarise

This chapter was devoted to the causes of economic imbalances. The introductory section offered a *model* according to which these causes may be analysed. We distinguished between structural mechanisms and volitional factors. We also distinguished between factors inherent in centres and peripheries respectively, and factors which are due to the asymmetrical interaction between the two. All these factors reinforce each other in a cumulative historical process.

In section II we discussed a few *non-economic factors* which are inherent in centres and peripheries respectively: the development, integration and efficiency of social institutions, the degree of economic freedom, the availability of social securities, the orientation of a culture either towards the past or the future, the educational approach, socio-psychological dependency, the quality and availability of health services, nutritional standards, and the growth rate of the population. All these factors contribute to the great discrepancies in economic potency found between the centre and the periphery.

Section III dealt with economic causes which are located *in centres and peripheries* respectively. We emphasised the differences in availability and efficiency of the factors of production in both the centre and the periphery: space, time, infrastructure, initiative, technology, organisation, raw materials, labour and capital. Production again depends on market demand. Market demand depends on distribution of income. Distribution of income is determined at least partly by the power of interest groups. In all these dimensions the centre has considerable advantages over the periphery.

In section IV we turned to the second type of economic factors: *asymmetrical interaction*. We mentioned a few crude and visible examples of how differences in power impact unequal relationships: colonial plunder, though historically severe, has made way in modern times for the *suction* of peripheral factors of production into the vortex of the centre, the *penetration* of the periphery by centre offshoots, unequal *trade relations*, the cumulative problem of international *debt* and the arms trade. We also mentioned examples of the abuse of power by those who wield it.

These crude indications will be augmented in chapter 12 with a new analytical approach to the economic interaction between centre and periphery. In chapter 13 we place this discussion into a more encompassing framework provided by the natural sciences - entropy, evolution and acceleration. But before we come to that we must deal with the asymmetrical relations caused by the difference between collective mindsets.

Notes

[1] For that see Nürnberger 1995.
[2] For the flawed concept of comparative advantage see Daly & Cobb 1989:209ff; Elliott 1994:52ff; Faber 1990:11ff, Johns 1985:149ff,229; Caporaso 1987:37, 80.
[3] Myrdal 1957; cf Thirlwall 1994:129. This factor is currently being rediscovered by main line economics: "the detailed pattern of advantage reflects the self-reinforcing virtuous circles, set in motion by the vagaries of history" (Krugman 1994:233f).
[4] Meadows *et al* 1992.
[5] Meadows et al 1992:14ff.
[6] For instance, Kindleberger 1958.
[7] For a discussion of different approaches see Faber 1990:12ff; Caporaso 1987:22ff, 92ff; Johns 1985:127ff, 225ff; Seligson 1984:105ff; Lovett 1987:21ff.
[8] For more detail see Nürnberger 1988 part II.
[9] Wogaman 1986:1f. See also the books of Küng, Duchrow, Stackhouse, Northcott, Meeks, McFague, Storkey, and others.
[10] Some economic historians believe that it was the mechanical routine of the conveyer belt which ingrained punctuality, constancy and precision in the Western mind. But that was only after efficiency had become an obsession among the organisers of the modern economy.

11 When asked why Africa could not follow the model of the "Asian Tigers", economists in South East Asia told me that China has a 4000 year old imperial culture which implied an age old experience of large scale organisation, while African society was composed of small autonomous tribes throughout the same period.
12 See K Nürnberger: Democracy in Africa - the raped tradition. In: Nürnberger, K (ed): *A democratic vision for South Africa*. Pietermaritzburg: Encounter Publications, 1991, pp 304-318. Also Andreski 1968; Klitgaard 1988 and 1991; Kotecha & Adams 1981.
13 H M McFerson: Focus on democracy and development in Africa. *Journal of Peace Research* 29/1992:243.
14 I have noticed the strange phenomenon that black students from various parts of South Africa leave out "rather" in the construction "rather than"; they also say "other ... other" instead of "some ... others". These construction may be caused by the structure of African languages, but the point is that it is not being corrected by the teachers.
15 Taylor 1963.
16 I have witnessed how the brother of a man who went to the city in search of work abandoned his home and his fields and moved into the latter's homestead with his family, ostensibly to look after the wife and children of the absent brother, in reality to live from the remissions of money sent by the brother to his wife. Due to traditional conventions there was no escape from such parasitism.
17 There are economists, however, who believe that population growth "can act both as a stimulus and an impediment to growth and development" (Thirlwall 1994:152). If population growth implies growing market demand, and if there is a lot of potential capacity which needs an incentive to unfold, population growth may constitute a stimulus. Population pressures may also lead to innovation. If linked to urbanisation it may lead to more rapid modernisation. It is also claimed that there is no empirical evidence to suggest a correlation between population growth and economic growth (for the lively debate see Thirlwall 1994:155,157ff). But this may be a short term impression. The growing pressure on resources and the rapid rise of unemployment where the industrial sector cannot absorb the work force must eventually tip the balance against development and cause a "low-level equilibrium trap": "First, rapid population growth may not permit a rise in per capita incomes sufficient to provide savings necessary for the required amount of capital formation for growth. Second, if population growth outstrips the capacity of industry to absorb new labour, either urban unemployment will develop or rural underdevelopment will be exacerbated, depressing productivity in the agricultural sector" (ibid 163).
18 Classical economics distinguished only three factors of production: land, capital and labour. Heilbroner & Thurow 1975:19ff; Ruffin & Gregory 1983:25ff. Some add techology. A comparison with our scheme shows that this is not satisfactory. Moreover, "land" which includes all resources, was sorely neglected as a factor of production (Daly & Cobb 1989:109ff). For the classical approach to the production function see Thirlwall 1994:66ff.
19 For a detailed treatment of space refer to the economic geographers, for instance Smith 1977 and 1979, Broadley 1992, etc.
20 Krieger Mytelk in Caporaso 1987:43.
21 For global "megatrends" to information, decentralisation and networking see Naisbitt 1982.
22 Krieger Mytelk in Caporaso 1987:70.
23 "Individual consumers are relatively weak; they are individuals, while suppliers and manufacturers of goods and services these days more often than not are large corporations. ... The obstacles facing them are immense. To talk of a 'free market' of equal individuals is absolute nonsense" (Goldring et al 1993:3).
24 *The American Vision*, p 6.
25 It is usually called entrepreneurship in economics. Heilbroner & Thurow 1975:136f; Ruffin & Gregory 1983:26, 426, 649f. Schumpeter (1943) has placed great emphasis on entrepreneurship and innovation. Also refer to "innovation" in economics text books.
26 For detail see chapters 2-4 of Nürnberger 1998.

Chapter 5 - Causes of economic imbalances | 137

[27] Cf Thirlwall 1994:119f.
[28] For the importance of "embodied technical progress", that is, technological advance translated into capital, see Thirlwall 1994:74f.
[29] For improvements in the quality of resources see Thirlwall 1994:77.
[30] For improvements in the quality of labour see Thirlwall 1994:76.
[31] Cf Ruffin 1983:25, 32f.
[32] Thirlwall 1994:114.
[33] Daly & Cobb 1989:72.
[34] Ruffin 1983:305.
[35] See the work of Schumacher (1973) and his Intermediate Technology Institute.
[36] Thirlwall 1994:104f; Nurkse 1955.
[37] Johns quotes Hirschman (1977): Samuelson put up (in 1948/9) a "brilliant theoretical capstone" saying that trade is "a potential force toward the equalization of incomes around the world" while consciousness of the "persisting and widening international inequality of incomes was becoming acute" (Johns 1985:128). Kaldor speaks of the "irrelevance of equilibrium economics" and the "vision of a dynamic world driven by cumulative processes" (Krugman 1993:9f). Of course, Gunnar Myrdal pointed out these phenomena much earlier. For a critical discussion of the theory of comparative advantage see Daly & Cobb 1989:209ff; Thirlwall 1994:366f, Johns 1985:149ff.
[38] For a summary of the debate on the international division of labour as an example of economic power relationships - including Emmanuel, Wallerstein, the Marxian approach, dependency theory etc. - see Caporaso 1987:1-42.
[39] Krugman 1993:9.
[40] Krugman 1993:26f. For the theory of self-fulfilling prophesy see Krugman 1993:101.
[41] Fröbel 1981.
[42] Junne in Caporaso 1987:72.
[43] For an overview of the debate see Lovett 1987:21ff and Faber 1990:11ff: the "comparative advantage" school (Ricardo, Heckscher, Ohlin, Samuelson) is contradicted by "trade pessimism" (Singer, Myrdal, Prebisch, dependency theory, Marxism, Barrett Brown), which is again countered by the neo-orthodox school with its insistence on free trade. Faber argues that the pessimists are wrong, but he takes the newly industrialising states and the oil rich countries as his example, which is misleading.
[44] Cf also Coote 1992.
[45] "Most of the $15 billion in technical assistance is spent on equipment, technology and experts from industrial countries - rather than on national capacity building in developing countries." *United Nations Development Programme* (UNDP) 1993:7.
[46] *United Nations Development Programme* (UNDP) 1993:7.
[47] Sanderson 1990:70. Agricultural subsidies in OECD countries rose from 28% in 1980 to 47% in 1986, in Japan they rose from 54% to 76% (Sanderson 1990:2). Even within rich countries these measures benefit the most powerful producers. In the US 14% of the farms account for 75% of the agricultural output and 75% of the benefits go to 25% of the farms (Sanderson 1990:2, 322).
[48] Sanderson 1990:327.
[49] For the current exchange rate system see Goldstein 1995.
[50] Say Germany can produce both wheat and motor cars more cost-effectively than South Africa. Then it would still be advantageous for South Africa to trade wheat against cars if it can produce wheat more economically relative to cars.
[51] "Because comparative advantage is a beautiful idea that it seems only economists understand, economists cling to the idea even more strongly, as a kind of badge that defines their professional identity and ratifies their intellectual superiority ... part of the economist's credo" (Krugman 1994:241). Imperfect competition leads to increasing returns, rather than to the expected decline in marginal returns. "In international economics ... from Ricardo until the 1980s (there) was an almost exclusive emphasis on comparative advantage, rather than increasing returns, as an explanation for trade."
[52] Krugman 1994:235.
[53] (Krugman 1993:74). "... over some range closer integration actually leads production to move perversely from the point of view of comparative cost." (Krugman 1993:97).

138 | Part I - The structure of the global system

54 "By 1990 ... 76% of advanced country export went to other advanced nations" (Krugman 1994:231).
55 The most famous of these speculators is George Soros, who ironically uses some of his massive wealth to "do good".
56 Donna Block in *Mail and Guardian,* July 3, 1998, p 6.
57 See Griffith-Jones 1988; Child in Paul, Miller & Paul 1992:114ff; Klay 1986:183ff; de Paiva Abreu in Tussie et al 1993:137ff (for Brazil); Lovett 1987:55ff, 137ff; World Bank (world debt tables 1988-89); Batista 1992 (Brazil).
58 For a balanced overview of the role of these two institutions, see *The Economist,* October 12, 1991, pp 1-54.
59 What are the facts? Within the single decade between 1980-1989 the debt of poor countries rose by 1800% (Elliott 1994:54). The debt of Kenya, for instance, amounted to 71.7% of its Gross Domestic Product and 272% of its exports in 1992 (Lehman 1993:122ff. Between 1982 and 1987, the total foreign debt as percentage of exports went from 85% to 278% in Venezuela (Child in Paul, Miller & Paul 1992:115). According to the World Bank there are 19 "severely indebted middle-income countries" Argentina, Bolivia, Brazil, Chile, Congo, Costa Rica, Ivory Coast, Ecuador, Honduras, Hungary, Mexico, Morocco, Nicaragua, Peru, Philippines, Poland, Senegal, Uruguay, Venezuela (World Bank 1990:x; 13ff).
60 Korten 1990:57ff.
61 World Bank 1990:19. For the relation between debt and poverty see page 126f.
62 Korten 1990:20.
63 Korten 1990:57, based on World Bank: *World Development Report* 1989:164f and 198f.
64 Child in Paul, Miller & Paul 1992:137ff. Taking military spending as his point of departure, Klay traces the following sequence of events (Klay 1986:183ff; for another discussion of the budget deficit under Reagan see Klugman 1994:151ff). President Reagan adopted a get-tough policy against the Soviet Union; this led to rising US federal defence spending, which was financed by rising budget deficits; this led to rising interest rates, which drew savings out of other nations and pushed up the dollar's exchange value. As a result, US export sales dropped and imports soared. In response, the US raised trade barriers while pushing for lower restrictions against US exports. This led to retaliatory action from its industrialised trading partners, but the Third World had no clout to do the same. One could add that the policy neatly fitted into the fad of supply-side economics described by Krugman 1994:82ff.
65 Tussie & Glover 1993:1ff, de Castro in Fontaine 1992:166.
66 We cannot spell out the detail of a very complex process. A useful "Survey of the IMF and the World Bank", including a history of their relationships with the private banks, is found in *The Economist,* October 12, 1991, pp 1-54.
67 Korten 1990:19.
68 This does not imply that insistence on greater efficiency and sound financial management, as enshrined in Structural Adjustment Packages, should be rejected. For an overview of the SAPs see Mosley in Fontaine 1992:27ff.
69 M Hirsh in *Newsweek,* March 20, 1995, p 13.
70 Broadley 1992:153
71 Elliott 1994:54. See the graph in Von Weizsäcker 1990:119; or the charts showing net resource flows and net transfers in indebted LDCs from 1980-1989 in World Bank 1991:126
72 The economist K W Kapp summarises the weaknesses of the market rather neatly: (a) it does not respond to needs but to purchasing power, (b) elitist and monopolist markets distort relative shortages and question consumer sovereignty, (c) the market does not take into account the distribution of resources over various generations and (d) the market ignores social costs (quoted by Duchrow 1987:153).
73 See Korten 1995 for detail.
74 Examples of genocide are the destruction of the Inca culture in Peru by the Spaniards, the virtual eradication of the San in the Cape by Dutch settlers and the devastation of the Hereros by the Germans in Namibia.

Chapter 5 - Causes of economic imbalances | 139

[75] Stackhouse 1987:129ff.
[76] All wars during the second half of the century have been fought, or are being fought, in the Third World. And only a tiny minority of these have been fought between countries.
[77] Korten 1990:22.
[78] Cappelen et al: Military spending and economic growth in the OECD countries. *Journal of Peace Research* 21/1984:361-373.
[79] Findings of the Brookings Institute reported by M Kettle in *Mail & Guardian,* July 3-9, 1998, page 17 and G Dyer in *The Natal Witness,* July 9, 1998, page 8.
[80] Apostolakis 1992:93. According to this author Rusett finds that in 1970 an increase of $ 1 in Milex implied a reduction of 42c in personal consumption, 29c in investment, 10c in exports and 13c in government expenditure (p 87).
[81] H M McFerson 1992: Focus on democracy and development in Africa. *Journal of Peace Research* 29/1992: 245.
[82] K Gyimah-Brempong: Do African governments favor defense in budgeting? *Journal of Peace Research* 29/1992 191-206.

PART II

The transformation of collective consciousness

# The social reality of convictions

The rationale of Part II

In Part I we discussed *social structures and processes*. Our frame of reference was the relation between material needs and the capacity to fulfil those needs. In Part II we analyse what happens *in our minds*. The two levels of the problem are closely interlinked. A social system can be described as a conglomeration of incarnate collective mindsets. It is collective mindsets which provide social systems and processes with coherence, legitimacy and stability. Without internalised mindsets (that is, shared assumptions, beliefs, values, norms, conventions, statuses, roles, procedures, goals and so forth), no social system could operate. By the same token, a change of collective consciousness is critical for social transformation.

To be real and effective, mindsets must entrench themselves in social systems. Once entrenched, mindsets determine the kind of mental structure (or culture) into which individuals must be socialised if they are to operate successfully within the social system. Therefore some materialists have assumed that social structures are primary, mental structures secondary; when social structures change, collective consciousness follows suit. For the bulk of an unconscientised and submissive population this may be true. But if there was nobody with a dissenting set of beliefs, values and norms and the determination to impact social reality, nobody would bother to attempt structural change in the first place.

It is also true that a social structure can generate its own dynamic, often in contradiction to the intentions of the mindset which it represents. Then it functions like a heavy vehicle getting out of control due to its inherent dynamics. But to forestal or redress such a development again demands the conviction and determination of the human agent.

So it seems as if the *transformation of collective mindsets* is the one indispensable ingredient of deliberate structural change. The motivation for change can be based on convictions or on interests. As we shall see in chapter 10, interests are adulterated convictions. Without mental transformation, social transformation follows its own structural dynamics, unhindered and undirected by human responsibility and control.

But humans are not supposed to be animals, who merely react to external circumstances guided by their instincts. For better or for worse, we have to make sense of a situation, become aware of deficiencies in the structures of reality, design visions of what ought to be, and aspire towards their realisations. Collective human will, following a particular vision, is the only hope we have to reach a just and sustainable future. That is why we deal with the transformation of collective consciousness first (Parts II and III) and with the transformation of social structures thereafter (Part IV).

In the *present chapter* we analyse the relation between various components of collective consciousness (convictions, interests and "popular wisdom"). In *chapter 7* we deal with a particular kind of conviction, the biblical faith, and its potential social impact. In chapters 8 - 10 we analyse the encounter between unequal partners at the level of collective consciousness. Indeed, even at the level of collective consciousness, power relations are pivotal. In *chapter 8* we look at the specific cultural contents of two mindsets which have interacted ever since the emergence of Western colonialism, namely traditionalism and modernity. *Chapter 9* highlights the psychological interaction between dominance and dependency and its socio-economic impact. *Chapter 10* deals with the distortion of perceptions by the ideological legitimation of collective interests. This will bring us to the end of Part II. Part III will then move from the popular to the academic level.

Throughout Part II, social reality will be confronted with *Christian assumptions*. The choice of the biblical faith as the major point of reference in Part II will be discredited by some readers as the outgrowth of sectarian preoccupations. But the idea of a non-sectarian approach is an illusion. What is given out as "our common humanity" is nothing but another conviction, namely the secular humanism of modernity. Among academics, and elites in general, modernity is currently in a dominant position. Dominant mindsets can afford to take themselves for granted and consider their rivals as deviant, flawed, or superstitious.

It will soon become clear, however, that modernity itself must be subjected to critique. Critique and reconstruction presuppose a particular system of meaning. Those who have a desire to float in limitless relativity will not be in a position to take a stand or make an impact. Since I am a Christian theologian, my choice of the Christian faith for an alternative system of meaning may seem obvious. And indeed, it is my theology which has led me to embark on the vast and secular project of coming to grips with the modern economy. Conversely, my economic and ecological studies have forced me to rethink my theology.

But the choice of Christianity makes sense apart from personal reasons. It is not by accident that modernity emerged and evolved in a cultural context which, by the time of the Enlightenment, had been co-determined by Christianity and classical antiquity for at least a millennium and a half. Christian theology must assume co-responsibility for the modernist revolution. It does not only have a *historical mandate*, but also a *historical obligation* to be concerned with the forces which led to the current situation.[1] The same cannot be said of other major convictions, such as Hinduism or Buddhism. As one of the architects of the modern world, Christianity is itself under critique; it has to review its past role; it has to rediscover its underlying assumptions; it has to develop its potential to arouse new motivations and give new directions.

Immanent and transcendent needs

Human consciousness is formed in response to experienced needs.[2] To understand the social reality of collective consciousness, let us begin with the crude differentiation between two levels of needs, that is, immanent and transcendent needs:

(a) *Immanent needs* belong to the sphere of reality which is, in principle, under human control, thus accessible to human understanding, investigation and manipulation. They include physical, psychological, intellectual, communal, cultural, social, political, economic, and ecological needs. So far we have been moving within this realm. Immanent needs depend on one's cultural peculiarities as much as on one's location in the social power structure. Bow and arrow, prerequisites of survival among the San, can at best form an ornament on the wall of a modern industrialist.

(b) *Transcendent needs* belong to the sphere of human reality which goes beyond human accessibility and control. They include (i) an authoritative system of meaning, (ii) the assurance of an individual's or a group's right of existence (acceptability and belonging) and the legitimacy of its social structures, and (iii) the authority to use the powers at an individual's or a group's disposal to achieve their goals. Humans cannot live without meaning, assurance and authority. In all cultures, world views and religions, these three entities are perceived to be derived from sources beyond humanity's own resources. Truth cannot be invented; it has to disclose itself. Assurance cannot be generated; it has to be granted. Authority cannot be usurped; it has to be allocated.

The relation between immanent and transcendent needs is close. The dualism between matter and spirit, which Western civilisation has inherited from the ancient Greeks, is no longer plausible. Transcendent needs do not occur in a spiritual sphere of their own, independent of immanent needs; they form the "depth dimension" of immanent needs.[3] It is precisely the experience of immanent needs, such as deprivation, misery, rejection, disease, inexplicable fate, vulnerability and mortality, which forms the mainspring of the human quest for meaning, reassurance and authority, not some abstract spiritual concerns. When in need, people consult oracles, bring sacrifices, or do research.

Moreover, the three transcendent needs follow a logical sequence: a system of meaning generates its own values and norms. Values and norms again constitute the criteria according to which the right of existence is either confirmed or questioned. Authority in turn is based on one's right of existence. Affirmation of one's right of existence is linked to a particular status and role in society. As such it is the source of authority to take over responsibility within allotted social positions and tasks.

Through its normative character, the system of meaning determines both individual behaviour and social structure. In some cultures, for example, honour demands that a leader be followed unquestioningly, that the death of a kinsman be avenged, that an adversary be challenged in a duel.[4] In other cultures honour is linked to critical autonomy, revenge is a sign of moral immaturity and a duel is ridiculous.

The particular way in which these three needs are fulfilled in a specific culture is internalised as a relatively stable set of mental structures and processes, which we call a *mindset*. This mindset interprets, legitimates and directs the relevant social structure. While it is true that a change of social structures inevitably leads to a change of assumptions and attitudes, especially among the masses, it is also true that without a change of assumptions, attitudes and motivations leaders and their followers will not bother to initiate social change and face the costs. Therefore mental and social structures and processes emerge, evolve, and deteriorate together.

Truth and conviction

Systems of meaning integrate experiences and observations into a picture of reality which appears to be more or less consistent and plausible. The structure of the system is determined by certain assumptions which are taken for granted. These "premises" are ultimate in the sense that they cannot be transcended towards, or derived from, something more profound. Nor can they be questioned without endangering the system of meaning which they sustain. That is why we call them "ultimates".

Convictions are accepted as truth, unless and until they are challenged or dethroned by rival truth claims with a greater power of conviction. They can also be dethroned by the weight of evidence. Because truth claims abound, especially in a pluralistic society, there is a constant struggle between convictions both in the individual and the collective consciousness.

The relative power of convictions varies, partly according to their ability to integrate existential experiences into a plausible system of meaning, partly according to their ability to establish emotional security by granting acceptance, belonging and motivation, that is, right of existence and authority. The current system of meaning is always in flux; it always reflects the latest outcome of the struggle for validity; it is never a purely rational construction.

The quest for truth is indispensable. Without a structured perception of reality human life falls apart. But truth is not the same as a set of doctrines which one has to accept, whether they make sense or not. Truth cannot be imposed; it has to establish its validity in human consciousness. In a plural society, this can only happen in an open dialogue between partners with different backgrounds who are committed to the quest for truth and, precisely for this reason, expose themselves to truth claims other than their own. An indifferent kind of tolerance or a naive kind of goodwill is not sufficient.

Nor is the fad of deconstructionist *post-modernism*. The latter can afford to yield to the lure of radical relativity only because the social system and its ideological foundations are still largely intact. Moreover, the dissolution of validity is suspiciously akin to the deliberate dismantling of inhibitions and the creation of wants by the capitalist drive to push up sales. One learns to fancy a gadget, a sexual partner, or a truth claim, acquire it, and discard it soon after. Nothing of value can be built on such foundations.

Similarly, the assumption that all convictions aim at the same transcendent truth is a convenient but dangerous illusion. It is an *illusion,* because it is not just the articulation (the imagery, vocabulary and symbolism) of convictions which differ, but the contents. If you want to reach Nirvana, the Christian faith is the wrong bus to board. Moreover, convictions often contradict each other fundamentally, not just superficially. It is *dangerous* because it conceals detrimental aspects of particular commitments. Convictions such as fascism, nationalism, Stalinism, some forms of fundamentalism and various cults can create havoc in human minds and human communities. All convictions must be required to expose themselves to mutual critique. A humanity in crisis can least afford to abandon its quest for valid, liberating and empowering truth. It is *convenient,* because it seems to obviate unpleasant confrontations and suggests an atmosphere of tolerance. But tolerance is not the same as indifference. The word tolerance is derived from a Latin word for "to bear a burden"; to be tolerant means to suffer differences, not to ignore them.

While we can never possess ultimate truth, we cannot do without truth altogether, thus without a struggle for the truth. What we need, therefore, is a committed and critical interaction between different convictions. All partners must recognise the difference between ultimate truth and their current perceptions of the truth. They must allow aspirant truth claims to try and prove their plausibility and validity in confrontation with rival truth claims. A truth which cannot hold its own in the face of critique is not the truth. What we espouse, therefore, is a confrontation not between human beings, but between truth claims, that is, a "battle between the gods".

Theoretically, confrontations of different truth claims under the umbrella of mutual acceptance can happen in any democratic society where the values of tolerance and human rights are firmly entrenched. Religions have often opposed free interaction because it subverts and relativises not only biased interpretations of social reality, but also the truth claims of specific convictions with severe social repercussions. Social identity is based on conviction and the breakdown of established social identities leads to anomie. The de-absolutisation and relativisation of truth claims have also lead to the privatisation of convictions, that is, to their public disempowerment. Pluralistic relativity again leads to value deterioration, the dissolution of the moral fabric of society and the degeneration of public wisdom to the level of immediate desire satisfaction. As a result, religions have often become fanatic. Witness the spread of fundamentalism in many parts of the world!

Therefore mutual acceptance cannot be taken for granted. Many religions and worldviews are not accidentally, but essentially intolerant. Democratic assumptions belong to a particular mindset among others. As such they challenge other truth claims. It takes considerable detachment to realise that a truth claim which needs propagandistic, institutional and military crutches to survive cannot be the truth. In the case of the Christian faith, it is the core assumption of the New Testament that God accepts us in spite of being unacceptable that leads to mutual tolerance.

Interests and ideology

Because they are foundational, convictions organise experienced reality as a whole and assign specific statuses and roles to individuals and groups within that whole.[5] With that they form a set of values and norms according to which the right of existence (acceptability, legitimacy, belonging) is granted or withheld.

But there is another phenomenon which vies with convictions in this respect, namely *interests*. Interests are not identical with needs; they are interpreted and prioritised needs. In most cultures the satisfaction of the need for food, for instance, is surrounded by a wealth of symbolism and ritual, which can include the extremes of abstinence, on the one hand, and lavish orgies on the other. This interpretation and prioritisation of needs again depends on the system of meaning. In African traditional cultures, for instance, cattle counterbalance human fertility, play an instrumental role in marriage transactions and help to structure the community; on modern commercial farms they are nothing but a profit-generating asset.

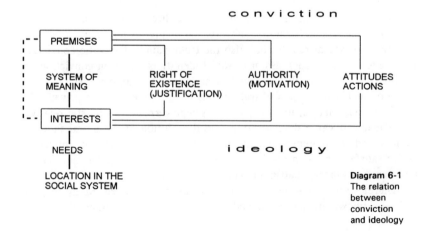

Diagram 6-1
The relation between conviction and ideology

While interests have a greater immediacy and vitality than convictions (hunger bites, whatever we believe), they must be justified before the forum of the prevalent conviction, otherwise one lacks the authority to satisfy them. A monk, for instance, may have sexual needs, but sexual gratification is excluded from legitimate interests by the monk's conviction. Diagram 6-1 shows how these entities are interlinked with each other.

Convictions ideally cover the whole of one's life world, while interests are inherently self-centred. Often they focus on the immediate gratification of desires. Therefore there is a constant tug of war between convictions and interests. While convictions have jurisdiction over the legitimacy of interests, interests manipulate convictions in their favour.[67] Imperialism, racism, capitalism and Marxism have all been justified, for instance, in the name of the Christian faith. We should not be surprised, therefore, that when the social structure changes, and interests begin to shift, convictions are reinterpreted and the system of meaning is adapted to the new situation. We call the product of the manipulation of a system of meaning and its system of values and norms to benefit particular interests an *ideology*.[8] We shall deal with the impact of this phenomenon on socio-economic realities in chapter 10.

The use of power in society presupposes vertical relationships, that is relationships between dominant and dependent sections of society. Karl Marx taught us that dominant convictions in society are the convictions of dominant groups in society. Or, to be more precise, convictions that appear to be the convictions of dominant groups! Elites can keep their true convictions secret and project "official" convictions for public consumption. Not all functionaries in Marxist-Leninist societies have been convinced Marxists! Dominated groups can also keep their true convictions secret and pander to dominant observances.[9] Hermeneutics speaks of "hidden transcripts" in such cases. Whatever the case may be, dominant convictions are widely utilised to legitimate the interests of the most powerful sections of society.[10]

The impact of a legitimating ideology corresponds, therefore, with the social power at the disposal of the groups whose interests are being legitimated. This can happen through overt propaganda or quiet diffusion. The latter happens when elites become reference groups for the rest of the population. To be "smart" one has to think, argue and act the way the public idols do. Modern marketing is, to a large extent, the art of manipulating the perceptions of "what is done", the art of engineering fashions and fads, sometimes adding an air of seriousness and profundity, more often simply exploiting base desires.

But the ambitious among underdog population groups can also utilise dominant convictions to defeat elites with their own weapons. In England the middle class claimed rights which aristocrats had previously claimed for themselves against royalty. Subsequent waves of emancipation, emanating from the proletariat, the colonial elites, the women, and the youth, have each in their turn utilised the same rhetoric. Another possibility is that oppressed groups keep their more fundamental convictions alive as the mainspring of latent resistance and revolution. The ancient Jews are the best example.

Popular wisdom

In heterogeneous or pluralistic societies there is an open market on which all kinds of beliefs and values can be traded. One would expect such a society to be inherently unstable, because there is no common frame of reference. Experience in such societies shows, however, that this is not necessarily the case.[11] Through constant social interaction between people of different persuasions a minimal public consensus emerges which reduces friction and facilitates communication and cooperation. We call it *popular wisdom* (diagram 6-2).[12] It can be expected that the dominant set of convictions in a society will acquire the greatest share in the composition of popular wisdom.

Diagram 6-2
The genesis of popular wisdom

The implication of this phenomenon is that more profound systems of meaning lose much of their impact on the public realm to relatively shallow and adulterated forms of collective consciousness.[13] The former tend to withdraw to the private devotion of individuals, or the piety of like-minded communities.[14] "No politics from the pulpit!" Popular wisdom, in contrast, takes over the role of legitimating social structures and processes. Because of its lack of foundation and cohesion, it can be manipulated much more easily by powerful interests in society. "Civil religion" often becomes the hand maiden of populist regimes and dominant classes. Certainly it is used as a quarry by profit seeking advertisers.

The privatisation of religion

The privatisation of religion is an anomaly. Religion tends to be either the legitimating ideology of a political order, or the determinant of personal relationships in primary groups. Usually it is both. The secularist idea that one can ignore religious conviction in *politics* with impunity is naive, to say the least.[15] Most great conflicts - think of Bosnia, Rwanda, the Indian subcontinent, the Sudan - are fuelled by religious or pseudo-religious sentiments. As mentioned above, secular modernity is only the currently dominant set of convictions among other potential candidates.

Primary group solidarity, on the other hand, has proved to be the most formidable motivating or demotivating power in human society. It operates on all levels from ruling elites to grass roots formations. Depending on the collective power of the agents of change, structural transformation can ensue from the top, or from the bottom, or from anywhere in between. We shall utilise these insights in chapter 14.

Religion as the legitimating ideology of a social order has a vested interest in neutralising its rivals. It wants to be taken for granted as the self-evident set of assumptions on which public consensus is built. It claims to be the "natural" and "rational" and "responsible" way of looking at reality. Other assumptions are tolerated as far as they do not lay claim to public relevance. It is well known that this "imperialist" attitude has been adopted, among others, by Judaism, Christianity, Islam, Fascism and Marxism. Today it is the liberal-capitalist and scientific-technological mindset of modernity which imposes its hegemony most pervasively upon collective consciousness.

Once a conviction has been relegated to the private sphere, transpersonal structures and processes do not seem to fall within its sphere of competence - even where the particular conviction is shared by large collectives. It is a queer fact that a religion may accept its subservient and socially irrelevant role without much protest.[16] This is not simply due to a lack of personal integrity. The might of the social power structure seems to leave believers with little choice. Social cohesion and social control always emanate from some sort of collective mindset which has incarnated itself in the social fabric. Private faith cannot compete with such a giant mental and social formation.

To make an impact, therefore, faith has to pose a serious challenge to the dominant set of certainties within as many primary groups as possible.[17] The goal must be to reach the "critical mass" which is needed to make a social process change direction.[18] For this to happen, faith must offer a cosmic vision which is able to generate a cosmic mission. This is how early Christianity, Islam, capitalist liberalism and Marxism have conquered large parts of the world. There is no other way. This does not mean that we should offer a particularly Christian economics or politics. There is no such thing! There is only a Christian vision, a Christian responsibility, a Christian dedication, a Christian way of serving the interests of the whole community rather than self-interest. But the economic detail must be worked out on the basis of observation and reason.[19]

Theoretically dominance can be achieved by force. But a revolution is only legitimate to the extent that it is supported, at least tolerated, by the majority. If a new system of meaning is imposed on an unconvinced population, it becomes counterproductive, regardless of its merits. This is one of the reasons why Marxism has faltered.

But that is only the structural aspect of the matter. More important is the fact that imposed faith is a contradiction in terms. There is really only one way of challenging a dominant system of meaning: the new mindset must develop a greater power of conviction than the incumbent one. There must be a genuine paradigm shift - due to the greater plausibility of the challenging system of meaning. This also means that a system of meaning should not strive for hegemony. It should fall back on nothing but the power of its truth. It must make sense in the face of all its possible rivals. If not, it becomes fanatic, repressive and ultimately self-destructive.

Science and religion

What is the relation between science and conviction? At face value it seems as if empirical science were simply engaged in the exploration of reality. Science is deemed to be based on fact, convictions on assumptions. But it is not as simple as that. Both science and convictions follow traditions. Science also makes assumptions; conviction also interprets experience. There is an element of *fact* and an element of *faith* in both.[20] Serious social analysts have claimed that science has become the religion of modern humankind.[21] Science is the authority in which they invest their trust, on which they base their hope, from which they expect their miracles.

The task of science is to understand reality. The task of religion is to transcend reality. Assuming that there is only one reality to which they both refer, science and conviction should be able to complement each other. If

Chapter 6 - The social reality of convictions | 153

they cannot, there must be something wrong with either, probably with both. Both science and religion can become ideological; both can be subject to cultural lag; both can become bogged down by internalised thought structures; both can become removed from experience; both can become misleading or irrelevant. They must both be called to task; they must both be accountable to the needs of the situation, otherwise they lose their legitimacy.

Revision: Describe the different elements which make up mindsets and their relation to social structures and processes.

Application: Try to specify the ultimate premises, the systems of meaning, the values, and the norms which give shape to capitalist liberalism and Marxism-Leninism respectively.

Critique: (a) "This description of mindsets is abstract, complicated and useless for any practical purposes." (b) "Your attempt to save the relevance of religion for public concerns is quite futile. People act in self-interest and no amount of philosophising and moralising will change that." How would you react to these statements?

Let us summarise

This short chapter introduces a series of discussions on the transformation of collective consciousness. There is a close interaction between collective consciousness and social structures. Mindsets tend to entrench themselves in a social system. Deliberate social-structural change comes about when *alternative mindsets challenge entrenched mindsets*. Therefore the transformation of collective consciousness is critical for social transformation.

Human consciousness evolves in response to experienced needs. We have distinguished between immanent needs and transcendent needs. *Immanent needs* include needs whose fulfilment is within human reach, such as physical, social, economic, political and ecological needs. *Transcendent needs* include the needs for meaning, for acceptability, and for authority - in that logical order - all of which are derived from a transcendent source.

Convictions respond to transcendent needs. These form the "depth dimension" of immanent needs. On the other side of the spectrum we find collective interests. *Interests* are interpreted and prioritised needs. So they contain an element of meaning, which is usually derived from adulterated convictions. The distortion of meaning to legitimate the pursuit of interests is called *ideology*. Generally, convictions have greater validity, interests have greater power. In a plural society interaction and dialogue lead to a minimal consensus which we called *popular wisdom*. While it enables the society to function, it is shallow and subject to abuse. Convictions, if they are serious, cannot abandon the quest for truth, but they have to struggle for truth together.

A dominant system of meaning entrenches itself in social structures and processes and relegates its rivals to *the private sphere*. To gain power and change the system, a marginalised conviction has to achieve a *critical mass* within the population which is able to challenge, redirect or oust the dominant set of assumptions. Finally, we drew attention to the fact that neither science nor religion is immune against ideological distortion.

Notes

[1] Cf Duchrow 1994:266.
[2] Cf the theological reflections of Meeks (1989:157ff) on needs.
[3] The term "depth dimension" goes back to Paul Tillich.
[4] Honour as a value in Western society has very deep historical roots, going back to the Germanic culture. See Russell 1994:119ff.
[5] The Israelite prophet, for instance, delivered divine messages to the "people of God", who in turn were supposed to witness to the moral order established by the Creator of the universe. The Marxist avantgarde was to assume leadership in the proletarian struggle for a classless society which was to be the fulfilment of history.
[6] Cf Sleeman (1976:141).
[7] For a deeper analysis of these processes see Part II of Nürnberger 1988.
[8] "Ideologies are systems of thought and belief by which dominant classes explain to themselves how their social system operates and what principles it exemplifies ... Ideological systems are ... not only evidential truths but moral truths" (Heilbroner 1985:107). But this they can only do because they link up with current public systems of meaning.
[9] "The goal of slaves and other subordinate groups, as they conduct their ideological and material resistance, is precisely to escape detection" (Scott 1990:193). Among the former slaves in Latin America the gods of popular religion, imported from Africa and adapted to the local situation, have survived and prospered under the guise of Catholic saints.
[10] CF Heilbroner 1985:107ff. For the whole phenomenon see Larraine 1979.
[11] For different interpretations of this phenomenon see Küng 1997:131ff.
[12] See the grandiose attempt of Küng to utilise the potential of popular wisdom to build a new global ethic for politics and economics (1997:91ff).
[13] Küng, who believes in the possibility of a global ethics, does not believe that ethics can be a substitute for religion. In particular he states that "No universal ethic, but only religion, can communicate a specific depth dimension, a comprehensive horizon of interpretation in the face of the positive ... and the negative ..." Religion provides ultimate meaning, guarantees unconditional values, norms, motivations and ideals, creates a home of spiritual security, and mobilises protest and resistance against unjust condi-

tions, even when these seem to be fruitless (1997:142f).
[14] Cf the interpretation of secularization by Berger 1973:111ff.
[15] Cf Küng 1997:115ff.
[16] Cf Meeks 1989:19ff.
[17] "Christians of necessity must be radical nonconformists" and must "refuse to be squeezed into the mould of our affluent, sinful culture" (Sider 1977:32).*
[18] Marwell & Oliver 1993.
[19] Storkey (1986:67 et passim) can be misunderstood as if his proposals were "Christian" in themselves. This would be problematic, however noteworthy - and worthy of critique - they were in themselves. Storkey's intention is, however, to develop a paradigm which is only Christian "in inspiration" and concedes that the "detailed arguments may be amiss" (1986:pp 202ff).
[20] Storkey 1986:68: all knowledge involves faith commitments.
[21] See, for instance, the works of Jürgen Habermas.

# Activating the resources of faith

What is the task of this chapter?

Chapter 6 sketched the main aspects of the mental dimension of social reality. In this chapter we reflect on the role of *a particular conviction*, the Christian faith, in the transformation of mindsets at the popular level. In chapter 11 we shall do the same with *a particular social science,* namely economics. In human consciousness, fact and faith work in tandem. Fact without faith lacks direction and inspiration; faith without fact is illusory.

The reasons for my choice of the Christian faith have been spelt out in the last chapter. Jewish-Christian assumptions are, for better or for worse, woven into the history of Western thought and behaviour, including fascism, secular humanism, capitalist liberalism, and Marxist socialism - as well as their hedonist, utilitarian, individualistic, colonialist and imperialist implications.[1] The theses of Weber and Tawney on the origins of capitalism in the "Protestant Ethic" are well known and Stackhouse traces the history of the modern corporation to the New Testament church and the development of technology to faith in a Creator God.[2]

I have argued, therefore, that Christianity has a historical obligation to review its role in the generation of the international order and a historical mandate to press for its transformation. On the basis of its faith, the Christian community must revisit its heritage, account for its motivations, face its responsibility, abandon its baggage, unearth its redemptive resources, and make its particular contribution. I hope that representatives of other religions will take up the challenge and do the same for their own convictions.

But my analyses are not meant for Christians alone. In a world growing together it is important that we come to know each other's commitments and confront them with the needs arising from our common predicaments. In view of the gravity and urgency of the situation we should be brave enough to subject each other's convictions to challenge and critique. If we remain sensitive to the common goal, the wellbeing of humankind, confrontations between convictions *can be* loving, pertinent, honest and liberating. They do not *have* to be arrogant, hateful and deprecating. Even atheists, agnostics and skeptics are invited to check whether faith has to be the great obstacle on the path of progress, the root of irrational behaviour, the repository of an obsolete worldview.

This chapter will not enter into the vast inner-theological discussions of the last three decades,[3] but respond creatively to the challenges posed by the economic and ecological syndrome analysed in this book. It concentrates on the single most important contribution the biblical faith can make to human existence, that is, *the capacity to transcend given reality*. We shall argue that transcendence is indispensable for human freedom and human responsibility. The definition of "economic man" as a profit, utility and pleasure maximiser closes human beings in upon themselves.[4] Without transcendence, reality imprisons and enslaves us. Transcendence shows that reality could be quite different; it discloses unrealised potentials; it opens up the future.

Section I of this chapter deals with an important aspect of the current situation, namely the lure of despair and despondency. Section II offers a brief description of the core assumptions of the Christian faith. Section III shows how faith helps us to transcend the limitations imposed by time, space, power, ethnicity and privilege. This discussion will continue in the following three chapters, where the response of faith to the assumptions of traditionalism and modernity (chapter 8), its capacity to horizontalise vertical relationships (chapter 9) and its role in overcoming ideological distortions (chapter 10) will be spelt out.

Section I: The lure of hopelessness

When we look back over the structures and processes analysed in previous chapters we cannot help but be haunted by the feeling of profound powerlessness, hopelessness and futility. The present situation is the cumulative result of collective decisions and actions over long periods of history.

Complicated networks of cause and effect seem to push us towards a future which a sane person cannot want to desire. Large power constellations struggle with each other for dominance and crush the efforts of well-meaning groups and individuals. Revolutions claim untold sacrifices and end up in disillusionment. Collective mindsets underpin these structural developments and ideological enslavements prevent them from being challenged effectively.[5]

But the immensity and complexity of social structures and processes is only one side of the problem. The other is the plurality of systems of meaning on offer. The current explosion of scientific insight inundates us with an avalanche of information which we can no longer digest. Reasonably coherent sets of assumptions find themselves surrounded by an increasing number of equally impressive alternatives. Hard hitting superficialities promise fulfilment on a noisy market of beliefs. A maze of detached and drifting assumptions, values and norms combine into makeshift worldviews with extremely short life times.

Postmodernism wants us to believe that every assumption under the sun has an equal claim to validity. There is no truth, there is only an entropic proliferation of mental constructs. Under these circumstances humanity cannot be expected to get its act together. It is also no answer to the problem of truth in a pluralistic society.[6] There is a parallel between capitalist commodification of everything under the sun and the post-modernist relativisation and atomisation of the truth.[7] Postmodern deconstructionism treats convictions, values and norms like commodities on the market place which have to sell themselves to overfed customers and which, once acquired, have already become stale and obsolete in view of more tempting alternatives. Moreover, the advertising and entertainment industries deliberately target value assumptions, dismantle internalised taboos and psychological barriers and reinforce primitive craving, irrational acquisition, over-consumption and wastage.

The third problem is the power of shortsighted and superficial interests which swamps any collective sense of responsibility. Politicians who reach the levers of power play the game which grants them their privileges. In economics global welfare considerations are ignored as much as in commerce and industry. The majority of people prefer to live their private lives as best they can under given circumstances. The poor struggle to survive and have no time for the niceties of ecological awareness. The wealthy do not like to be disturbed. Devotees use religion as an escape route from a reality they cannot master. Irresponsibility manifests itself in the form of a lack of commitment and reliability, imprecision, inefficiency, crime, and corruption.

Faith and fatalism

So is the attempt to "do something about the situation" simply futile? Most people seem to think so. The appeals of social analysts and futurologists seem to fall on deaf ears just as much as the inciting sermons of politically engaged theologians. Where do we figure on the global stage with six billion largely ignorant, self-interested and enslaved actors!

Realism is indispensable to dispel wrong expectations, but it cannot have the last word.[8] It is precisely the role of faith to prevent realism from turning into fatalism.[9] The human being cannot live without a hope that transcends current limitations. Hopelessness leads to paralysis. But faith is not a kind of make-believe utopianism. Faith is protest against apparent inevitabilities. It is faith which detects that fatalism is not realistic; fatalism buries sensitivity for the possibilities of the future under the dead weight of despair.

Faith keeps the future open. It is God who is in charge, not automatic mechanisms.[10] If God is in charge, we are in charge - at least potentially. At any point in time the chain reactions of cause and effect do not simply predetermine the future cause of events but open up a vast array of choices. It is our human mental fixations which obstruct the realisation of these possibilities. And it is here where the battle lines must be drawn.

Collective mindsets, which underpin social structures and processes, are the outcome of ideological power struggles. They are not immutable. The decisive questions are (a) which mindset can muster the greatest power of conviction in response to pressing needs and (b) which mindset is capable of displacing its rivals as the formative assumptions of social power structures? Wiener, the founder of cybernetics, said that a low energy level with high resolve can control a high energy level without direction, just as the eye of the driver controls the muscles that operates the brakes that bring a great truck to a sudden standstill.[11] People of faith must learn to match the determination of those who are committed to the pursuit of self-interest.[12] Great movements have small beginnings. If a movement gathers momentum a "critical mass" can be reached which can make a substantial impact. As the demise of Marxism-Leninism in the Soviet Union and the apartheid system in South Africa has shown, entrenched social systems can change.

Faith and relativity

So the lure of hopelessness must be dispelled. But we must also get out of the swamp of relativity. Relativity saps our resolve. It is true that reality is in historical flux, that perspectives vary, that the equality of dignity implies the right to one's convictions, that integration demands tolerance, that there is no platform out there from which we could establish ultimate truth, that research outcomes depend on initial assumptions.

However, that does not mean that there is no truth. Truth is like a system of sign posts. While sign posts point in all kinds of directions, they are not arbitrary. Placed at a particular junction, a sign post can be absolutely reliable, though the same sign post placed at another junction may be misleading. Mental constructs have a shape which can be described, processes have a direction which can be determined, particular inputs lead to particular outputs, the application of certain assumptions have certain consequences.

The immense power of the industrial and commercial economy, for instance, is rooted in the basic assumptions of modernity. That the assumptions of traditionalism could lead to a similar kind of development is wishful thinking. Buddhist withdrawal from reality, Muslim fundamentalism, African ancestor veneration, Christian spiritualisation of salvation, Marxist usurpation of low class interests, the fascist will to power - all these mindsets have the most profound impact on social and natural developments.

To be conscious of historical flux and endless variability does not imply, therefore, that some values are not more conducive than others to the healthy evolutionary process, to creating an environment for life to flourish, to equity, to concern and to responsibility.

Because the Christian faith is geared to what ought to be, it has to indicate directions which are universally valid. Universally valid directions may be informed and checked by religious assumptions, but they must be in line with observation, analysis, visionary intuition and construction. Other faiths must check whether they are able to agree with these directions and enter into the debate. Instead of indulging in perpetual doubt and cynicism we should get on with the job.

The privatisation of the Christian faith

For reasons stated in the previous chapter, many Christians have restricted their faith to the private realm of church and family. They cannot have a good conscience for doing so. After all, their creed postulates the cosmic lordship of Christ - a lordship which is defined as the dedicated service of the authentic human being to the wellbeing of God's creation. Two traditions help Christians to evade this issue. The first is the apocalyptic notion of a divine future which relegates the Lordship of Christ to an imagined end of times. The other is the Hellenistic notion that the lordship of Christ is applicable only to the spiritual, not to the physical and social realms.

The statement attributed to Jesus, "My kingdom is not of this world!" (John 18:36) has been misquoted endless times to legitimate this dual evasion.

Chapter 7 - Activating the resources of faith | 161

The kingdom of God will come when God wills it to come, it is said, and humans can do nothing about the evil world as long as it lasts. The only valid response to the depravity of the world is to accept one's personal Saviour, gain "peace with God", and love each other. The peace of the rest of creation is not part of the Christian agenda.

With due respect to my own spiritual kith and kin, this kind of piety is nothing but a (probably unconscious) attempt to rationalise oneself out of responsibility.[13] It is true that humans cannot create the ideal conditions envisioned by the concept of the kingdom of God. But that does not mean that there are no approximations, or that a retreating horizon cannot lure us forward to new vistas. A vision is not meant to be reached, but to provide direction and inspiration. If such Christians were serious, they would go into the desert and wait for the kingdom, as the ancient Qumran sect, some hermits and some monastic orders have done. Instead they seem to enjoy this world as much as everybody else.

Genuine eschatological hope does not imply that our earth will be discarded by its Creator because it is worthless.[14] If that were the case, it could indeed be rendered uninhabitable by us with impunity. The same would be true if we followed the Greek idea that the material and temporal world was evil and that authenticity could only lie in a spiritual and eternal realm - whatever that may mean. Genuine eschatological hope looks forward to God's redemption of his creation, a creation which is precious in his eyes.

Human beings are part of God's creation. We are mammals and belong to the animal kingdom; our "souls" are embedded in our bodies; our bodies are part of communities; communities are part of society; society is dependent for its very survival on the rest of the natural world.[15] A spiritual realm, isolated from the concreteness of physical and social reality, is an idealistic delusion. It cannot claim to be biblical. In the Old Testament the concreteness of human existence is taken for granted. The New Testament authors fight a relentless battle against Gnostic tendencies which deprecate the body and emphasise a bodiless spirit. Similarly, early Christians did not believe in the immortality of the soul, but in the resurrection of the body and the earthly reality of the kingdom of God.[16]

If that is the case, there can be no "peace with God" unless God is at peace with his creation. While the world is not what God intends it to be, God cannot be at rest. God can rest when he finds his world to be "very good" (Gen 2:1f). Then the people of God can enter with God into his rest (Heb 4:1-11). Until such time *the restlessness of God translates into the restlessness of his people*. True faith is giving both oneself and God no rest until the world is whole (Is 62:6-7). If it is true that in the Kingdom of God there will be no tears any more, then tears must cease to flow even now as far as we can help it.[17]

The spiritual dimension

Demythologised, the concept of God represents the ultimate source of reality *as it ought to be* in personal, communal, natural and cosmic terms. Rightly understood, God is not only the power underlying the evolutionary process, but also the authority who establishes the goal of this process. As such God is the guarantor of the priority of the evolution of an integrated whole over the urge of the part to actualise its own potentials at the expense of other parts. The concept of the will of God articulates the demand that the self-interest of the part be defined by, attuned to, and taken up into, the dynamics of the interests of the whole.

These are entirely rational assumptions to make, while the idea of the autonomy of the individual human being is irrational.[18] That the part depends upon a constellation of forces greater than itself is self-evident. The autonomous development of the part at the expense of the whole is cancerous and, because it destroys the preconditions of its existence, fatal.

The whole referred to above is the reality we experience. Faith does not substitute an imaginary heaven for a transformed world.[19] Because God, the Source of reality, has made us part of this reality, there can only be one reality relevant to us, that is the one we experience. Visions of another world, whether conceptualised as "heaven above", a perfect "beginning", or fulfilment in an "age to come", are expressions of what this world ought to be according to God's intentions, in contrast to what it currently is. They are also expressions of what the world is expected to become, because the intentions of a powerful and benevolent God cannot be thwarted. This reality-transcending hope evokes a sense of direction. It also generates confidence and courage in times of frustration, adversity, suffering and death. It does not lead to acquiescence and despondency.

Faith does not mean that we hold unproven assertions or superstitions. Faith is trust as opposed to anxiety, responsibility as opposed to autonomy, commitment as opposed to self-satisfaction.[20] Trust leads to hope. Hope is directed towards a vision of what ought to be. A genuine vision activates. In contrast, religious fatalism obstructs the transformative power of faith. When faith turns into fatalism it abandons itself. If the Christian faith cannot overcome escapist and fatalistic attitudes, it will remain irrelevant and useless in terms of the problems discussed in this book. It will also remain out of step with its own authentic heritage. Faith must again become faith, that is, empowering trust and activating hope.

Faith is also not a lazy way of shifting responsibility to God. True faith is a conscious decision to go with God against a situation which is out of tune with his purposes. Faith is like a fox restlessly running up and down its cage seeking a way out. Faith does not try to escape from an intractable situation, but to rise above it, to protest against it, to claim mastery over it. Faith does not take "no" for an answer. The assurance that human limitations are not God's limitations continues to motivate believers even when there seems to be no way out any more. The biblical witness abounds with instances of prophetic agony, spiritual warfare and tenacious struggle against apparent inevitabilities.

Being at peace with God, therefore, can mean nothing else but sharing God's creative authority, God's redemptive concern and God's comprehensive vision for his world - thus sharing his restlessness.[21] Similarly, being an "enemy of God" can mean nothing but obstructing or ignoring his redemptive purposes in this world.[22]

Revision: "Faith and fatalism do not match." Explain this statement.
Application: A militant youth leader ridicules the Christian faith for offering a "pie in the sky when you die." Does he have a point?
Critique: (a) "Enthusiasm to transform the world belongs to the pathology of faith and has always led to frustration and disillusionment. Sober believers know that the redemption of the world must be left to God." (b) "Faith has always obstructed involvement in social change. Activists should not allow faith to undermine their resolve." How would you react to these statements?

Section II: The inner dynamics of the Christian faith

Constitutive elements of the Christian "story"

We have talked about a faith that is designed to conquer fatalism and despondency. What is the secret of this faith? The biblical God is believed to have disclosed himself in a historical process covering the better part of one and a half millennia. But God's self-disclosure has always been a personal experience. So its tradition has taken the form of a narrative.[23] This narrative can be summarised as follows:

1. At the root of the biblical faith in all its forms lies the manifestation of the intentions of God, the Source of reality, in *a redemptive event*. The liberation of Israel from Egyptian slavery is an example which has been seminal in the history of this faith. Here a redemptive experience was believed to be due to the intervention of Yahweh, the God of Israel, on behalf of his people. Other examples are the occupation of the land, the institution of the kingship, and the gift of a moral and social order. The religious interpretation of such redemptive events implied two basic faith assumptions: the *mastery* of God and the *benevolence* of God.[24]
2. Both the experience of need and the redemptive response of God lead to an awareness of the discrepancy between what reality *is* and what it *ought to be*. This includes the difference between unauthentic and authentic human life. Authentic life is life in fellowship with God, a life which forms a living expression of the intentions of God. That is the existential dimension of faith. While the church can only be "in the world", in fact "part of the world", it will feel the urge to be "different from the world".[25]
3. But the biblical faith cannot be expressed appropriately in terms of the authenticity of isolated individuals. It is healthy relationships which matter most.[26] And if God is the creator of the world, faith must develop cosmic horizons. The *ultimate intention* of God for the world as a whole is the comprehensive wellbeing of all his creatures.[27] In the Old Testament this goal is called *shalom*,[28] in the New Testament it is called the "age to come", or the "kingdom of God".
4. The "mission of God"[29] is derived from the vision of God. If God's ultimate intention is the comprehensive wellbeing of his entire creation, then the target of God's *specific concern* is any deficiency in wellbeing found in any dimension of life. To be authentic means to enact God's specific redemptive concerns on the basis of God's comprehensive vision and in the power of God's creative authority.
5. Christians conceptualise human authenticity as the new life of Christ. The roots of this concept go back to the ministry of Jesus of Nazareth. God's vision was spelt out in the *preaching* of Jesus. Jesus announced that the realisation of God's vision, the "kingdom of God", was near. A sense of proximity and urgency is essential because a vision does not challenge and motivate if it is conceptualised as being remote in space or time. However, proximity and urgency indicate a shifting horizon. As a goal is being achieved, a subsequent goal comes within sight.

6. God's concern for any deficiency in wellbeing in any dimension of life is symbolised by the *ministry* of Jesus: healing of lepers; acceptance of outcasts; feeding the hungry; forgiveness of sinners; according people in need priority over principles such as the Sabbath law; redefining the status of women, children and outsiders, and so on.
7. Redemptive concern implies *sacrifice and suffering.*[30] The most characteristic faith statement of Christianity is that "Christ died for us". He "became sin for us so that we might become righteousness in him";[31] he "humiliated himself" so that we might become elevated;[32] he "became poor to enrich us".[33] In the terminology of this book we could say that Jesus went from the centre to the periphery so that we might move from the periphery towards the centre - but only to join Christ in his movement from the centre to the periphery.[34] It is the *reversal* of marginalisation *through* marginalisation.[35]
8. We are *meant to share* in the new life of Christ through the transformative power of God's Spirit. In contrast to the new life of Christ, our own lives are not authentic. So our right of existence is constantly being questioned. We are accused, judged, condemned. The same is true for our world. A reality which does not conform to the intentions of its Creator is doomed.
9. To share his authority, love and vision, God *accepts us into his fellowship in spite of the fact that we are not acceptable.* The classical expression of the gospel is formulated in *legal* terms, namely that we are justified by grace accepted in faith. It can be translated into *communal* terms by saying that God accepts the unacceptable into his fellowship - suffering their unacceptability - so as to transform their unauthentic existence into an authentic existence. Acceptance implies belonging and participation.[36] This is what constitutes the "good news", as Christians understand it.
10. According to the Christian "story", God identified himself with Christ, the authentic human being, as his representative and plenipotentiary on earth. In Christ he suffers our unacceptability. He also bears the present condition of our world in his patience. The ultimate expression of the willingness of God to accept the unacceptable and to suffer its present condition is the *crucifixion* of Christ. It is here that he exposed himself to human enmity - though only to overcome it from within. The proclamation of the crucified Christ as the *risen Lord* elevates an event which happened once upon a time in a particular locality as God's present, valid and accessible act of salvation for all times and places.[37]
11. Of course, God accepts us and our world not to condone our unacceptability but to overcome it. The *Spirit of Christ*, that is, the reality of the risen Christ, is present in the *Body of Christ*, that is, the community

of believers.[38] In faith the community identifies with the death of Christ, the symbol of the end of inauthenticity, and with the resurrection of Christ, the symbol of authenticity.[39] Identifying with Christ, they anticipate their own deaths as the termination of sin, and their own resurrection into a life in fellowship with God. This is an activating, invigorating, transforming identification. Their depravity is superseded, in the power of the Spirit, by the authenticity of the new life of Christ. This is an ongoing struggle.[40] Faith also anticipates the transformation of the world according to God's intentions.[41] In this sense, the church is "missionary by its very nature".[42]

12. As a consequence, believers begin to *share God's creative authority*. They are liberated from other loyalties and slave masters, including God's own law, other religious, ideological and social authorities, collective interests and individual desires and addictions. They become sovereign masters over their lives and their life worlds.
13. They also begin to *share God's redemptive concern* for his world, that is, his concern for nature, for other humans and for their own lives. They share in God's suffering acceptance of an unacceptable reality and feel the urge to do something about it. That is what it means to "take up the cross." Responsibility for other humans and our entire life world includes economically relevant attitudes such as commitment, reliability, efficiency, precision, frugality and so forth.
14. They also begin to *share God's comprehensive vision* for the world, for others and for themselves. Their own lives are seen in the context of the future of the community, the society, the natural environment, the cosmic whole.[43] A vision is not reached; it is like a receding horizon. But it motivates, activates and gives direction.[44]
15. All this is, at the same time, a personal, a communal, and a cosmic experience. The Christian faith is the faith of the *community of believers,* the instrument of God's mission in the world. There is no private Christian faith. There is also no stagnant faith. The mission of God is holistic, comprehensive and dynamic.

The mission of God

These fundamentals of the Christian "story" have to be spelt out in historical terms. We shall argue in chapter 8 that Christian mission has been part of the Western dynamic. This does not imply that Christianity has always been a dynamic type of religious faith. In its earlier forms it shared the traditionalist worldviews of its environments. More often than not, it has entrenched itself in social institutions, imprisoned the minds of people in mental structures, and obstructed human progress. Rigid forms of faith are not very helpful. If dynamic Christianity has been part of the problem, only dynamic Christianity can be part of the solution. But then it must be rediscovered as a faith with a cosmic mission.

Mission in general is a process in which people are activated by the urge of a vision.[45] Human freedom is enlisted by a higher cause which demands dedication. That is also true for other convictions such as Islam, liberalism and Marxism. Recently, secular scientists have discovered the importance of such a vision: "A sustainable world can never come into being if it cannot be envisioned."[46] The theological task of today is to establish a critical link between the vision of the biblical faith and the visions of secular scientists.

As mentioned above, the Christian faith is future-directed by definition, because God himself is believed to have a vision.[47] The biblical past is retrieved only so as to get an impression of the direction in which history is supposed to be moving under the direction of God. The ultimate intention of the biblical God is the *comprehensive wellbeing* of all human beings within the context of the comprehensive wellbeing of their entire social and natural environments.

The term "comprehensive wellbeing" is not meant to indicate a utopian never-never land. As long as we live in this world, there are trade-offs between the needs of the parts and the needs of the whole. Human beings cannot survive without consuming the substance of other creatures. Death is built into the system. To be realistic we should, perhaps, rather speak of *balanced*, or *optimal wellbeing*. The concept "comprehensive wellbeing" is here used merely to emphasise the fact that God's redemptive intentions cannot be reduced to the salvation of a disembodied soul in a spiritual heaven. It can also not be reduced to the interests of one group at the expense of others. It can also not be reduced to the interests of humanity at the expense of the rest of the natural world. Comprehensive wellbeing is meant to cover all dimensions of reality: body and mind; individual, community, society and humankind as a whole; the present and all future generations; social structures and natural environments.

Obviously the overall goal of God has not been reached. There is a yawning discrepancy between what is and what ought to be. This implies that reality is not fixed; history is not closed; the unexpected can happen; the unlikely is possible. The anticipation of God's ultimate intention activates Christians to move forward into the future. It determines the direction of their hope and their action. As we shall see in the following chapters, the capacity to transcend the particular towards the whole in terms of time, space and power is one of the most fundamental prerequisites for a socially and ecologically responsible economy.

If the ultimate intention of God is the comprehensive wellbeing of his creation, the specific targets of God's redemptive action are *particular deficiencies* in wellbeing. Such deficiencies can occur in any dimension of life. To understand God's mission in its total context, therefore, we have to analyse the human need structure. In chapter 6 we have distinguished between immanent and transcendent needs. Our analyses in Part I have laid bare the most pressing immanent (economic and ecological) needs of humankind at present. In Part II we focus on transcendent needs, that is, the need for *meaning, acceptability and authority.* At the roots of the biblical faith lies the conviction that God responds redemptively to this whole range of human needs.[48]

A further observation is of critical importance for our purposes. The biblical witness maintains that God's redemptive intention and God's creative mastery are mediated through historical events and human actions. God does not act "straight from heaven" as it were. According to the biblical witness, he acts through his creatures. God also works according to the laws of nature, which are, after all, the laws of God's own creation. God's redemptive action in the life, death and resurrection of Christ, the authentic human being, is the supreme manifestation of human mediation. Those who belong to Christ are privileged to participate in Christ's life of suffering and service in the power of the Spirit. It is they who form the "Body" of the "risen Christ".

A body is the concrete manifestation of a person, the material basis of human communication and action in this world. By sharing the new life of Christ in fellowship with God, Christians participate in God's creative authority, God's redemptive concern and God's comprehensive vision for the world. In other words, they are involved in his redemptive action in the world. This action targets specific human needs. Where people do not feel this urge, they are not engaged in God's mission; they only theorise and moralise. The prerequisite for mission to get off the ground is a spiritual transformation with social manifestations.

Once again: mission targets the entire spectrum of human needs. Needs are historically and contextually specific. To be relevant, mission has to establish where the needs of its social environment are located. Traditionally, mission has concentrated on transcendent needs. Often these were seen in complete isolation from immanent needs.[49] To the detriment of its mission, Christianity assimilated the Platonic dualism between a material and a spiritual world and between body and soul. This dualism can no longer be upheld, whether in terms of modern scientific insights, or in terms of biblical assumptions. The need for meaning, right of existence and authority becomes pressing precisely when one is struck by sudden fate, or when one is cast into powerlessness and misery. Transcendent needs form the "depth dimension" of immanent needs, not some spiritual concerns abstracted from life.[50]

There are two necessary and complementary ways of meeting transcendent needs. The first is to fulfil the immanent needs which gave rise to transcendent needs in the first place. To regain their sense of meaning, assurance and authority, outcasts must be accepted into communities, famine victims must be fed, warring factions must be reconciled. When feeding the hungry, we affirm their dignity, that is, their right to exist. The assertion of their right to exist, in spite of their limitations and failures, is the essence of the gospel.

The second way is to transcend the limitations placed on our abilities to fulfil these needs. This happens through prayer, proclamation and symbolic enactment. A patient suffering from terminal cancer is reassured of God's purposes (meaning), God's love (assurance) and God's calling (authority) in spite of her illness. This happens through prayer, a pastoral word, a sacrament, or an act of love. The capacity to transcend implies that we do not have to accept the ultimacy of experienced reality. Human and earthly limits are not the limits of God. Only superficiality can assume that transcending a seemingly hopeless situation is meaningless, even in economic terms.

Under no circumstances may the second act, the act of transcending our limits, bypass the first. Transcendence indicates that our limits are not God's limits. Therefore transcendence is legitimate only beyond the point where humans have reached their limits, not before. To tell a patient that her disease is immaterial, because her soul is healed in Christ, or to tell an oppressed person that slavery is of no consequence, because we are set free in Christ, are fraudulent versions of the Christian message.

Revision: How can a personal relationship with God lead to involvement in social reconstruction?

Application: In which sense is the plight of street children an economic as well as a spiritual problem?

Critique: (a) "Visions have led to the catastrophes of fascism and Marxism. In economics we need hardnosed pragmatism." (b) "Christian mission has to do with preaching the gospel of Christ, not with involvement in politics and economics." How would you respond?

Section III: The capacity of faith to transcend reality

An outdated evolution theory, based on the principle of the "survival of the fittest", seemed to suggest that control of individual selfishness was an ethical imperative at odds with the evolutionary conditioning of human nature. Recently the natural sciences have discovered the importance for living organisms of going beyond individual self-interests. A more adequate concept of evolution is based on "a systems theory that focuses on the dynamics of self-transcendence".[51] It is not individual organisms that strive for survival but relationships in contexts. "An organism that thinks only in terms of its own survival will invariably destroy its environment, and, as we are learning from bitter experience, will thus destroy itself."[52]

This is true for animals. Human beings are more than animals. For them the ability to transcend is not just an instinct, but a cultural potential which can be activated. This potential is the result of specifically human evolution. One does not need to be a believer to discern that the imaginative capacity to go beyond individual selfishness, even beyond empirical and historical limitations, is one of the most remarkable characteristics of the human species.[53] It is the mainspring of meaning, assurance, self-esteem, ambition, courage, trust, concern and hope. None of these qualities are subject to empirical verification. But without the transcending capacity of faith and morality, the great achievements of humankind would have been unthinkable.

Today we must learn to apply this potential to the economic-ecological sphere. In chapter 8 we shall see that the narrow horizons of traditionalist communalism and the self-centred individualism of modernity must both be transcended towards an awareness of social and natural reality as a whole. The problem is that conventional Christianity is, to a large extent, itself both traditionalist and individualist. We need a mindset which opens up our horizons in terms of space, time and power. It must create a sense of responsibility for the six billion people living on our globe today, for all future generations, and for the life of all creatures with whom we share this planet.

There is a long Christian tradition which considers the human being as the "crown of creation" and the rest of the world as created for human consumption or utilisation. This is *scientifically* misleading because any entity is part of an intricate system of relationships on which it depends; the context is logically and temporally prior to the individual. It is *ethically* untenable because self-centredness in all its forms cannot be justified, even if it was the collective selfishness of humanity as a whole in relation to the rest of creation.[54] It is *theologically* illegitimate because it is anthropocentric, rather than theocentric.⁵⁵ Human suffering does not include animal suffering; divine suffering does.⁵⁶ As Moltmann pointed out, it is not the human being who is the crown of creation, according to Gen 1, but the Sabbath, that is, the situation in which God can stand back from reality and say "it is very good".⁵⁷

Obviously, from the vantage point of the individual, reality looks like a system of concentric circles or spheres going from the most immediate to the most remote. But to be ecologically responsible, we need to begin with the outermost sphere and work ourselves backward from greater to smaller contexts until we find our own place in the system.⁵⁸ Science is beginning to become aware of this fact. As the cosmologist Brian Swimme argues, we must internalise the "new story" of the generation of the universe.⁵⁹

But it takes "the mind of God" to acquire such vast horizons.⁶⁰ The Christian faith claims that, by sharing Christ's new life in fellowship with God, humans do share the mind of God (1 Cor 2:6-16). It is significant that in its long history the biblical faith has developed a concept of "the world". This concept encompasses, but far surpasses, what has always been experienced as the human "life world". The biblical faith has also developed a notion of the difference between the "passing age" and the "age to come", between the world as it is, and the world as it is intended to be. Moreover, the biblical faith assumes that the human being is responsible for the world, but not to the world; instead humans are responsible to the transcendent Source, Owner and Sustainer of the world. This assumption is the precondition of their freedom from the world and their freedom for the world.

As mentioned above, these abstract terms must be filled with the substance of historical reality. Our present task is to show how the biblical faith has the potential of "breaking through the inner limits", as ecologists call it.⁶¹

Why should the biblical faith have a built-in tendency to go beyond the given? In its search for redemption, faith locates every new experience within the realm of divine mastery and divine benevolence. As faith moves through history from one constellation of needs to the other, its awareness of the power sphere of a benevolent God constantly widens to include ever new situations. Moreover, because it does not build on human possibilities, it cannot accept human limitations to be final. As a result, it is always one step ahead of given experiences in the direction of unknown possibilities. And it is precisely the experiences of need, suffering and powerlessness which evoke such "spiritual conquests". There is nothing our age needs more than such an explosion of our narrow horizons.

Transcending the limitations of space and time

The biblical faith transcended the limitations of *space*. As the Jews learnt to their horror, the "promised land" could be lost and its inhabitants scattered. During the Babylonian exile, at the latest, they learnt that the boundaries of their fatherland were not the limits of the presence of their God. Even less could the presence of God be confined to a sanctuary in the "holy city". He could be approached anywhere on earth, even among pagans and on enemy territory. In the New Testament the spatial determination of the presence of God was abandoned altogether (Mat 28:20; John 4:19-24).

As a result the world became one. There were no holy places. Or rather, the entire world became a holy place, the stage of God's presence and action. God's unconditional acceptance was, in this way, prefigured in geographical terms. No-go areas lost their awe-inspiring secrets and powers. Already during the exile the bold commandment was formulated: "fill the earth and subject it to your rule".[62] As we shall see below, to rule meant "to care for". In contrast to the land, however, the limits of the earth as a whole were seen to be absolute. They were confined to what is accessible by what is inaccessible, that is, God's own space beyond, "heaven above". According to the biblical witness it is what we do on earth that determines our authenticity and integrity. By implication, the earth is all we have. If we destroy it, we impoverish the race.

In similar fashion the biblical faith transcended the limitations of *time*. In early times the hope was directed at Yahweh's intervention some time soon in the interest of his chosen people and in the confines of the "promised land". Thus a particular time was envisaged which would bring a specific redemption to a specific nation. But suffering intensified. Consecutive empires (Assyrians, Babylonians, Persians, Hellenists, Romans) ravaged increasing chunks of the known world. Promises for a restoration of the davidic kingdom were not fulfilled, let alone the expectations of a universal empire ruled from Jerusalem by the Israelite king.

This forced the Jewish faith to go beyond current history. It broke down the barriers of "this age" towards a vision of an "age to come". The "age to come" is juxtaposed to all times as the time of universal redemption. So God's unconditional acceptance was prefigured in terms of time. This vision also broke down ethnic frontiers towards a cosmic "kingdom of the Most High", in which all people would ultimately be ruled by the representative of Yahweh, the God of justice, peace and prosperity *(shalom* in Hebrew). It is quite likely that the Parsist religion, which believed world history to be moving towards a giant showdown between the forces of good and evil, opened up the Jewish mind for the vistas of cosmic history. What could be more apt at a time when everybody is bent on satisfying immediate cravings!

Transcending human limitations

The biblical faith is not built on speculation. The driving force which made the Jews transcend historical time was *faith in Yahweh's incorruptible justice*. The national catastrophes (the destruction of Samaria in 721 BC and of Jerusalem in 586 BC) were interpreted as Yahweh's punishment for Israel's lack of faithfulness and righteousness. With the growing complexity of a scattered and increasingly heterogeneous "people of God", collective accountability made way for individual responsibility (Jer 31:29ff; Ez 18:2ff). Those who fulfilled his law had to be rewarded, those who did not, had to be punished.

This is the origin of the post-exilic concept of the last judgment. Again it implies that this life is all we have; our authenticity and integrity are established between birth and death. But the insistence on the incorruptibility of God's justice made the Jews think the unthinkable: if God's justice would not be fulfilled during this life, it would certainly be fulfilled some time after, even if the dead had to rise for the purpose. The implication is that we can never say "after me the deluge". Though we may no longer live, our irresponsible behaviour will continue to be on our heads. All this presupposes that God's acceptance is conditional. To become acceptable, norms have to be fulfilled. We shall soon come to the concept of unconditional acceptance. The point, at this juncture, is that the limits of human life were transcended.

But humans are embedded in vast, historically evolving contexts. Unrighteous actions by individuals and communities could no longer make the occurrence of great natural and political disasters plausible. The insight began to dawn that humans cannot even attain their own authenticity on their own accord. After all, they are human, not divine. They are creatures of God, not creators of themselves. Yahweh was God, the Source and Master of reality. So it was he who had to be deemed responsible for the current state of the world, including human failure, and it was he who would overcome these aberrations. So not only human limitations, but also human failures were transcended. There is no more powerful antidote against fatalism and despondency than that.

Transcending the ambiguity of reality

Of course, this had to lead to the agonising question why a God who is credited with mastery and benevolence would allow evil to emerge and persist in the first place, a problem called "theodicy" (the justification of God) in theology.[63] The biblical faith did not produce a satisfactory answer to this question. Indeed there can be no such answer because the question reflects the inescapable dialectic between what reality *is* and what reality *ought to be*. As long as this discrepancy exists, the urge to overcome it cannot subside, and no amount of metaphysical speculation can solve the problem. The question is, rather, whether we are able to forge the contradiction into the concept of a single and sovereign God, or whether we succumb to the temptation of dualistic escape routes - which also offer no solutions.[64]

Once the ultimate Source of reality is perceived to be one, the tension within this concept is translated into a forward looking dynamic. Only God can resolve the conflict by transforming what is into what ought to be. This is the root of the radical future-orientation of the biblical faith: God will make good his promises. He will reconstruct reality as a whole. There will be a new heaven, a new earth, a new social order, and a new human being. He will take over responsibility for human failures. He will accept the unacceptable into his fellowship and transform them in this fellowship. It is only in conjunction with this radical kind of assumption that the concept of unconditional acceptance could mature.

This universalisation and radicalisation enabled the biblical faith to cover ever more complex experiences of reality. In nomadic times the Hebrews expected progeny, a long and healthy life, fertility of their flock, grazing, water and protection against hostile clans from their God. During the time of national consolidation, human needs included agriculture, urban fortifications, war, politics, justice, liberty, national prestige, central government, and so on. In other words, the system of meaning was constantly broadened and updated.[65] Insights from Israel's religious environment were incorporated. In each case, the initial wave of syncretism led to a prophetic reaction, which restored the integrity of the biblical faith, but on a more inclusive basis. The awareness of human needs also grew in depth. It was the realisation of the pervasiveness of evil, whether in the human heart, in society, in nature, or in the cosmic order, which led to the expectation of a transformation of reality as a whole.

The coming transformation did not materialise as expected. Unfortunately, at this juncture the Christian faith bought into the Hellenistic alternative. The Jewish concept of the coming kingdom of God was translated into the Hellenistic concept of a timeless and immaterial sphere of authenticity. This eternal and spiritual world was removed from, and untouched by, the corrupt and fleeting world of earthly matter and human history. Thus the evolution of the biblical faith culminated in a spirituality which tended to leave the physical, social and natural dimensions of reality behind. But a spiritual world without a material substratum is an illusion. There is no doubt that the earthbound intention of the biblical faith must today be retrieved from this kind of "enculturation".

Transcending human suffering

The perception of human suffering was also transformed in this process. In early times we find much "wailing before God", and this is only natural (e.g. Ps 22). The problem was that Israel could not understand why Yahweh allowed his faithful to suffer. Often they were worse off than indifferent and unscrupulous evil doers (Ps 37:1ff). By the time of the exile of the Jews in Babylon, suffering began to be understood as a sacrificial and redemptive act (Is 52:13 - 53:12). This attitude grew in strength when Jews suffered martyrdom. Slowly the insight dawned that God himself suffers and that God's servants are bound to share in God's suffering. These insights helped early Christians to understand the torture and execution of their leader. In short, suffering had acquired a redemptive meaning.

Using an apocalyptic frame of reference, the apostle Paul developed a profound "theology of the cross". Suffering prefigures death, that is the end of the "flesh" (corrupt human nature). But death is only the human dimension of the imminent end of "this age". Believers rise above their suffering by anticipating both their death and their transformation into a new authentic life (Rom 6). They actively "crucify their flesh" (unauthentic humanity) and "walk in the spirit" (authentic humanity). They do so in anticipation of the transformation of "this age" into the "age to come".

Hope beyond death has often been decried as "a pie in the sky when you die". However, in times of a seemingly irredeemable situation, it is precisely this kind of hope which empowers the suffering to rise above their circumstances, not only to endure but also to confront their misery. For millions of people the transcendence of the limits of time is the only source of hope - and hope is a prerequisite of survival.

Transcending ethnic and social barriers

One of the most pervasive characteristics of social reality is the distinction between "us" and "them", between "ingroup" and "outgroup". The ingroup can be very narrowly defined. The patriarchs of Israel, Abraham, Isaac and Jacob, seem to have believed in a God who had chosen their clan and who was concerned, primarily, with this clan's survival and prosperity. The dictates of justice and the concern for the weak and vulnerable were confined to the limits of the ingroup.

The clan developed into a tribe, then a nation, but ingroup-outgroup attitudes remained intact. It is the experience of political oppression by their neighbours which led to the awareness that Yahweh, the God of Israel, was the God of all nations. The implication was that the divinities who ruled over these nations were also subject to Yahweh (Ps 82:1f). So Yahweh became the God of all cosmic powers. The continuing temptation to acknowledge the authority of these seemingly more powerful gods led the Israelite prophets to the revolutionary statement that other gods have no existence beyond the earthly symbols of their carved or cast images.[66]

As a result, a worldview emerged which unified humankind in a single system of meaning. This system of meaning contained a normative system in the form of Yahweh's law. The most central aspects of this law were social equity, public justice and concern for the weak. The ritual overgrowth of the law was attacked in no uncertain terms by the Israelite prophets.[67] The needy had to be accorded priority. The idea that human need was only relevant if it was felt by members of the ingroup could not persist. Eventually all authority - not only Israelite but also pagan, not only human but also divine - were subjected to this central criterion of acceptability.

Human kings were disparaged because of their lack of justice, and Yahweh was praised as the cosmic king because he was the God of justice (Ps 146). The other gods were deemed to have been placed into positions of authority by Yahweh to maintain justice (Ps 82:6,3f). Other gods were not denied their existence on the strength of metaphysical speculations, therefore, but deemed to be demoted because they did not appear to be fulfilling Yahweh's mandate (Ps 82:2ff). In the same vein, the risen Christ was deemed to have been placed in authority over all cosmic powers so that the universe would be united under the principle of his self-giving concern for the redemption of others.[68]

Rightly understood, therefore, the explicit and sometimes ferocious exclusiveness of the biblical faith was not rooted in religious arrogance and intolerance, but in the concern for the rights and the wellbeing of the oppressed, the suffering and the lost. The Jews themselves had experienced the religious legitimation of arbitrariness and oppression, whether by their own kings or by foreign rulers. Gradually they came to the conclusion that divinities which demanded the sacrifice of children, legitimated oppression and exploitation, and supported the subjugation and eradication of foreign nations, had to be "removed from office."

This is as valid a concern today as ever. Idolatry has to be overcome because it elevates the temporal and partial to the level of the eternal and absolute. It enslaves people from within. It confines their vision and their living space to limitations that can be and should be transcended.

The most drastic dissolution of ethnic and religious barriers emerged, however, where the insight dawned that God in Christ accepts the unacceptable into his fellowship, suffering their unacceptability.[69] If God suspends his conditions of acceptance, we cannot want to make our acceptance of each other conditional. As the conflict between Peter and Paul (Gal 2) demonstrates, the biblical faith had to learn to transform even its own concept of God to conform to this criterion - and to do so drastically. "Canaanites" can no longer be driven out of the "promised land" to make way for the "people of God" in the name of the biblical faith.[70] The "near" and the "far" are to be forged into a new humanity.[71] There is neither Jew nor Greek, slave nor free, male nor female "in Christ" (authentic humanity).[72] In short, God's acceptance had become unconditional, and unconditionally redemptive.[73]

Transcending special privileges

If all people matter, every person matters. The combination of the concept of a single humanity with the concept of equity yielded an awareness of universal human dignity. Yahweh, enthroned beyond the heavens, stoops down to pick up the abject from the "rubbish dump" and seats them with the elite (Ps 113:4-8). Paul scorns Jewish privileges to gain Christ, the authentic human universal (Phil 3).

But dignity includes accountability. In early times accountability had been collective. Children had to suffer because of the sins of their ancestors. By the time of the Babylonian exile collective accountability was abandoned in favour of individual accountability (Jer 31:29ff; Ez 18:2ff). The idea of a "last judgment" beyond death was based on the premise that individuals would have to justify their actions before the ultimate Judge.

This concentration on individual dignity and individual accountability should not be confused with the selfishness of greed and piety. Paul's personalised theology of "flesh" and "spirit" was still embedded in his universalist theology of "this age" and the "age to come". For Paul the individual "gifts of the Spirit" had no other function but to "build up the Body of Christ" (1 Cor 12 - 14). The same can be said of theology up to the Reformation.

During the so-called Enlightenment, awareness of the vastness and relativity of the universe grew beyond proportions. The result was a profound metaphysical insecurity. During this time, philosophy, theology and piety sought new foundations in the subjective experience of the individual. Secular humanism claimed autonomy for the individual. Utilitarianism, hedonism and pragmatism reduced the goals of life to the immediate satisfaction of the immanent needs and desires of the individual.

In the course of these developments the natural and social contexts of the individual disappeared from sight. This loss of context manifested itself in the preoccupation of pietism with linking the soul to its personal saviour, the preoccupation of existentialism with the authenticity of the self, and the avarice of the "economic man" found in economic theory. All these are relatively recent Western phenomena. In political theology, liberation theology and ecological theology, believers have recently begun to retrieve the natural and social dimensions of their faith. If humankind is to have a future, these dimensions must be appropriated by elites and masses alike.

Human rights

The emphasis of the biblical faith on equity and justice also led to enshrined human rights.[74] Christians have often argued that, being sinners, humans have lost all rights and depend solely on grace. But that is a theological misunderstanding. Rights are necessary precisely because of human depravity. Having been denied the privilege of the "presence of the Lord", Kain, the primeval murderer, received a mark so that he would not be killed himself.[75] In fact, the entire law was deemed necessary to serve as a bulwark against the ravages of sin.[76] It goes without saying that in the biblical context there were no rights without responsibilities. Two narratives may serve to illustrate this fact:

(a) The story of Naboth's vineyard in 1 Ki 31. According to the legal perceptions of the Phoenician princess, Jezebel, the rights of the king were absolute. Therefore the commoner Naboth, who refused to sell his vineyard to king Ahab, had "cursed God and the king". In contrast, Yahweh's prophet proclaimed Naboth's right to his inheritance. No Israelite king was above the law of Yahweh.

(b) The Jubilee law was meant to restore any property alienated from a family after 49 years, that is after two generations at the latest. At a time when economic sufficiency depended on access to agricultural land, such a legal provision was of incredible significance, even though we are not sure when it was actually implemented.

Thus property rights were indeed protected in Israel. However, these were not the property rights of the affluent against the claims of the poor, but the property rights of the poor and vulnerable against the encroachments of the rich and powerful (cf Is 5:8). Translated into modern terms, not the individual profit maximiser was protected, but the resource base of the extended family. This has implications for ecological concerns as well: if God is the creator of the universe, not only humans but all creatures have rights.[77] It also has consequences for the issue of power concentrations in society to which we turn now.

Transcending power concentrations

In the Ancient Near East, the king was believed to be the representative and plenipotentiary of the deity on earth.[78] Whoever disobeyed the king rebelled against God. The king was the adopted "son of God" or the "image of God". He presided over the earthly realm of a divinely instituted cosmic order. All blessings flowed through the king to the people. This idea found its way from Egypt via the Canaanite city state of Jerusalem into the court of the davidic dynasty. Its original intention was positive: the king was regarded as the mainspring of collective wellbeing based on justice, peace and stability. But the motive was extensively abused to legitimate authoritarian rule and imperial ambitions - including David's own despotism and his subjugation of neighbouring kingdoms (Ps 2).

Due to continued experiences of royal abuses, the prophetic movement began to project a genuine king, a "prince of peace", a champion of justice, into the future (Is 9; 11; cf Dan 7:13ff, 26ff). That is the root of the messianic tradition. Because of his active concern for the outcasts and the suffering, Jesus was identified with the expected prince of peace. The idea of the king as an oppressive tyrant acting in the name of an oppressive God, as found in Ps 2, was identified as being "pagan" and turned on its head: the "Son of Man" had come not to be served but to serve and to give his life for many. That was to be the pattern of authority in the new and authentic community (Mat 20:25ff; cf John 13). By implication there can be no legitimate authority which is not a servant of the people over whom it rules. Extrapolated to present circumstances this ideal can only materialise in a democracy. As management science begins to discover, for corporate life it is stewardship and partnership rather than leadership which matters.[79]

This does not mean that concentrations of power have to be rejected as a matter of principle. The idea that power concentrations can be avoided or abolished is either romanticism or, as in the case of Marxism-Leninism, ideological deviousness to gain absolute power.[80] As our reflections on entropy and evolution in chapter 13 will show, reality is, by definition, composed of power concentrations. But legitimate power is mandated, not usurped power.[81] As a result, the centre has to relinquish its position of dominance and utilise its power to empower the periphery. The periphery has to relinquish its position of dependency and assume responsibility for its own life. Thus vertical relationships are horizontalised. We shall come back to these insights in chapter 9.

Revision: Sketch the potential impact of the capacity to transcend on social reality.
Application: Which role did the Christian faith play in breaking down ethnic and racial barriers during the apartheid era in South Africa?
Critique: (a) "The achievements of modern society which you have enumerated are the achievements of secular humanism. They have been rejected and obstructed by the church for centuries. For Christians to claim credit for them is truly preposterous." (b) "African traditions had more respect for human dignity and the sanctity of nature than the rapacious Christian culture of the West." Comment.

Let us summarise

Human perceptions, interpretations and directions all rest upon irreducible assumptions. Assumptions come in sets; we call them "mindsets". Assumptions are not just working hypotheses; they are commitments linked to the human need for meaning, right of existence and authority which arise from immanent needs and predicaments. As such they have the power to channel our analysis, decision making and action in particular directions.

All this is true for both science and religion. In chapters 11 - 13 we shall challenge and reformulate a particular science, namely economics, in response to the needs of the emerging economic-ecological crisis. In this chapter we have challenged and reformulated a *particular religion,* the Christian faith, in response to these needs. Apart from the personal faith of the author, the choice of this religion is appropriate in view of the immense role the Christian faith has played in the emergence and evolution of modernity, which is the dominant mindset of our time.

In section I we attacked *religious fatalism,* arguing that a fatalistic faith in the biblical God is a contradiction in terms. In section II we sketched the *exis-*

tential core and the *missionary dynamic* of the Christian faith. Participation in the creative authority, the redemptive concern and the comprehensive vision of God, the Source of reality as a whole, implies an acute awareness of the discrepancy between what is and what ought to be. The ultimate intention of the biblical God is the comprehensive wellbeing of his creatures. The target of his specific redemptive action is any particular deficiency in wellbeing, which can occur in any dimension of life. Divine action is mediated by human involvement in its dynamics.

In section III we traced a number of historical trajectories which show how the biblical faith has *transcended the limits of space, time and power.* This has a number of momentous consequences. Forbidden places, sacred times and uncanny forces are demythologised. God is Lord of the universe as a whole and of history as a whole. Natural laws are God's laws. The networks of cause and effect do not predetermine the future, but open up new opportunities and choices. The human being is called to participate in God's authority, love and vision - and do so in freedom and responsibility.

It is here on earth, in our history, with the powers at our disposal, that our integrity and our authenticity are established. But human limitations cannot limit God. On the one hand God's justice reaches beyond earthly reality and human history. Therefore our deaths cannot prevent God from taking us to task. On the other hand, our failures themselves can be transcended. *God takes over responsibility* for the universal discrepancy between what is and what ought to be. He accepts the unacceptable in view of its transformation. And he invites us to participate in his suffering, transformative acceptance of the unacceptable.

The biblical faith looks forward to *a new and redeemed reality*. This vision generates inspiration and provides direction for human action. It leads to a redemptive interpretation of human suffering. It transcends the boundaries of ethnic groups. It redefines the legitimacy of power concentrations in terms of service to the weak and vulnerable. It postulates human dignity and enshrines human rights. All this has infinite implications for economic and ecological concerns.

Notes

[1] See the epilogue of Nürnberger 1998.
[2] 1987:125ff and 141ff. That science and technology can only emerge where the world is demythologised by faith in a transcendent God has already been argued powerfully half a century ago by the German theologian F Gogarten in *Der Mensch zwischen Gott und Welt*. As Stackhouse points out, A van Leeuwen in *Christianity in World History* went further by maintaining that technology is a new evangelism (based on biblical foundations) which spreads a genuinely cosmopolitan faith, which eventually has to lead to a new faith in the Creator. In *The technological society* Jacques Ellul, in contrast, argued

that technology is a fallen angel which has emancipated itself from divine and human control and causes havoc. However that may be, these authors agree that science and technology can be traced back to the biblical heritage.

[3] Cf Daly and Cobb, Boff, Ruether, Moltmann, Duchrow, Stackhouse, Goudzwaard and de Lange, Küng, Granberg-Michaelson, Wilkinson, Block, Milbank, Villa-Vicencio, Issar, Birch, Eakin and McDaniel, and many others.

[4] Luther called this frame of mind *incurvatus in se ipse"* (curved into oneself).

[5] Cf also Küng 1997:247.

[6] "(It) is no answer to the divisive forces of religion to offer the insipid soup of indifference as the 'postmodern' consensus of society" (Küng 1997:136).

[7] Cf Pieterse in JTSA 94/1996.

[8] Cf Cobb 1992:11ff. for the following.

[9] Cf Storkey 1986:196ff.

[10] Cf Meeks 1989:9ff.

[11] Stackhouse 1987:152.

[12] From a secular point of view, Block 1993:221ff distinguishes the cynic, the victim and the bystander as unhelpful.

[13] After Rauschenbusch and his "Social Gospel" movement, European Political Theology (Moltmann, Metz) and Latin American Liberation Theology (Gutierrez and many others) have linked concept of the "Kingdom of God" and even the concept of personal holiness with socio-political involvement. See, as one example among many others, Sobrino 1985:80ff, 124ff. See Bosch 1992:376f for the new "turn to the world". Northcott provides a useful overview of what he terms the "flowering of ecotheology" (1996:124ff) and distinguishes between emphases on God, the human being and the Earth.

[14] Classical theology has discarded this idea with its doctrine of creation (Wogaman 1986:34ff). But this is hardly sufficient.

[15] Concern for animals, especially, has a long, though intermittent tradition. In biblical times it has been expressed in passages such as Gen 1:29f (humans spill not blood!), Jonah 4:11, and Rom 8:19ff. Names such as Francis of Assisi, Arthur Broome (initiator of the SPCA), Albert Schweitzer with his "reverence for life" ethics, etc. come to mind. For detail see Linzey 1987.

[16] In an unusual breadth of holism, Sallie McFague (1993:13ff) extends the model of the body to the planet Earth and its ecological needs, and even ventures the thought of the "body of God". For the panentheistic connotations see page 149. Note also the juxtaposition of organism and mechanism pp 91ff.

[17] Is 65:17-25; Rev 21:1-4.

[18] It is true that "the protestant view of the Church, which understands it as an association of individual believers who possess, outside the social context, their own direct relationship to God, articulates more fully what was always latent within Christian self-understanding" (Milbank 1993:399). But does this imply that our relationship to God bypasses innerworldly mediation? Does one's primary relationship to God, thus one's independence from social control and primary group pressure, lead to individualism and spiritualism? That precisely is the misunderstanding of liberalism. The new community must be built on emancipation *from* social fetters, but *for* social responsibility.

[19] According to the Lord's prayer, "heaven" is the sphere where God's will is done, the model of what should happen on earth, thus an expression of what ought to be as opposed to what is.

[20] I owe the insight that faith is primal trust to Martin Luther (Large Catechism, first article of the creed). Cf Wolfhart Pannenberg: *Anthropologie in theologischer Perspektive*. Göttingen: Vandenhoeck & Ruprecht, 1983, 68ff.

[21] Bosch rightly says that the church cannot boast in a "realized eschatology" for itself over against the unredeemed world (1992:150).

[22] Cf Müller-Fahrenholz 1995.

[23] I continue to use the male form for God without intending its discriminatory implications. The biblical perception of God is neither inherently gender specific nor necessarily singular. The most common Hebrew word for God, elohim, is a plural noun.

Due to the personal nature of this God, expressed in the context of the Ancient Near Eastern patriarchal and hierarchical social order, the tradition of his appellation has assumed a distinctly male form. In my view no satisfying inclusive appellation has been found so far.

[24] See Jon Levenson: *Creation and the persistence of evil*. New York: Harper & Row 1988.
[25] Bosch 1992:386.
[26] Cf Wogaman 1986:35.
[27] Cf Bosch 1992:399f.
[28] Cf Northcott 1996:193ff. It can also be referred to the sabbath rest of God, when God had looked at everything he had made and behold, it was very good (Gen 1:1-2:3).
[29] Cf Bosch 1992:389ff. See also Newbigin, Sider, Nicholls and others.
[30] Cf Cobb 1992:16ff.
[31] 2 Cor 5:21; cf Rom 4:25.
[32] Phil 2:6.
[33] 2 Cor 8:9.
[34] Nürnberger 1988:303f.
[35] Waves in a pond move from the centre to the margin and are deflected back to the centre (Lee 1995:30f). This also means that Jesus is not simply the marginalised human being per se, which would deny the symbol of incarnation (Lee 1995:78ff).
[36] Wogaman mentions this as his first "general priority" (1986:40).
[37] Son of God, Son of David, Son of Man, the Annointed (Hebrew: Messiah, Greek: Christ) are all titles of the king who was believed to be representative and plenipotentiary of God on earth. For the royal frame of reference implied in these assumptions see K Nürnberger: The royal-imperial paradigm in the Bible and the modern demand for democracy. *Journal of Theology for Southern Africa* No 81, Dec 1992, pp 16-34.
[38] Cf Bosch 1992:172.
[39] Rom 6.
[40] This struggle is expressed in the formulation *simul iustus et peccator*, meaning that believers are sinners and righteous at the same time.
[41] 2 Pet 3:13; Rev 21:1. Cf Meeks 1989:23ff regarding the implications for the church.
[42] Bosch 1992:372f.
[43] Cf Cobb 1992:17ff, 20ff. See Northcott 1996:124ff for a survey of recent "ecotheology".
[44] This is how I formulate the classical dialectic between the "ultimate" and the "penultimate", between "hope-in-action" and "action-in-hope", between eschatological hope and ethics. Cf Bosch 1992:510.
[45] Genuine mission is not motivated by a dogmatic system or ideology which requires to be imposed upon non-believers, thus by imperialist or hegemonic pursuits, but by an active anticipation of the authenticity of the other and of the world and the urge to act as facilitators to let it materialise. This is not spelt out clearly enough in Bosch's "interim definition" of mission (1992:8ff).
[46] Meadows, Meadows & Randers 1992:224f.
[47] Cf Cobb 1992:12ff.
[48] Cf Meeks 1989:157ff.
[49] Cf Northcott 1996:211ff.
[50] I owe this term to Paul Tillich.
[51] Capra 1983:310f.
[52] Capra 1983:313 and 1992:72ff.
[53] Cf Kieffer 1979:12ff for a secular view of the human capacity and necessity to transcend, based on the theory of evolution.
[54] In this regard Küng's "core of a global ethic" (1997:110) is deficient because its horizon includes only humanity.
[55] Linzey 1987:16ff.
[56] Cf Campolo 1992:65ff.
[57] Moltmann 1985:276-296.

[58] Cf Nürnberger 1987a.
[59] Swimme 1996.
[60] This is not necessarily restricted to the Christian faith. Naess distinguishes between "ultimate premises" (based in religious conviction or philosophy) and "platform principles" which can be derived from various faiths. A platform principle is that "the wellbeing and flourishing of human and nonhuman Life on Earth have value in themselves ... independent of the usefulness of the nonhuman world for human purposes" (Drengson & Inoue 1995:11 and 49).
[61] Laszlo 1977:367.
[62] Gen 1:28.
[63] Cf Levenson, op cit.
[64] Religions have found various ways of dealing with this experience. (a) *African religions,* for instance, ascribe the realm which humans can understand and influence to the sphere of a powerful tradition, mediated by the ancestors, and leave the rest to a mysterious being which cannot be understood or approached. In other words, the powerful Source of life is distinguished from a redemptive agency with limited power. As we shall see this leads to anxiety and fatalism. For the Hebrew faith the Redeemer had to be identical with the Creator otherwise he was not God. (b) The ancient *Parsist religion* of Zoroaster, with which the Jews came into contact when they became subjects of the Persian empire, conceptualised a good god (Ahuramazda) and an evil god (Angramainyu). World history was perceived to be the scene of a cosmic struggle between the two divine principles. That is by far the easiest way of defining the world in religious and ethical terms. When some Christians perceive the "devil" as something like a counter-god, they are closer to Parsist dualism than to the biblical faith. (c) But the *Jewish religion* has always rejected this way out. Satan is the "public prosecutor" at the heavenly court, whose job it is to test and accuse people who are unfaithful to God. In his ambition he tries to deceive people into sinning so that he has something to show (Job 1:6ff; Mat 4:10; Lk 10:18; 1 Cor 5:5; 1 Tim 1:20; Rev 12:10f). But he is in no way an autonomous counter-god. (d) Ancient *Greek dualism,* notably that of the Platonic school, assigned good and evil to spirit and matter. Our task in life is to disentangle ourselves from the historical and concrete sphere of life and to strive for the eternal and spiritual. Where that philosophy was followed consistently, not only the outside world but also our own bodies were of no consequence. There were "gnostic" currents in early Christianity which indeed arrived at such conclusions. Less extreme forms could be found throughout the history of the church. Many Christians believe that this kind of spiritualism is a genuine expression of the biblical faith even today. It is not. The gnostic heresy was relentlessly attacked by the most prominent authors of the New Testament documents, notably Paul and John. (e) The *Christian faith* insists that God loves this world; he lays claim on this world; he came into this world; he wants to redeem this world. God has entered bodily human reality in Christ; Christ rose bodily from the dead and constituted his church as his body; he lays claim not just on our spirits but on our bodies; he wants to redeem us as bodies and souls. Faith in the resurrection of the body means that our earthly bodies belong to Christ and may not be abused (John 1:14; 3:16; 1 John 1:1ff; 4:2; 1 Cor 6:13ff; Rom 12:1ff). In a similar vein God created this world and found it to be good; it belongs to him; it is most precious to him; he will reconstitute it. By implication it may not be abused. Therefore the Jewish-Christian faith, in its genuine form, rejects all forms of dualism and maintains the dialectic between the mastery and the benevolence of the one and only God.
[65] Confrontation with the Canaanite fertility cult led to the integration of nature; confrontation with Assyrian and Babylonian mythology to the integration of celestial bodies and cosmic powers; confrontation with Persian speculations to the integration of global history; confrontation with Hellenistic philosophy to the integration of a spiritual world.
[66] Is 40:18-20; 44:12-17.
[67] E.g. Is 58.
[68] Most clearly formulated in Eph 1:7-10; 20-23.
[69] Nürnberger K 1983: Socio-political ideologies and church unity. *Missionalia* 10/1983,

47-57.
[70] Ex 23; Deut 7.
[71] Eph 2:11ff.
[72] Gal 3:28. Unfortunately Paul himself was not consistent in his views concerning women. In 1 Cor 12:13 gender is omitted and in 1 Cor 11 and 14 discrimination against women is justified with the old arguments.
[73] McFague (1993:207) extends this insight to non-human nature.
[74] Cf Villa-Vicencio 1992:117ff.
[75] Gn 4:13-16.
[76] Gal 3:19.
[77] Linzey 1987:68ff.
[78] For the following see Nürnberger K: The royal-imperial paradigm in the Bible and the modern demand for democracy. *Journal of Theology in SA*, No 81, Dec 1992, pp 16-34.
[79] Block 1993:6ff, 27ff, 41ff.
[80] For detail see Nürnberger 1998, chapter 4.
[81] Wogaman 1986:26ff. See my reflections on the distribution of power in Nürnberger 1998:197ff.

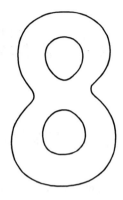

Traditionalism and modernity

What is the task of this chapter?

Part II deals with the transformation of collective consciousness. We began with the basic aspects of this dimension of social reality. Then we discussed the potentials for change of a particular mindset, namely the biblical faith. In the present chapter we deal with power discrepancies at the level of collective consciousness by comparing two unequal and radically opposed mindsets, traditionalism and modernity.

In section I we describe the revolution modernity has brought about during the last few centuries. In section II we compare its assumptions with those of traditionalism. We also argue that the impact of modernity on traditionalism has been disruptive and debilitating for the latter. In section III we sketch various reactions to conquest. Finally we reflect on the potential challenge of the Christian faith to these two mindsets.

Section I: The revolution of modernity

A new dynamic

Centuries back a new dynamic was born among the civilisations of humankind. It began in the countries surrounding the Mediterranean Sea and shifted to North Western Europe. Building on older civilisations in Egypt,

Mesopotamia and Asia Minor, it combined Greek and Hebrew perceptions of reality into a unique synthesis. The Greeks contributed the insight that reality was structured according to dependable principles, which can be explored by empirical research and rational thought. The Hebrews contributed the vision of a linear, goal-directed history and the power of a supreme will. Asian, Oriental, Latin and Germanic influences added further dimensions which we do not need to pursue.[1]

The origins of the new mentality were embedded in metaphysical thought and religious conviction, but during the last 500 years it became progressively human-centred, individualistic and secular. Its view of reality became mechanistic and optimistic, evolutionary and progress-oriented.[2] Its ethic demanded the active subjugation and transformation of all aspects of reality to the advantage of the human being. It had a pioneering and missionary spirit in which idealist zeal to help humankind solve its problems combined with ruthless pursuit of individual and collective self-interest.[3] It released the individual from metaphysical constraints, traditional authorities and hierarchical structures and gave personal initiative and enterprise free reign. It placed a premium on the optimum utilisation of all available resources to obtain maximum private satisfaction or corporate gain. The overall effect was a spirit of fierce competition in which everybody attempted to outwit and outperform everybody else.

Like most historical processes, the evolution of this new way of thinking began to unfold very slowly and gradually picked up speed. For a long time its potential seems to have been arrested by legal and hierarchical institutions, a static metaphysics[4] and a superstitious religion.[5] Beginning with the "Renaissance" and culminating in the "Enlightenment", it progressively discarded these shackles. Set free, it began to evolve very rapidly.[6] In our times it is accelerating at a breath-taking pace. In fact, not only the process as such, but even its acceleration has become institutionalised and entrenched as the dominant determinant of human life.[7]

The spiritual content of this mentality diversified into various ideological movements. *Liberalism,* the philosophy of freedom, justified the pursuit of individual self-interests. It found its counterpart in *nationalism,* which justified the pursuit of collective, but ethnically particularist self-interests. The belief in *progress* justified the pursuit of the self-interest of humankind as a whole at the expense of the rest of creation. The urge for freedom and power led to the dismantling of patriarchal, religious and aristocratic structures of authority and the emergence of the ideas of *democracy and human rights.* Because there were losers in the game of achievement and competition, a powerful demand for equality arose, at least in the form of equality of opportunity. This was the root of *socialist* thought.

When seen in the greater contexts of history, very divergent phenomena appear to be aspects of the same historical movement.[8]

(a) The voyages of discovery and conquest led to the establishment of *global empires* and the subjugation of virtually the whole of humankind under the rule of a handful of European nations: Spain, Portugal, the Netherlands, Britain, France, Russia, Belgium and - for a short time - Germany and Italy. The ideological thrust is highlighted by the following description of the British Empire: "How marvellous it all is! Built not by saints and angels, but the work of man's hands; cemented with man's honest blood and with a world of tears, welded by the best brains of the centuries past; not without the taint and reproach incidental to all human work, but constructed on the whole with pure and splendid purpose. Human and yet not fully human, for the most heedless and the most cynical must see the finger of the divine."[9]

(b) Drawing from much older roots in Egypt, Greece, China, India and the Islamic world, the development of *science and technology* experienced a series of breakthroughs with the achievements of great scientists such as Copernicus, Kepler, Galileo and Newton. These early beginnings led to the marvels of our century: space travel, genetic manipulation, organ transplants, nuclear power, computers, and so on.[10] Its missionary zeal was as conspicuous as the absence of its ethical self-critique.[11] According to Habermas science and technology have come to form the fundamental legitimating ideology of the late capitalist civilisation.[12] With equal justification one can say that they have become the religion of masses of secularised people because it provides them with their system of meaning, their right of existence, their authority to act, their sustenance, their miracles, their hope for the future, but also the ambiguity of the "hidden God" which can devour us all in a nuclear holocaust.

(c) The emergence of the *capitalist system,* built first on commerce and colonial exploitation, then on the industrial revolution, and finally on information technology, led to an accumulation and concentration of wealth and power unheard-of in prior phases of human history. The capitalist system has a powerful ideological base,[13] and produced its own "antithesis" in the form of the socialist movement.[14]

(d) Competing Christian and Islamic *missionary movements* of global proportions[15] were followed by similar waves of secular, materialist, liberal and socialist thought which swept through the elites of the entire world.

Chapter 8 - Traditionalism and modernity | 189

The evolution of the scientific-technological mindset

The "scientific mentality" did not arrive overnight; it developed gradually. It can be characterised as an impulse of the human mind which *longs for autonomy,* that is, for the freedom to see for itself, to think for itself, to act according to its own judgment. It wants to understand reality so as to control it. It progressively discarded other-worldly or superhuman authorities and powers as fabrications of the human imagination. It scorned the authority of the caretakers of old beliefs and wisdoms. It turned its back on the past and looked ahead into the future. Given circumstances were no longer accepted but questioned. Not what was real, but what was possible caught the imagination. And the possible was to be realised by using human potential, not by appealing to divine intervention. Humankind became the centre of interest and the source of confidence.

So the new "conquest of reality" was based not on trust, nor on authority, but on method. Two philosophical approaches took the lead, empiricism and rationalism. According to *empiricism* only things that can be perceived by the senses, or demonstrated by experiment, are taken to be real. According to *rationalism,* only what makes sense to human reason is accorded significance. Everything else must be ignored. Modern science is a combination of empiricism and rationalism. It begins with observed facts, from which logical conclusions are drawn, which are again subjected to experimental tests.

Once one understands how reality operates, one can work out how it can be controlled and changed to suit human purposes. So science (= the attempt to understand reality) flows naturally into technology (= the method used to change reality). *Technology* is applied science. Planning is an essential part of the technological approach. It is the attempt to obtain maximum output with the minimum input of effort, time and raw materials. Science helps technology to become more efficient, while technology subjects science to planning. We no longer wait for chance discoveries or inventions. Instead, we conduct planned research on a large and institutionalised scale.

Social developments are not balanced. As we have seen, some processes lag behind, others surge forward, even leap far ahead of others. Those who gain a competitive edge are able to use their power in their own interest at the expense of others. Vast discrepancies in resources, capabilities and life chances open up within human societies. While some become affluent beyond measure, others are marginalised. These discrepancies heighten the potential for conflict in society. With the development of ever more powerful and sophisticated weaponry, these conflicts have become forbiddingly destructive.

Obviously the power of the human race as a whole over *other creatures* also grew beyond measure. Humankind can now oppress and exploit nature at will. Since the Enlightenment, the world has lost its enchantment. Other creatures are without soul, feeling and emotion. They can be hunted down for pleasure, utilised for consumption, or tortured in scientific experiments.[16] This rapaciousness is bound to lead to serious consequences for the life chances of future generations. On the other hand, these two imbalances, the economic and the ecological, have led to a longing for universal prosperity, justice and peace. This longing opens up opportunities for redemptive action.

Not only outward structures like houses, factories or highways, not only the interaction between people within society, not only human dealings with nature, but *the human being* itself, in body and mind, became the object of analysis, manipulation, programming and planning. Education, propaganda and advertising mould collective consciousness. Biological sciences transplant organs and manipulate genes. Sexuality and the thrills of violence have become prime targets for economic exploitation through mass media and entertainment industries.

All this is extremely hazardous and calls for ethical alertness. Today we battle not so much to subdue nature as to keep our own creations under control. But the empirical sciences do not have the tools to deal with ethical questions. For a considerable time, a "positivistic" creed dominated the academic world. It was believed that science should not be prescriptive but descriptive. Normative statements were considered to be subjective and arbitrary. In the mean time the dangers of this stance have become apparent within the sciences. You can generate nuclear power, but what do you do with it? You can manipulate the genetic material of plants, animals and humans - but what will be the outcome if you do? But is every deployment of technology ethically justified? Hitler's death camps worked like clock-work, as did the American attack on Iraq. Scientific knowledge and technological efficiency cannot make people discern and do what is right and beneficial.

One result of specialisation is that experts lose their capacity to see reality as a whole. They concentrate on a tiny part of it, though they know more detail within that tiny part than previous generations dreamt of. Specialists have been called "professional idiots". This is not a very kind expression, but it does indicate the problem: the technical civilisation has lost an all-embracing and integrated comprehension of reality. You might be a Nobel prize winner in nuclear physics and be entirely ignorant of the forces that make society tick - not to mention religion or morality, where the ignorance and helplessness of professionals is most alarming.

This problem has been taken up by some branches of the scientific enterprise. The close interaction between social relationships, the mind and the body is attracting attention, for instance in the diagnosis of psycho-somatic ailments. There are specialisations which cut across conventional boundaries, such as economic geography and ecological economics. To regain a more comprehensive vision of the interaction between various dimensions of the social and the natural world the discipline of systems analysis has been developed. It is often utilised by new academic disciplines such as ecology, peace research and futures research (futurology). Here empirical data, derived from different disciplines, are fed into computer simulations under various sets of assumptions. The alternative outcomes of these computer calculations are called *scenarios,* that is, possible futures.

A new understanding of time

Traditional cultures distinguished times and occasions according to their religious and social significance. The time of a marriage or a funeral was different from the time when you built a hut or plowed a field. The technological civilisation has taught us to measure time in equal portions: seconds, minutes, hours and so on. Human life has been fitted into time grids controlled by clocks, calendars, stop watches, deadlines and appointments. Human beings have discovered that time is a precious asset which is constantly running out. To lose time is to lose production, opportunity, money, and fulfilment. Time management has become a part of industrial organisation. Punctuality and efficiency signify the time dimension of technological precision. Having "no time", we have become the slaves of time. For previous cultures time had been no problem.

Time is not only an abstract flow measured by watches and calenders. The sequences of time also mark chain-reactions of cause and effect. Complicated networks of such chain-reactions make up the sequence of historical events. The new understanding of time has led to an appreciation of the importance of historical processes which lead us from the past into the future. When science and technology began to flourish during the last three centuries, historical research also flourished.

Science and historical awareness have a common root, namely the spirit of discovery and conquest described above. Whereas scientific research tries to discover how things are, historical research tries to discover how they have become what they are. The motive is the same: once you know which forces determine reality, you can use them to manipulate reality. Historical consciousness is, therefore, not only interested in past events but also fascinated by the possibilities of the future. The present has come to be seen as the decisive transition between what has been and what may come to be. In part this new type of historical consciousness is also the result of the breakdown of meaning and the loss of identity which we shall presently discuss.

When you become uncertain of who you are and where you are, you look back to see where you came from, and forward to where you may be going.

A huge amount of knowledge has been unearthed through this research. Whole philosophies and ideologies have been based on theories about the flow of history and the possibilities of progress. Undoubtedly the most influential of these was Marxism. But nationalism of all kinds almost equalled Marxism in vigour. In the scientific field, theories of evolution and progress have stimulated experts to self-sacrificing effort and demanding norms of efficiency. For quite some time the future and its "unlimited possibilities" captured the imagination of the Western world. But the catastrophes of World Wars I and II, the dangers of a nuclear holocaust during the Cold War, the breakdown of cultural norms, and the growing awareness of ecological destruction have changed initial optimism to a more sober, if not pessimistic assessment of modern progress.

The reconstruction of the social order

Technology has led to production by machines, thus to *industrialisation.* Industrialisation drew people from the land into cities. We call that *urbanisation.* Society ceased to be based on small scale communities, intimate relationships and simple institutions, such as the family, clan or tribe. Large conglomerations of people called for *large scale organisations:* transport and traffic, communications networks, supply lines, distribution channels for raw materials and consumer products, credit facilities, etc. A large administrative machinery developed a dynamic of its own. We call that *bureaucracy.* In all these instances the technological method was applied to make the organisation of society ever more efficient. We no longer live in a natural environment but in a "man-made", artificial world, especially in the cities.

In Europe some groups advanced more rapidly than others along the way of emancipation, scientific insight and economic progress. They soon outperformed and ousted the hierarchical leadership of the feudal system and formed a new elite. This elite was not organised *upward,* but *forward.* In metaphoric language, it did not aspire to be on top of a pyramid, but to be the engine of a train. The last coaches of the train were occupied by masses of people who did not share the vision, the drive, the education, the economic resources and the political power of the elite. They may even have resented and resisted the new spirit. But they were, in spite of themselves, dragged along and forced into the new age. The conservative reaction had no future.

However, sooner or later underdog population groups began to acquire collective counterpower in the form of trade unions and socialist parties in a democratic state. This is how the "class structure" of Western society, or, in our terminology, the centre-periphery structure within the centre, came about. A more radical revolutionary movement, Marxism-Leninism, was only regionally and temporarily successful.

The evolution of individualism in the West

Although the individual was seen as part of a redeemed community, the responsibility of the individual before God, and the acceptance of the individual by God, was one of the outstanding characteristics of faith in *New Testament times*. During the Middle Ages, in contrast, the individual was perceived to be embedded in a great hierarchical order which encompassed church, society and the cosmic whole. Offices and rituals mediated between God and the believer.

The Renaissance began to loosen up this perception, but it was the *Reformation* which powerfully re-stated the Christian conviction that every person was immediate to God, both in terms of human accountability and divine redemption. The Catholic church was widely perceived to be an enslaving and illegitimate institution and the state a necessary evil.

Due to a series of startling *scientific discoveries,* the mediaeval worldview slowly but surely disintegrated. The earth was no longer the centre of the universe, but a tiny speck of dust lost among the galaxies; the human being was no longer the "crown of creation", but an insignificant accident in a long process of evolution on this tiny speck in space. There was no longer a heaven above where God ruled, an earth below where humans had to prove themselves, and an underworld which they were to escape at all costs.

During the same time the *feudal system* disintegrated. The initiative passed on to a commercial middle class intent on gaining individual liberties and privileges, particularly in business. The structured social system with the nobility on top and the serfs at the bottom came to an end. The *journeys of discovery* revealed that Western religion and culture were not unique. There was no longer one truth, guaranteed by the church, but a host of different convictions, all with their own claims to validity.

There was nothing to stand on, nothing to believe in, no above and below, only an endless sea of relativity. Imagine an astronaut who, during a space walk, suddenly finds himself with the supply line from the space craft cut off. There he drifts among the billions of stars with nothing to cling to, no external support, no direction, entirely dependent on the diminishing resources in his space suit, left alone with his inability to manoeuvre in space, the brevity of his time, and his lack of control over the immensity of cosmic reality surrounding him.

That was the feeling which led to modern individualism. Deprived of "objective" clues, the Western mind turned inward into the subjectivity of the individual. The only certainty I could ever have was my own experience; the only goal I could have was my own self-interest.

(a) In Philosophy, Descartes subjected everything to methodological doubt. The only thing he could not doubt was that he was doubting. Hence his famous conclusion: *cogito ergo sum* (I think therefore I must exist). The approach which builds on reason and nothing but reason is called *rationalism*. Francis Bacon, in contrast, accepted as real only what can established with our senses or their extensions. We call this *empiricism*.

(b) In religion, believers turned to the internal experiences of the individual. I feel guilty; I have to repent; I have to accept Christ as my personal Saviour. We call that *pietism*.

(c) In politics, kings and emperors claimed absolute power for themselves. They believed that their countries and their subjects were their personal possessions with whom they could do what they wished. We call that *absolutism*. When individual subjects claimed their rights against the kings in popular revolutions, it became the nation state which became paramount. Again self-interest ruled, though it was now a collective self-interest. We call that *nationalism*.

(d) In economics Adam Smith came up with the idea that, if every individual worked for personal gain, an invisible hand would coordinate their efforts in such a way that the society would prosper. We call *free enterprise*. Soon the world was considered to be nothing but material to be exploited at will. Whether minerals or fossil fuels, animals or plants, workers or beautiful women - you treat them as factors of production or objects of consumption. They have no value of their own apart from the utility they have for you.

One does not have to be a Christian to notice where this kind of worldview has brought us. The *extended family* has made way for the core family, the core family for living together, and living together for the short pleasure of sexual relationships without commitment. In *politics* our century has seen the most horrendous wars in history. The second World War cost 50 million lives. The self-interest of nationalist groups lies behind the bloodshed in Sri Lanka, Bosnia, Rwanda, Campodia, Israel, and so many other places. In *economics* self-interest has led to unbelievable accumulations of wealth for a minority and the impoverishment of vast sections of the population. We are in the process of depleting our natural resources and destroying our *natural environment,* thereby undermining our own survival as a human species. We have sufficient nuclear weapons to blow up the earth several times over.

In our *private lives* we have come to believe that we have a right to the immediate satisfaction of all our desires. And if we cannot afford more, we still believe that we are entitled to more, so we take it by force. That is the root of modern criminality. In *religion* this kind of self-interest has led to the irrelevance of the Christian faith in social issues. Christians, who are concerned about nothing but their personal salvation, have no contribution to make to the church as a living community, to society at large, to the natural world as a whole. Religious selfishness is still selfishness.

In all these cases humans resemble cancer cells in a body, cells which live off the body, but which are not integrated into the system. Being a law unto themselves, they endanger and kill the body. And by slowly killing the body on which they prey, they undermine the prerequisites of their own survival.

Revision: Characterise the modernist revolution briefly.
Application: If it is true that technology has given us the tools to exploit, pollute and destroy the natural environment on which all future generations depend, should we not go back to the pre-scientific culture of our ancestors?
Critique: (a) "You deprecate both the achievements of science and colonialism. Where would South Africa be today if its original inhabitants had been left to their own devices?" (b) "The Western mentality is brutal, dominating, selfish and arrogant. If we do not retrieve the African culture of ubuntu we are doomed." Comment.

Section II: Traditionalism and modernity

The effects of colonial expansion

The new dynamic was not confined to Europe. Through exploration, conquest, colonialism, imperialism and large scale European migration to the Americas and to Australia, it poured over the rest of the world. On the one hand, this brought modern development to the conquered populations, but it also caused havoc among their religious certainties, cultural traditions and social institutions. All cultures are, to a certain extent, in motion. They assimilate new ideas, adjust to new circumstance and face new challenges. Some Third World populations, which managed to remain in control of their own affairs, were able to transform their cultural heritage to such an extent that they became competitive. Japan, Turkey, Russia and even Africa yielded early examples of such internal and autonomous transformations.[17]

But colonial subjugation was widely experienced by the victims like a tornado leaving disorganisation and destruction in its wake. Their sovereignty was annulled, their leadership utilised, their economic potentials exploited. Traditional patterns of conviction were rendered obsolete by new systems of meaning, new values, norms and goals. Established social hierarchies were thrown into disarray and disintegrated. New roles and status definitions emerged. Responsible adults who had run the affairs of their communities for ages with self-confidence and skill, became minors, primitives, uneducated and unskilled in terms of the new value system. They were ridiculed not only by their new masters, but also by their own children who acquired the new ideas at school. The dissolution of social cohesion led to widespread anomie and breakdown of morality. Pervasive experiences of failure and dependency led to a loss of morale.

Countless volumes have been written on the effects of the colonial encounter on the partners concerned. It is impossible to do justice to the complexity of the problem within a single chapter. For our purposes it suffices to consider the effects of the confrontation between two general types of mindset, traditionalism and modernity. They are situated at opposite extremes of the spectrum, one at the outer periphery, the other at the inner core. What follows below are ideal type descriptions. Neither of them exists in pure form; they are inextricably mixed up in the giant cultural supermarket and there are countless variations and stages between them. Coca Cola and transistor radios have entered into a symbiosis with magic spells and burial rituals. Yet the contours of these two types can be recognised readily enough in the real world. By juxtaposing the extremes, we simply offer a model which can be varied and applied according to particular circumstances.

Characteristics of traditionalism

The traditionalist system of meaning is characterised by three aspects.[18] In the first place the *fabric of reality as such* is perceived to be constituted by "dynamistic" power. This is a type of power which cannot be defined in terms of natural law. It is rather like the movement of winds in the atmosphere or the currents in the sea. It can be calm or stormy; it can go in any direction; it is basically unpredictable; it is neither good or bad in itself.

Yet by means of appropriate rituals its currents can be channelled in directions which are profitable or detrimental to humankind. Diviners and community leaders use *public rituals* to establish the direction of present currents, channel them to the advantage of the community and protect the latter against detrimental flows of power. Detrimental flows of power are set in motion by sorcerers and evil spirits. Sorcerers manipulate the flow of power in secret, for private gain and against the interests of the community. It is not the magical means which differ between legitimate and illegitimate use of dynastic power, but the purpose.

Secondly, the kind of dynamistic power which gives life and health to the community flows from past generations to future generations through the current generation. The channel of this flow is the *male lineage*. This means that it moves from male ancestors to the grandfather, to the father, to the son, to the grandson, and from there to the not yet born. For this reason the community is patriarchal. Female members are attached to the male genealogical "spine" like "ribs". For this reason too, it is imperative that a family has male progeny. A childless person cannot become an ancestor, is not remembered after death and thereby drops out of existence. In cases of barrenness there are socially sanctioned ways to obtain substitutes.

The regulated flow of life-supporting power is the most fundamental and thus the most precious asset a community has. For this reason disturbances in social relationships are avoided at all costs. *The status and the role of every person* is meticulously defined according to sex, age and seniority. To go beyond one's socially sanctioned sphere of competence is a serious offence. Transitions from one status to another, such as birth, puberty, marriage, parenthood and death, are executed and legitimated by elaborate rituals called "passage rites" in cultural anthropology. All patterns of behaviour are strictly circumscribed and jealously guarded by ancestors and elders alike. Because disturbed relationships can upset the entire social system, any transgression is punished severely. Reconciliation with the ancestors and between the living members of the community is effected by means of elaborate ceremonies.[19] Traditionalism is hierarchical by definition.

The third aspect is the conception of the *Supreme Being,* or High God. The realm of power which is understood and controlled by the community is limited. Ancestors have more power than the living, whether to bless or to curse, but they are not omnipotent. Chiefs and fathers can channel power to their dependents in general and diviners can reveal and manipulate special powers. But there comes a point where a drought persists, a disease cannot be healed, a woman will not conceive although all relevant rituals have been performed. This shows that universal power goes infinitely beyond the realm accessible to human beings. This universal power is conceptualised in the Supreme Being.

In many traditional societies the Supreme Being is personified and a number of myths are spun around this figure. But, in contrast with the Hebrew Divinity, the traditionalist Supreme Being is not generally experienced as a person - though it is generally personified in traditionalist mythology.[20] It does not speak, make demands, hear prayers, or accept sacrifices. In contrast with the Greek divinity, it also does not represent rational principles which can be explored and utilised. Its basic function is to give conceptual expression to the vast context in which organised communal existence is embedded and which goes beyond human understanding and manipulation. The nearest Western equivalent is the concept of "fate", that is the overall flow of reality which is beyond human prediction or control. When fate strikes, there is nothing you can do but wait and suffer until the storm is over.

Particular *values and norms* emanate from this system of meaning. The cultural tradition is more than a social convention. It is the communally harnessed power of life in its flow. Any interruption of this flow is life threatening. There is no authentic life that operates outside, or in conflict with this flow. Death is an unavoidable rupture, which potentially threatens the continuity. This explains the importance of rituals which are meant to heal the rift. In the first place, death itself is believed to be contagious. In the second place, inappropriate behaviour during the funeral rituals may endanger the peace between the deceased and the living. In the third place death always represents a severe disruption of the protective social canopy. As a result, when death occurs, all activities cease and everybody concerned hurries to the funeral rituals. Expenses can be enormous. In the Western counterpart, death has only a nuisance value and people try to discount the "loss" as quickly as possible.

But death rituals and ancestor veneration also have something to do with overcoming a potential loss of authority and status. If authority and status could be lost through death, they would also be relativised in life.[21] As a result social relationships are organised in such a way that this will not happen: the deceased, who have less life than the living, are charged with more authority and power by means of appropriate rituals. Various phenomena even seem to suggest that the ancestors have no life at all, they only have authority, which in turn is located in conscious or subconscious memory. When they are forgotten, they cease to exert any influence. And if one has no progeny to remember one's existence one cannot become an ancestor.

The social power factor also comes into play in other ways. A development project which seems to give power to foreign influences, for instance, can be wrecked clandestinely but deliberately. A pump may be in constant disrepair until the representative of the donor agency has left. It may also stay in disrepair as a symbol of the powerlessness of outsiders who, in their demonstration of technical superiority, have questioned the competence and authority of the traditional leaders and their deceased guardians.

Possible economic consequences

We have mentioned the importance of progeny, the patriarchal and hierarchical order, the closely defined realms of competence, the strong orientation towards past generations and the traditions they handed down to the living, and the low threshold beyond which human self-confidence and initiative revert to fatalistic acquiescence. Authority is concentrated in the ancestors, the chiefs, the elders and fathers, the males in general, and the diviners. The right of existence is confirmed or questioned on the basis of these values and norms and authority is allocated accordingly. With due circumspection one can venture a few economic consequences of such a mindset.

Before we continue, it must be understood that there is nothing determinative about these consequences.[22] Whether they indeed come into operation, and to what extent, depends on a whole set of external and internal circumstances, most prominently the way the leaders of the community perceive their interests.[23]

(a) The willingness to *share resources* and bear the socially weak and vulnerable is proverbial. Unless the entire community starves, no insider is ever exposed to destitution. Under the enormous strain of growing peripheral misery the importance of this trait cannot be overestimated.

(b) Traditional societies have their own kind of "democracy" in that decision making processes are based on *social consensus*. After consensus has been reached, no opposition can be tolerated. As a result, individuals sacrifice their own opinions and initiatives to social harmony. Consensus includes the deceased and may not be disrupted.

(c) Closed spheres of competence and the priority of the community over the individual militate against the activation of *personal initiative* and the development of personal potentials. Any deviation from the norm is viewed with suspicion. The tightly-knit social network offers belonging and security to every member of the community, but it also frustrates the development of alertness and astuteness. The accumulation and utilisation of economic resources for private gain, in particular, is made virtually impossible; in fact, it would be tantamount to sorcery.

(d) The orientation towards the past militates against the exploration of the potentials of the future. *Change* is mentally and institutionally resisted.

Cultural evolution cannot be stopped, but its pace is reduced to a minimum. Some traditional cultures have remained virtually unchanged for long periods of time.
(e) The acceptance of human powerlessness over reality as a whole, represented by the Supreme Being, leads to a relatively low threshold where boldness and resistance turns into *fatalism and acquiescence.*
(f) Because *nature is uncanny*, it must be handled with circumspection and respect. This may seem to be ecologically advantageous and, in comparison with the rapaciousness of the technological civilisation, it undoubtedly is. But fear of nature is not the same as responsibility for nature. In the age of technology, ecological consciousness is not based on powerlessness, ignorance and fear, but on freedom and responsibility.
(g) Orientation towards biological ancestry is designed to underpin *small scale cognate communities,* rather than large social formations. Large scale social organisations require complex administrative structures and technological developments. In China, for instance, ancestor veneration is superseded by a universalist mindset which was able to sustain a vast empire for millennia. In most parts of Africa, in contrast, the traditional structures of families, clans and tribes persisted.

Yet all this is culture, not nature. A change of circumstances can lead to redefinitions which can modify accepted patterns of behaviour or explode the entire system. This is what happened with the impact of Western civilisation on traditionalist societies.

Characteristics of modernity

As mentioned above, the modernist system of meaning is characterised by two fundamental aspects. First, and in line with the Greek heritage, reality is perceived to be composed of physical energy which follows *natural laws.* These laws are explored by scientific research and utilised by technology. Power is channelled through knowledge expressed in tools, procedures and organisations which are able to facilitate its flow.

Second, and in line with the Hebrew heritage, reality is perceived to be subject to the mastery of a *supreme will.* Originally this was believed to be the will of an omnipotent God. Paradoxically, humans who subjected themselves to this will shared in its authority. In recent times, however, divine sovereignty was scrapped and human mastery remained. More than that, it was the *individual* who became sovereign. In principle, the modern human being recognises no authority but its own - whether natural, social, or divine.[24]

Chapter 8 - Traditionalism and modernity | 201

In principle too, nothing is inaccessible, forbidden or taboo; everything is at the disposal of the human being. "The aspect of science that capitalism seizes upon is the reduction of the universe to an array of units of energy that can be legitimately used for any purpose whatsoever."[25] The world is composed, not of uncanny constellations of power, which must be kept in a healthy balance, but of so much *material,* which is available for deconstruction and reconstruction.[26] Moreover, the difference between natural resources and human resources is blurred; both can be exploited at will, except where society lays down rules to control abuse.[27]

Potential calls for realisation. The lure of increasing mastery translates into a *forward thrust.* While traditionalism is geared to preservation of the past, modernity is geared to conquest of the future. Not the wisdom of the elders count, but the ingenuity of the youth. Those who cannot keep pace are left behind. In the place of the dignified institutions of traditionalism, modernity has institutionalised progress as such.

Personal ingenuity can only emerge when individuals are free to develop their potential to the full. Thus, while traditionalism is based on subordination, modernity is based on *emancipation.* Its goals have become shifting horizons; the future is open and indefinite; space is no longer circumscribed, at least not in principle. There is no sky which could set the limits.

The need to become acceptable (thus the right of existence) is basic in both cultural types. But in traditionalism it is based on integration into the community, in modernity on *personal achievement.*[28] Personal achievement is a norm which severs the individual from communal solidarity. In modernity not accommodation but competitiveness reaps rewards. Prestige is established by upward mobility, not by the steady growth of one's progeny. Acceptability is undermined by downward mobility, not by childlessness. Belonging must be earned by performance, not preserved by obedience. The worst thing that can happen to the individual is not to become an outcast, as in traditionalism, but to become a dropout. Because achievement leads to income, conspicuous consumption is a measure of status.

It is not just desire, but ambition and the craving for excellence that was given free reign in capitalism.[29] They assume such enormous power because they define acceptability. Mediocrity, which Fukuyama believes to be part of liberal democracy, may characterise the proverbial common man and woman on the street, but not the leaders in science and industry.[30] Liberals may clamour for equality of opportunity, but not for an equal share of the masses in the fruits of the excellence of the few. Everybody should enjoy burgeoning creativity and productivity, but not security, belonging and peace of mind.

Authority follows from this value system. In principle, all human beings are deemed to have the same right, even the obligation, to make the best of their potential. If they do not, they have only themselves to blame. In traditionalism, ambition is deliberately curtailed; in modernity, it is highly rewarded. The same is true for curiosity. In childhood people are taught, not to shut up and obey, but to explore and construct. Competition for knowledge, resources and power is not deemed to be dangerous, but beneficial for society. The accumulation of economic potentials and their utilisation for private gain, the most horrendous vices in traditionalism, are the most glittering virtues in modernity.[31]

Authority is granted to competence and efficiency, not to seniority. Therefore the society is not ordered in a patriarchal hierarchy, but in a staggered *class system*. It is not based on sex, seniority and age but on differences in income and power. Within this society *cooperation* can assume highly complex forms of organisation. However, it is not motivated by communal interdependencies and loyalties, as in traditionalism, but by functional advantages. Even when cooperation is extended to collectives, such as professional organisations, employers organisations, trade unions, social classes, ethnic groups, nation states, international economic organisations, and so on, the basic motives remain self-interest, utility and functionality.

When modern humans seek solutions for daily problems, or when they dream of future states of prosperity, they turn to science and technology, rather than to deities. Capitalist liberalism has its *own system of meaning*. It also has its own set of criteria for granting acceptability and authority. So we are here confronted with a new kind of conviction which competes with older convictions, including Christianity. Because of the economic powers it has unleashed, the gratifications it continues to promise, and the means of communication it can muster we should not be surprised if it is capable of outcompeting its religious rivals. That is the root of the mass secularisation we are witnessing today.

But there is also *a systemic compulsion* to perform and consume.[32] In a competitive system one either outcompetes one's competitors or one is outcompeted by them. There is no stability, only growth or decline. One has to cut down on costs, rationalise production, extend the resource base, improve efficiency, gain information, innovate, find niches, realise potentials, develop new markets, create new needs - or go under! Performance and growth are not only values and virtues but necessities for survival and prosperity.

Economic consequences of modernity

It can be expected that the *socio-economic consequences* of this mindset are diametrically opposed to those of traditionalism. Let us summarise them once again (diagram 8-1):

(a) While the evolution of traditional society marks time, due to its addiction to the past, the thrust towards the future in modernity leads to accelerating *scientific and technological advancement.*

(b) While traditionalists are content with reasonable levels of security, food and fertility, modernity pushes forward to growing *accumulations of productive capacity and wealth.*

(c) While living standards in traditionalism remain relatively equal, yawning *discrepancies in potential and wealth* open up in modernity.

(d) While traditional structures of authority are bound into fixed roles, statuses and procedures, the power generated by the modern system *can easily be abused.* This again makes it necessary to institutionalise countervailing processes, such as democratic checks and balances.

(e) While traditionalism grants belonging and security even to the weakest members of the community, in modernity individuals have to *fend for themselves.* It is precisely because of this lack of safety nets that social securities have to be institutionalised.

(f) On the consumer side all stops have been pulled. While traditionalists take from nature what they need, modernity opens the *sluice gates of acquisitiveness,* leading to the ruthless domination, exploitation and destruction of the natural world.[33] While not so long ago saving, mending and making do had been virtues, people are now ceaselessly induced to spend, enjoy and discard.

(g) *Equity and sustainability* are sacrificed.

It is, by definition, a ruthlessly selfish, volatile and exploitative system. It is mellowed only where its disruptions cut into one's own flesh, or where concessions are in one's own best interest, for instance where better wages make for more loyal employees. Profit, utility and pleasure maximisation are not only allowed, but have become normative. In capitalist liberalism freedom has assumed the character of self-interest and tolerance has assumed the character of non-accountability. Live and let live; perish and let perish!

The utilitarian and hedonistic attitude is so ingrained in the modern mind that it is taken for granted as natural and normal. In fact it is everything but natural and normal. It is the specific cultural acquisition of a historically unique social formation. As such it must be subjected to scrutiny and critique. Once released, however, the new dynamic has begun to feed on itself and it is extremely difficult to turn it around.

But why should we even try? What is the point in getting indignant or excited about the prospects of humanity, nature, or anything at all except one's immediate private interest? Only a powerful faith with comprehensive horizons can answer this question. But is there still a place for strong convictions and wide horizons in the modern world? The erosion of authority, morality and accountability has not stopped with patterns of acquisition and consumption. "What is in it for me?" has become the overriding question not only in the economic sphere, but in all dimensions of life. In a "postmodernist" age, convictions are traded on the open market: acquired, enjoyed and discarded when more "exciting" alternatives present themselves. It seems as if the very fabric of earlier systems of meaning is falling to pieces.

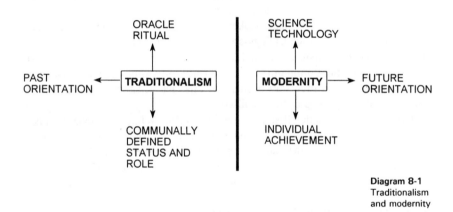

Diagram 8-1
Traditionalism and modernity

Traditionalism and modernity are unequal contenders

When the Western aggressive and competitive spirit interacts with the traditional spirit of social and ecological equilibrium, it is obvious who the winner will be. The South African situation is perhaps the clearest case in point. Dance, ecstasy, music and prowess, whether in play, hunting or battle, are evidence of the enormous vitality shored up in the traditional mindset. But it is not geared to the tactics of outwitting, outmanipulating and outperforming others economically. Traditionalists normally prefer to make peace and find their slot in the evolving social hierarchy. By the time a new Westernised elite had arisen among the colonised and had learnt to fight back on Western terms, Westerners had used their advantage to entrench themselves in unassailable positions of power - fully convinced that they were entitled to occupy these positions because they had won the competitive game.

Obviously the encounter did not remain on the level of "free competition". The destruction of Indian societies and cultures in Latin America; the imposition of the opium trade on China; the expulsion of native populations from land "opened up" for European settlement in Kenya, Zimbabwe, Namibia or South Africa;[34] the spatial organisation of the colony to suit the commercial interests of the motherland;[35] the massive slave trade from Africa, Indonesia and other places; the incorporation of chieftainships into the imperial administration through "indirect rule"; the introduction of poll tax to force Africans to sell their labour to farms and mines; the institutionalisation of racial discrimination in the colonies, with South Africa being the most prominent example; the acquisition of control over oil fields in the Middle East; the linking of "development aid" to imports from donor countries - all these phenomena witness to the ruthlessness with which Westerners have utilised their power for their own economic advantage.[36]

We do not want to suggest that there were no wars in traditional societies. Competition for scarce resources, the lure of power and dominance, conquest and subjugation, brutality and robbery, slavery and exploitation have been around for a very long time in human history. Nor do we want to suggest that there are no examples of very rapid adjustments of traditionalists to modern conditions. It is the magnitude of the power discrepancies caused by colonialism which we want to highlight.

Revision: Summarise the contrast between traditionalism and modernism and its economic consequences.

Application: A man, who works as a trained artisan in a factory, sacrifices a goat to his ancestors because his wife does not fall pregnant. How would you interpret this behaviour?

Critique: (a) "Social reality just does not match with your theory. Economic development has taken place in societies as diverse geographically, culturally and ideologically as Sweden, Japan and Botswana. It has nothing to do with world-view and religion." (b) "This 'analysis' is nothing but a manifestation of your prejudice against Third World cultures. The economy of African peoples was strong and healthy and they carried their heads high, until the white man stole their land and cattle, subjected them to wage-slavery, destroyed their human infrastructure and restricted them to ghettos, whether in South Africa or in the world at large." (c) "You betray hostility against progress. Not the

> achievements of science and technology, but the ravages of capitalist greed have led to the problems of the modern world." (d) "Why should traditional peoples adopt modernity in the first place? They have been much happier without it." With which of these statements do you sympathise? Are they tenable?

Section III: Reactions to conquest

It is hardly possible to imagine a greater contrast in mindsets than that between traditionalism and modernity. When they came into confrontation with each other, the outcome was predictable. What strikes the eye is the vast technological superiority of the modern civilisation over traditional societies, but it is not the only problem. Much more fundamental is the difference in power located in the structure of collective consciousness. The initiative and self-confidence of traditionalists have been subdued from childhood; they do not feel entitled to conquer as much of the world as they can for their own benefit. They have not learnt to use economic weapons to outwit and outperform their counterparts. In fact, ruthless economic competition is a totally foreign concept for them. There was no way they could withstand the onslaught of the Western civilisation. The consequences have been far-reaching (see diagram 8-2 for the following).[37]

The first reaction to this traumatic experience was *resistance*. When resistance proved to be of no avail, it was followed by *withdrawal* into the fortresses of the old tradition. Traditional cultures are not static, but they evolve very slowly. When traditionalism was confronted with a subversive, aggressive and destructive alternative, it tended to apply the brakes even more forcefully. The fear that the slightest concessions would cause a landslide, made it hesitant to allow any changes at all.[38] But ultimately there was no future in conservatism.[39] Overtaken by events, traditionalists tended to fall into fatalism and despondency.

The alternative was *dependency*. Dependency is inherent in traditional cultures. When the new masters had established their authority, it was natural for traditionalists to transfer their loyalty to them.[40] The new masters in turn were flattered by this unexpected respect and faithfulness. It seemed to legitimate their leadership and supremacy. Eager to escape the enslaving achievement norm at home, and traumatised by the feelings of inferiority intrinsic to a competitive society, colonial Europeans generated feudal attitudes and relished their patriarchal positions. According to some analysts, this is what led to the typical colonial syndrome of dominance and dependency with all its static and sterile implications.

But there are also Third World situations, notably in Latin America, which go back to the time when feudal structures were still in place in the colonial motherland. Traditionalists, whether slaves or indigenous people, were simply integrated on the lowest rungs of the system. Although there have been occasional rebellions in feudal contexts, they have usually questioned not the system as such, but only particularly harsh conditions or injustices which violated the social contract.[41] By and large a traditionalist mentality lacks the inner authority to claim equal dignity with feudal overlords. We shall come back to both forms of dependency in chapter 9.

Dependency can be of two kinds. The first is *acquiescence.* Here the dependent do not cherish the hope that they will ever play a role on par with the dominant group. They find security, acceptance and belonging in their subordinate position. They acquire pride and status by identifying with a great master. This attitude greatly helped to stabilise colonial regimes.

The alternative is *ambition.* Here the dependent become fascinated by the powers which their masters seem to control and try to penetrate their secrets. Particularly low ranking members of traditional communities, whose stake in the traditional heritage was negligible, must have recognised their chance to gain more power and status. Still conditioned by their old culture, they often learnt the new laws, patterns of behaviour and ideas by rote, without necessarily understanding the underlying rationale. They oscillated between two worlds of meaning and tried to fulfil the demands of both.[42] Their mimicry earned humiliating ridicule from their colonial masters. Colonial elites tended to despise an acculturating "dandy" more than a faithful servant who "knew his place".

On the other hand colonialists also used these acculturating locals as handy tools in their conquest, administration and exploitation of the colonies.[43] Obviously the colonisers were not interested in the upward mobility of the colonised and rejected all their attempts to enter the privileged class. Upstarts are never readily accepted as equals by members of a dominant group. That is the mainspring of colonial racism, which found a late and extreme development in apartheid South Africa. In its roots and manifestations it is close to aristocratic attitudes and quite different from anti-Semitism.

But the ambitious could not be detracted. There was no looking back; their new identity lay in the future.[44] It is in such acculturating groups that the self-confidence and the competence to *demand and fight for freedom* emerged. When the age of decolonisation dawned, it was the ambitious who succeeded in laying their hands on the levers of power abandoned by the colonial administrators. Not the traditional leadership, but a new, Westernised elite took control of the newly established nations.

Obviously not everybody could come out at the top. Those who had internalised a traditional culture, did not find it easy to compete with those who were at home in the urban-industrial civilisation. Most of the ambitious encountered *endless frustrations*. Acculturation is a ceaseless struggle between conflicting loyalties. At the best of times, catching up is an arduous process; when the dominant culture is engaged in accelerating change it becomes worse. Frustration can lead to *anomie*. One has turned one's back to the old culture in contempt, yet one fails to gain acceptance in the new. One literally has no place and becomes socially marginalised.[45]

Frustration can also lead to *regression*. Unnerved by its incompetence to operate in the modern setting, the remains of the traditional society were prone to respect those who "had made it". Even for westernised leaders, regression was the easiest way to entrench their positions.[46] Once in power, one gloried in the status of being a "very important person." The absence of a powerful and critical middle class created a power vacuum underneath these leaders. If one lost one's position, one could end up at the rock bottom of an impoverished society. So one had to make sure that this would not happen, and that one was well cushioned if it did. As a result, abuse of power and corruption became rife among these leaders. Life presidencies, nepotism and clienteles abounded.[47] "Predatory authoritarianism has been a central cause of Africa's underdevelopment."[48]

Frustration also breeds resentment. It can build up to a spirit of *rebellion* against the dominant civilisation. Among acculturating groups, rebellion and defiance do not manifest a desire to revert to ancestral traditions, which are no longer considered to be viable.[49] The rebellious grope for a new identity, which is equal to, yet distinct from, the cultural identity offered by the colonialists. The indigenous heritage is searched for elements which can be utilised to forge such a new identity. These traits are idealised, reinterpreted, transformed and integrated into a new world view, which is both modern and culture-specific.[50]

Deep down, however, the Western reference group retains its fascination and legitimating power. Most post-colonial leaders displayed a great need for recognition: the need to be represented on international platforms, to acquire modern prestige objects, such as airlines, fancy airports, five-star hotels, universities, presidential palaces, and to be driven by chauffeurs in luxury limousines.[51]

The need to discard the dominant paradigm, yet retain the modern spirit, also made one responsive to alternatives available within modernity. For quite some time, Marxism seemed to offer this critical alternative. Revolution against capitalist imperialism became a powerful symbol all over the Third World. A number of states experimented with Marxism-Leninism. But sooner or later disillusionment with this approach began to set in. Marxism was, after all, just another version of the Western dynamic. It was just as foreign to one's indigenous heritage, just as materialistic, technology-minded, and oppressive as its capitalist counterpart. It did not really fit African or Asian conditions, nor was it necessarily capable of pulling an underdeveloped country out of misery. Not able to deliver the goods, its oppressive character undermined its legitimacy. When it collapsed in its old strongholds, the Soviet Union and its former satellites, it lost its attractiveness in the Third World.

Rebellion is not a stable position. To overcome or replace the dominant culture is virtually impossible at this stage in history. If regression into traditionalism is no way out, one has to *adapt* and be *integrated*. One alternative is to become part of a train pulled by another locomotive. The majority of Westerners also find themselves in this kind of class dependency. But when there is no commensurate reward, carelessness and listlessness may result.

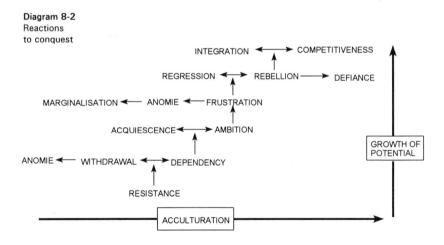

Diagram 8-2
Reactions to conquest

The other alternative is to press forward and upward to reach the commanding heights. Some of the most gifted have faced the challenges of the scientific and technological age and become *competitive* on the latter's terms. They are a small, but growing minority who, for all intents and purposes, become part of the global elite. In peripheral contexts their superior but underutilised qualifications cry out for fulfilment. Often they cannot withstand the lure of migration to the centre. It goes without saying that both alternatives have not helped the peripheral economy to get off the ground.

Population growth - the ailing society

The predicament of the marginalised mindset may be exemplified by the phenomenon of population growth. The problem as such is complex. Determinative factors are urbanisation, education, availability of contraceptives, standards of living, gender, culture and religion.[52] Here we simply want to highlight cultural disempowerment as *one* of the factors.

We have seen that growing numbers without increases in resources and productivity have devastating economic and ecological consequences. A healthy and self-confident society can be expected to invent ways and means to overcome a threat to its survival and prosperity. High growth rates were themselves a healthy reaction of pre-modern cultures to high demands for labour in traditional agriculture, crippling epidemics, recurring droughts, high rates of infant mortality, roaming beasts and raiding intruders. For Abraham and Sarah the promise that their descendants would be like the dust of the earth (Gen 13:16) constituted a message of salvation.

Times have changed. The natural environment no longer resembles a threatening monster but a beaten dog, clinging to its life. Fields have been eroded, grazing denuded, forests chopped down. In many areas cattle manure is the only fuel available. Water has become a problem. Traditional sources of income have become inadequate in a world where everything depends on money: food, housing, security, education, training, employment, status. It sounds like blasphemy, but in such a situation the avalanche of births has the same economic effect as swarms of locusts or plundering hordes in earlier times: they destroy the material preconditions of life.

A healthy population would have adjusted to the new situation by drastically reducing its birth rate. Even in traditional societies mechanisms existed which, though inadvertently, controlled population growth. Sexual relations, especially before marriage, were subject to severe taboos and social sanctions. Among the Basotho a pregnant girl would be married off to an elderly man. Sexual intercourse was not allowed during lactation, which could be extended to two years or more. Among the Amazulu there were regular virginity checks of young girls by their peers. Young men were not allowed to marry while in military service, which could extend to over thirty years of age. In some traditional cultures infants who did not meet the norm, such as twins, were put to death.

Apart from these traditions most cultures are imaginative when it comes to warding off a threat to their prosperity. But when a population is either culturally disempowered or economically exhausted it may lose its grip on life and become a victim of its circumstances. Normal social correctives cease to function. Examples from all over the world show that large families are linked to poverty. Once the standard of living and the quality of life rise, the growth rate begins to drop.[53] Under modern circumstances it can no longer be assumed to be a sign of robust health if a population grows rapidly, but a sign of social paralysis.

What could be the reason for this phenomenon? We have argued above that a dynamic civilisation, undergoing accelerating change, overwhelmed a relatively stable traditional culture and sucked it into its vortex. The victims not only lost their political self-determination but also the psycho-cultural means to control their lives under vastly changed circumstances. In a state of dependency, the overlords are seen to be in control of the situation they had created, whether militarily, medically, agriculturally or administratively. Weakened by poverty and frustration, one tended to abandon one's initiative and responsibility and allowed others to determine one's fate. In this sense Bedjaoui certainly has a point when he says that the Third World "is prolific because underdeveloped, not underdeveloped because prolific."[54]

Part of the dependency syndrome is that the dominated became passive recipients of modern medical services. While traditional healing was understood and performed by the community, modern medical science was beyond one's grasp and out of one's control. One did not want to do without it, because it was able to perform unheard-of miracles. But one did not "own" it. Otherwise one would not have continued to consult traditional diviners and herbalists alongside modern clinics. One also no longer felt responsible for the consequences; one did not even perceive the danger of population growth; and if one did, the means to counter it were out of reach. In a certain way passive acculturation may be worse than no acculturation at all.

The new rulers in turn had a truncated perception of the responsibility they had taken over. The application of modern medicine by colonial authorities and missionary hospitals catered for one side of the demographic process and neglected the other. While the humanitarian motivations for reducing infant mortality and raising life expectancy may have been most laudable, they were built on individualistic, rather than demographic assumptions. In the end the suffering caused by a population outgrowing its economic potential may be vastly more catastrophic than that caused by traditional causes of mortality.

Cultural integration

When trying to come to terms with the syndrome of dominance and dependency, one is faced with a dilemma. On the one hand all cultures must be accorded *equal dignity*. Dignity is fundamental to what it means to be a person. One's cultural identity defines one's collective identity within the context of an overarching system of meaning. This, in turn, supplies the norms according to which one is deemed acceptable or not. Acceptability again determines the measure of authority with which one feels entitled to take one's life into one's hands. Once an accepted system of meaning is questioned, people lose their bearings. The first prerequisite of cultural interaction is, therefore, acceptance of the equal dignity of all human cultures and religious convictions.

On the other hand we cannot ignore the fact that the modern mindset has unleashed powers unmatched by any other. Economic development, which has led to unprecedented prosperity, simply cannot be built on a traditionalist world-view. If one insists on shielding a traditional culture from modern influences, one simply has to *forfeit modern development* and the prosperity it generates. Enjoying modern patterns of consumption without commensurate increases in the quality and quantity of production will lead to dependency and misery.

Some populations, such as the Japanese,[55] have voluntarily allowed the scientific-technological world-view to *permeate and transform* their culture from within. They neither discarded their cultural heritage in favour of Western civilisation, nor did they allow the past to obstruct their way into the future. The word *voluntarily* is critical because it refers to the autonomy of the actors. Where cultural innovations have been imposed by colonial powers, the people concerned have not been in charge of their lives and we have discussed the results.

Viewing the ascendancy of modernity one is tempted to call for as rapid a transculturation of traditionalist societies as possible. But it is not as simple as that. In the first place, cultural change takes time. The main reason is that the most determinative cultural conditioning takes place in early childhood. In disrupted families, it is often the grannies who take care of the children of that age. By the time they reach school, the traditionalist foundations have already been laid.

More important, however, is the fact that modernity is not an unmitigated blessing. The industrial civilisation has undergone serious credibility crises.[56] Two world wars; political revolutions; a devastating depression; the rise and collapse of fascism; the crumbling of all the great empires, including the Soviet Union; racism in the US and in Europe; the ecological effects of indus-

Chapter 8 - Traditionalism and modernity | 213

trialisation; the oil crisis; the ideological conflict between West and East; the arms race; the prospects of a nuclear holocaust; the humiliation of the US in Vietnam and of the Soviet Union in Afghanistan; the failure of attempts to develop the Third World; growing unemployment, even in industrialised countries; rocketing budget deficits; the international financial crisis; the loss of credibility of socialist alternatives to the free market economy - all these factors and many others have undermined the optimism and self-confidence of the technological civilisation.

Uncertainty leads to fear, and fear leads to a conservative attitude. In the technological civilisation conservatism does not revert to traditionalist patterns, but tries to secure the advances of modernity. One *settles in with the accelerating process itself* and hopes for the best. Doubts are discredited as unwarranted pessimism. Freedom is absolutised; responsibility is suspended. Truth is no longer deemed fundamental for life. Convictions and values are drowned in post-modernist relativity; everything is permitted; all possibilities can be explored - from drugs to nuclear tests. Life is lived in the present; the future no longer matters. "After me the deluge!"

Theoretically the current crisis of the dominant worldview could open up *new opportunities* for the dominated. When a dominant culture crashes from its unassailable throne, the solidarity of a common pilgrimage could emerge. And indeed, the search for a new indigenous identity by Third World elites, a new self-confidence, a powerful drive for agency and self-sufficiency show that the spell of dependency may be disintegrating. But modernity, though hollow from within, has not relinquished its power, and is not likely to do so in the foreseeable future. The power game is by no means over. Therefore Third World leaders often react with great indignation when Westerners suggest that they should not follow the exploitative and destructive ways of the West: not chop down their forests, not poison their environment, not arm themselves with weapons of mass destruction. "It is now our time to pollute and become affluent!"

At a deeper level, uncertainty in the West can also devastate Westernised elites of peripheral countries for whom the West represents a living utopia. To a large extent their value system is derived from, and dependent on, the Western reference group. Cultural autonomy has not been reached as yet. Where such foundations begin to give way, *one may lose one's bearings* and begin to drift in any direction. To regain a foothold one shores up political, economic, social, or mental bulwarks. Massive abuse of power is one of the possible outcomes. Religious fundamentalism is another.

All this does not mean that mindsets *determine* social processes. Mental and social structures interact in a complex network of feedback loops, in which other influences, notably changes in the natural and social environment, also play a role.

Revision: Sketch the different ways in which traditional societies have tried to cope with the impact of modernity.
Application: Have you observed some of the phenomena described above in the relation between ethnic groups in your own country or on visits abroad?
Critique (a) "Your picture is one-sided and unfair. The colonialists have brought good administration, modern education, infrastructure, commerce and industrial development into much of the Third World. The majority of countries in Africa would still be in the iron age, if it were not for European colonialism!" (b) "You ride your hobby horse until you break its back. Every single factor you consider adds to the problem, none to the solution. Without doubt the real world is much more balanced than what your theories want to make us believe." (c) To reduce the birth rate by force, as the Chinese are doing, or to refrain from applying modern medicine to reduce the death rate is both immoral and impractical." With which of these statements do you sympathise? Argue out your own position.

Section IV: Challenges of faith to traditionalism and modernity

Let us look at a possible response of the biblical faith to the problems of traditionalism and modernity discussed in this chapter.[57] The system of meaning of this faith maintains that reality is not primarily determined by unstable and unpredictable forces, which have to be kept in a precarious balance, nor by unchangeable laws of nature which humans can explore and exploit at will, but by the mastery and benevolence of a personal God whose goal is the comprehensive wellbeing of his creation. Human participation in his redemptive action has two basic consequences, freedom and responsibility.

Freedom and mastery

Human beings are meant to be *below God* and *above creation*, including themselves. That is their unique dignity and status as human beings. Only when humans can transcend themselves and their life worlds towards a higher entity can they overcome their enslavements to uncanny forces and material idols. This higher entity must be a person, otherwise their own personhood could not be transcended. It is of critical importance that the human person should be transcended, because its absolutisation has led to some of the worst atrocities of human history. Humanity must be accountable to a higher authority without being enslaved by it. How can this paradox be solved?

Being persons, human beings are the only creatures who can relate personally to a personal God, who can enjoy his fellowship and participate in his authority, concern and vision. If the world is the creation of a personal God, there can be no uncanny forces which are not subject to this God. This implies that there is nothing divine or demonic in the human psyche, in society, or in nature to which humanity could be subject. "The earth is the Lord's and what is in it" (1 Cor 10:26). Where these assumptions are assumed to be true reality becomes secular and the human being becomes free. Human freedom means participation in the mastery of God.[58]

That *traditionalists* are not free seems to be self-evident to Westerners. The dynamistic composition of reality, the uncanny possibilities of diviners and sorcerers to manipulate dynamistic power, the unknown designs of roaming spirits, the vindictive authority of jealous ancestors and the fateful unpredictability and inaccessibility of the Supreme Being all contribute to this impression.[59] The concomitant assumption that emancipated *modernists* are free, however, is a dangerous illusion. While it is true that humans are not meant to be slaves of ancestral authorities, victims of sorcery, and targets of a blind fate, it is also true that they are not supposed to be prisoners of inescapable networks of cause and effect. Freedom from natural laws, social conventions and scientific paradigms is every bit as difficult to contemplate in the modern worldview as freedom from uncanny forces is in traditionalism.

Faith deliberately transcends the forces of reality towards the mastery of God. On their own, humans are part of the reality they perceive and experience, not its masters. As a result their desire to be autonomous can only lead to new enslavements. Autonomy means, in *traditionalism,* that you go it alone and become a sorcerer. The consequences are catastrophic, not only for the victims of their sorcery but for the sorcerers themselves. They drop out of the sphere of collective wholeness in which wellbeing is possible.

In *modernity* autonomy means, by and large, being cajoled by irrational desires and drives, lured by treacherous promises of fulfilment, whipped by the necessity to perform, fettered by primary group loyalties, pushed around by social currents, conventions and fads, straight-jacketed by scientific paradigms. Modernity generates the desire to be on a permanent high through constant novelty, drugs, sex, powerful machines, or deafening music. Only if we can transcend reality is it possible to rise above it. Christians are convinced, therefore, that *humans can only be sovereign masters of themselves and their worlds if they perceive themselves to be participants in the authority, concern and vision of God, the ultimate Source and Master of reality.*

Freedom and responsibility

According to Gen 1:26ff, humans have been created "in the image of God". The image of God, an ancient royal title, is the representative of God on earth to whom the creation of God is entrusted. Every human being has the right and the potential to realise this status. This means that reality as a whole is accessible to human understanding and utilisation. Here modern scientific assumptions are correct. It does not mean, however, that the human being, as the representative of God, is meant to be autonomous. When humans share God's authority, concern and vision, they feel a deep responsibility for God's creatures, both present and future. Being set free *from* the world, humans are free *for* the world, including nature, culture and society. The tension between what is and what ought to be cannot be avoided, but it can become a task. Within the range of specific powers and responsibilities humans have the authority to maintain, or to change the world.

This presupposes a criterion. In how far humans opt for stability or for change depends on an assessment of what *serves the intention of God* best in overcoming concrete deficiencies in wellbeing of the whole of reality. Other convictions have their own definitions of what ought to be. Because some of them correspond with the Christian criterion, while others do not, there is no way that confrontations on the religious level can be avoided. Neither the concern for a large progeny in traditionalism, nor the immediate satisfaction of fleeting desires in modernism are in line with the Christian criterion, namely what serves the wellbeing of social and natural reality as a whole.

Humans are supposed to be free and responsible. In fact they are not. Along with many other religions, the Christian faith is aware of human depravation and frailty. In other words, the designated agents of transformation are themselves in need of transformation. The demand to be what they ought to be *judges and condemns* them - and rightly so.

But condemnations do not lead people out of their predicament. The most remarkable contribution of the Christian faith is the conviction that our right of existence is a *divine gift*.[60] It does not depend on our capacity to fulfil a norm; it is granted by God, the one authority which transcends all social and natural authorities. God accepts his creatures *unconditionally*. He accepts them not because they are acceptable in terms of his norms, but in spite of the fact that they are not acceptable. Through the mediation of the "Christ event", God suffered their unacceptability (the death of Christ) *and* manifested his determination to overcome it (the resurrection of Christ). Symbolised by the death and resurrection of Christ, God's rejection of what is evil serves the establishment of what is good.

The implication of unconditional acceptance is that humans are liberated from enslaving norms.[61] Belonging and participation are, in almost all human communities, subject to the fulfilment of conditions; in the fellowship of Christ it is, at least in principle, unconditional.[62] On the one hand, *ancestral criteria of acceptance* found in traditionalist societies, which imprison people in the past and block their way into the future, are relativised and suspended as norms of acceptance. With that they lose their inhibiting and debilitating power. At the same time everything that is healthy and good in these traditions is retrieved.

On the other hand, the *achievement norm* of modern civilisation, which whips people into restless production and senseless consumption, loses its ultimacy and power. At the same time genuine advances in understanding and potency are retrieved. In other words, both the respect for ancestral traditions and the admiration of technological achievements are *superseded* by a sovereignty and purposefulness which, on the one hand, allows humans to leave the past behind and move into the future with confidence and authority, and, on the other hand, considers human work as a gift and feels free to stop the machine and celebrate the Sabbath.

Responsibility and community

Unconditional acceptance is acceptance into the fellowship of Christ. In concrete terms this fellowship is a caring community which transcends tribal, ethnic, national and class boundaries.[63] The tight confines of traditionalist solidarity are pried open and made more inclusive. For *uprooted traditionalists* it is vital to be integrated into a larger, non-ethnic community. We have seen that the breakdown of tribal structures under the impact of acculturation and urbanisation leaves a void which causes severe anomie and suffering. The emotional security granted by the community of believers offers a new foundation. The rapid rise of the African Independent Churches in South Africa, for instance, seems to replicate the growth of Christianity in the spiritual and social chaos of the pre-Constantinian Roman Empire.[64]

But the new community is, in principle, capable of combining belonging and security with freedom and future orientation - a feature which is not necessarily realised in all these churches. For those traumatised by racial and class discrimination, unconditional acceptance into the Christian community restores dignity and self-acceptance. As Liberation Theology and Black Theology have shown, it may reinforce their motivation to struggle for political liberation and economic advancement.

On the other side of the spectrum, the voluntary association of the Christian community is able to redeem *Westerners* from the loneliness and the meaninglessness caused by the technological civilisation. It gives them a new vision with a new mission - the ultimate antidote to meaninglessness. It offers them a communal cohesion which does not erode their freedom, yet evokes their communal responsibility. The fulfilment of the need for comfort and fellowship continues to attract Westerners to the church. It gives them resources to cope with the daily rat race, alienation and moral chaos encountered in the secular world. Only a superficial view can assume that this is economically irrelevant. The problem in Western churches is, however, that the expectations of what faith in Christ could offer have become confined to a private, otherworldly spirituality. To be relevant it must again become comprehensive and public.[65]

Progeny and profit

One consequence of unconditional acceptance is the demythologisation of progeny. It has seldom been recognised that progeny as a norm of acceptance, recognition and belonging has a profound impact on population growth in traditional societies. How did the biblical faith deal with this issue? In early times the Israelites, like other traditional cultures, depended on progeny for their survival. Accordingly, progeny became a sign of God's favour. There was no survival after death; life was maintained within the sequence of generations. How could Abraham believe in God's benevolence if God allowed his clan to die out? As mentioned above, salvation entailed the promise of an offspring as large as the "dust of the earth" and the "stars in the sky".

The individualisation of accountability referred to above signified the first great shift in this regard: it is *my* life which counts for God, not just the life of the clan, tribe or nation. Once this is clear, children are no longer decisive in one's relation to God; what counts is personal authenticity and integrity. Jesus did not marry, neither did Paul. But the final breakthrough came when acceptance became unconditional. Now progeny disappeared with all other conditions of acceptance. A person is acceptable merely on the strength of being a person, created and loved by God. This is the basis of emancipation from enslaving norms, including the expectation to perform as sires and childbearers.

In the technological civilisation progeny is of no consequence. Recognition and acceptability depend on personal or collective achievement. As producers people have to be able to compete. Dropouts are despised or pitied. Unsuccessful individuals and firms, even whole nations which cannot raise their Gross National Products, are considered to be "failures". As consumers people seem to miss authentic life if they cannot afford a television set or a car. People are whipped through their lives by the ceaseless urge to "better their positions". Aggressive marketing and the demonstration effect of elitist consumption are instrumental in enslaving whole populations from within.

The biblical faith confronts this set of obsessions head on. In the first place, it denies the validity of these norms. Neither proving one's worth in production, nor keeping up with the Jones' in consumption, are valid criteria of authenticity and acceptability. True, basic needs must be fulfilled, but petty motivations of profit and enjoyment are not worthy of the human being. Humans are meant to share God's creative authority, God's redemptive concern and God's comprehensive vision. Humans should be sovereign, constructive and responsible. These criteria of authentic life are valid for the great achievers and for the less gifted among us. They are also accessible to the superrich and to the miserable.

But much more fundamental than the replacement of a misleading norm is the fact that the biblical faith denies the right of *any* norm to enslave us and hunt us through our lives. Nobody can earn acceptance through any achievement whatsoever, because acceptance is a gift. The right of existence is granted unconditionally by the highest Judge thinkable. That is what the classical formulation "justification by grace, not by the works of the law" means.

Transcending the limits of time

The radical future orientation of the biblical faith is designed to overcome the inhibiting past-orientation of traditionalism. In *traditionalism* the power of life, the meaning of life and the directives for life flow through the patrilineal lineage. Therefore the past, represented by the ancestral spirits, plays a life-determining role. In the biblical faith, in contrast, meaning, authenticity and power are direct gifts of the personal Source of reality. The deceased cannot represent authenticity because they are just as much under the judgment of what ought to be as the living.

Therefore making contact with the deceased, a common feature of all traditional cultures, was prohibited in the Mosaic law. In fact, it was considered to be an offence so serious that the death penalty was imposed on the practice. The patriarchs were remembered because they marked the beginning of the relationship between Yahweh and Israel, but they were ascribed neither life, nor power, nor authority over the living. It was Yahweh, the "living God" and the "God of the living", who blessed and punished his people, who gave direction to life and demanded obedience, who was addressed in praise, sacrifice and supplication. And, as we have seen above, this God of life was increasingly perceived to be the God of the future. The modern future orientation may very well be a late consequence of this early breakthrough in human insight.

The New Testament built on the Old. Faith in Yahweh's incorruptible justice had, in the mean time, opened up the possibility of transcending death: God's intentions cannot be thwarted by human limitations. However, in Christianity the frightening prospect of a "last judgment", in which we were to account for our deeds, was superseded by the notion of a transition from inauthenticity to authenticity. The most radical symbolism was developed by the apostle Paul and his school: by his crucifixion Christ has left unauthentic human reality (the "flesh") behind and has risen into authentic human reality (the "spirit").[66] The authentic human being is enthroned above the powers that rule the universe. A comprehensive reconstruction of reality is envisaged. We, the living, are invited to anticipate our own transition by identifying with the death, the resurrection and the empowerment of Christ.

Thus authentic life is determined by a vision of a future which is qualitatively different from the past. Authentic life is seen within the context of the "age to come", which is a vision of authentic reality as a whole. This marks the fundamental difference between ancestor veneration and faith in the resurrection from the dead. Ancestors do not represent new life or a new reality; on the contrary, they reinforce the old. Resurrection must be understood as the gift of a new, authentic life.

The Christian faith also challenges the *modernist* worldview. Here it is not the past which enslaves, but the present. Due to the empiricist stance of modernity, the power to transcend death was progressively lost in the West. This earth and this life is all we have; it is only here where true life can be found. This is fair enough, but authentic existence has increasingly been defined, not in terms of justice and service, but in terms of utility, satisfaction and enjoyment. To reach true life we must exploit our senses, our possessions, our society and our natural environment as much as we can. Our bodies, society and nature are only means to an end. Authentic life is defined in terms of the sexy, the young, the pretty, the successful, the prosperous, the healthy. The particular blessings of old age are forgotten. The power and significance of death is suppressed.

This basic individualism and selfishness permeates the attitudes even of those who still adhere to the Christian faith. In concrete life it is business as usual. Faith is restricted to concern for the salvation of our "immortal souls" from eternal condemnation. This can be achieved by linking up with our "personal Saviour". This is a truncated and adulterated faith. We must regain a notion of authentic existence which defines what ought to be, not in terms of a disembodied soul at peace with a worldless God, but in terms of the Creator's peace with a reconstructed universe.

Transcendence has economic and ecological consequences. In the first place innerworldly enjoyment does not lead to authenticity; life for others is much more rewarding. Christ became poor so as to enrich others (2 Cor 8:9). Christians should know that, in practical terms, the "cross" means suffering for the sake of others. They should know the difference between faith and idolatry. Fulfilment of life does not depend on the possession of ever more gadgets. When they have reached the threshold where "enough is enough", they should be able to do without.[67] True riches consist in serving others in the power of Christ - a wealth which greed and longing will never attain. The lure of selfishness must be transcended towards the needs of others.

But we must go a step further. Apart from cutting down on consumption for the sake of others, it is hardly possible to overestimate the significance which transcendence could assume, if the global catastrophe predicted by some ecologists and economists cannot be averted. In fact, this catastrophe has already arrived for millions of people, notably in Africa. The vision of an authentic reality *empowers the suffering* to endure and confront their misery. It prevents those who have reached the limits of their tether, who seem to have become helpless victims of "superhuman" forces of history, from falling into despair. Human freedom and dignity can be upheld in situations of mass starvation and social chaos only by a powerful and world-transcending faith.

The question of truth

Postmodernist deconstructionism, like its Marxist counterpart, is a radical and highly deceptive form of modernity. It assumes that all mental constructs are oppressive and that emancipation implies consistent deconstruction. That all constructs possess inherent rigidities, and that these rigidities can obstruct rather than enhance life, is true. But the reverse is also true, namely that life depends on organisation. The reality on which life depends, and of which it forms a part, is an evolving system of constructs. Without constructs no life is possible.

As we shall see in chapter 13, the second thermodynamic law says that all construction (evolution) implies deconstruction (entropy). However, consistent deconstruction, without simultaneous reconstruction, does not lead to a life in freedom, but to entropy. And, as the technological revolution has shown, reconstruction without a set of criteria is irresponsible. In the field of collective consciousness, entropy is called anomie. We have more anomie at present than is good for our future.

Postmodernist deconstructionism is also, in spite of itself, in line with the capitalist ethos. According to the modernist creed the world is composed of nothing but material to be dismantled and put together again in the pursuit of profit, utility and pleasure. Throughput must be enhanced, obstacles must be removed. To remove marketing obstacles, the advertising and entertainment industries deliberately and methodologically dismantle inhibitions and commitments. All values are questioned. "Just do it!" This slogan found on car stickers can apply equally to extramarital sex, currency speculations, the use of drugs, the production of nuclear weapons, or the development of cults. Postmodernist relativity fits neatly into the ostensibly emancipatory drive of the capitalist system. The message is freedom, but the motive is the market.

Academics with well paid and secure jobs can afford the naiveness of emancipatory relativism, because they live off the proceeds of the capitalist economy. The captains in commerce and industry themselves, in contrast, follow a hard nosed agenda and a tight discipline firmly based on self-interest. There is absolutely no relativity and no tolerance here. The question is not, therefore, whether there should be mental constructs; the question is, which mental constructs are in line with the overall goal of comprehensive wellbeing, including a kind of freedom which is disciplined by commitment and responsibility. The Christian faith believes that there should be no compromise in this regard.

Revision: How can the Christian faith transform the motivations of traditionalism and modernity?

Application: A migrant labourer with a traditional cultural background encounters a computer technician in a modern factory for the first time. Can you imagine how Christian foreman could mediate in a conflict between the two?

Critique: (a) "Those who allowed themselves to become subject to an imaginary big boss in heaven cannot possibly be free." (b) "It is Western religious imperialism which has enslaved Africans, not their own cultural traditions." (c) "The religious obscurantism of a pre-scientific age will not be able to solve the problems of the future. What we need is hardnosed pragmatism and warm humanitarianism." How would you react to these statements?

Let us summarise

This chapter continued our series of reflections on the transformation of collective consciousness which forms the basis and corollary of social change. Its particular thrust was the existence of power differentials at the level of worldview assumptions. In section I we dealt with the *modernist revolution*. Human autonomy; the freedom to develop initiatives; a linear concept of time and a profound faith in progress; a utilitarian and hedonistic value system; the empirical, rational and pragmatic approach of science and technology; industrialisation, urbanisation, and large scale social organisation describe some of its manifestations.

In section II we compared modernity with traditionalism. In the *traditionalist* system of meaning reality is determined by dynamistic powers. Such powers can be activated by rituals which are beneficial, or sorcery which is detrimental to the community. Its blessings are channelled from the ancestors through the male lineage to the descendants. Right of existence and authority are determined by conformity to predefined statuses and roles in a hierarchical community. Where insight into, and control over, reality reach their limits, traditionalism tends to become fatalistic.

In the *modern world-view,* power is unleashed by scientific research into laws of nature and their technological utilisation. Right of existence and authority are tied to individual and collective achievement. In principle, human possibilities appear to be unlimited. The *impact* of the dynamic, future-oriented and ruthlessly competitive Western civilisation on a culture geared to stability, continuity with the past and social harmony was devastating. We closed the section with a discussion of the uprooting and debilitating effects modernity has had on traditionalism in colonial and post-colonial times.

In section III we enumerated some *reactions to conquest.* They range from various forms of resistance to various forms of accommodation. In each case the dominated found it hard to compete with the dominant culture, even though the latter was not without its crises. Population growth may be an example of the effects of mindset disempowerment.

In section IV we spelt out the potential impact of human participation in *God's vision and mission* on traditionalism and modernity. Humans can be liberated from uncanny forces operative in traditionalism and from the material obsessions characteristic of modernity. Genuine freedom *from* reality, based on benevolent mastery, is inextricably linked to responsibility *for* reality. The message of God's unconditional acceptance of the unacceptable suspends all norms of acceptance. Debilitating ancestral prescriptions in traditionalism and the inciting power of the achievement norm in modernity are overcome. But individual freedom is matched with responsibility in the context of a free and caring community.

Notes

[1] The phenomenon alluded to is very complex. I have been enlightened by C F von Weizsäcker's seminal work 1973.
[2] The genesis of the idea of progress is complex. Initially "development" was a philosophical concept, whose roots go back to Aristotle. It suggested that an idea will unfold in time like a seed. Applied to nature it became a tool for the interpretation of material reality. Applied to the human spirit it became the mainspring of idealist, historical and dialectical thought. Applied to the institutional realm it became a prime tool of the social sciences. The voyages of discovery, growing empires, new inventions, and the phenomenal success of science and technology triggered off the general progress optimism of Western civilisation. Darwin's concept of evolution gave a powerful impetus. Historical experiences such as the senseless sacrifices during World War I, the atrocities committed during World War II, the dangers of a nuclear holocaust, the global deterioration of the environment, and others, provided a pessimistic backlash. (Information gleaned from J Ritter (ed) 1972: *Historisches Wörterbuch der Philosophie*, vol II, Darmstadt: Wiss Buchgesellschaft, pp 550ff.
[3] The spirit is represented superbly by Rudyard Kipling's idea of the "whites man's burden". See Stokes 1960:27ff; Sandison 1967:64ff.
[4] The metaphysic was static because it did not do justice to the historical flux of reality, but posited timeless relationships.
[5] Roman law, a neo-Platonic hierarchical world-view and Germanic perceptions of fate, all appropriated by the Catholic church, may have been important stabilising factors. On the other hand the Roman idea of a pax Romana breathed a missionary spirit, and Germanic mentality was success-oriented. The high regard for personal honour and absolute loyalty, the restless striving for glory and heroic deeds, the mentality of self-assertion and conquest, which led to the explorations of the Vikingers and the historic migrations of Germanic into Central Europe were strongly dynamic cultural motives. (Information gleaned from Chantepie de la Saussaye (ed): *Lehrbuch der Religionsgeschichte*, vol II, Tübingen: Mohr, 1925, second edition, pp 540ff.)
[6] Cf Hagen 1962.
[7] For a good introduction to the phenomenon of acceleration see Toffler 1983.
[8] Cf Von Weizsäcker 1994:23f. Galtung distinguished economic, political, military, communicational and cultural imperialism; according to him the missionary movement belonged to the latter.
[9] Lord Rosebury, as quoted by Stokes 1960:17f.
[10] Fukuyama believes technology is the force that gives direction to history (1992:76ff).
[11] "In Japan (science) had even secured a missionary success as great as that of the Christian churches in Africa" (Knight 1986:206).
[12] Jürgen Habermas in *Technik und Wissenschaft als Ideologie*. Frankfurt/M: Suhrkamp, 1973. See also in Connerton (ed) 1976:330ff, 363ff.
[13] For a superb analysis see Heilbroner 1985.
[14] For a detailed description of the underlying ideologies see Nürnberger 1998.
[15] The ideological component of the missionary movement can be gleaned from statements such as the following: "I have no conceivable doubts that England has been privileged with the extraordinary success, power and grandeur of the last three centuries, solely for the purpose of sustaining true religion in the world." A theologian quoted by Sandison 1967:4. For South African examples see Nürnberger 1988, part II.
[16] Campolo (1992:25ff) argues that it was science, rather than Christianity which led to this objectivation of nature.
[17] For Japan see Morishima 1982; for Africa see Bundy 1988. Cf Fukuyama 1992:74f.
[18] For an analysis of a particular traditionalist mindset see Nürnberger 1975.

[19] It needs to be noted that not all traditional societies are communally organised. Billings has made an interesting comparison: "Tikana and Lavongai societies are shaped by cultures of contrasting style. Tikana culture is group-oriented, institutionalized, and egalitarian, while Lavongai culture is individualistic, non-institutionalized, and peck-ordered." (Dorothy Billings: Cultural style and solutions to conflict. *Journal of Peace Research* 28/1991:250). As she describes them, the Lavongai are almost a carricature of the notorious capitalist society, displaying a chaotic battle for shifting excellence over others and the exploitation of the weaknesses of the vulnerable, if they are not like an unstructured hostel community or uprooted urban youth.

[20] See Nürnberger 1975.

[21] The Egyptian pyramids were apparently meant to ensure that authority and status transcended death, thus stabilising life. The same with the early Jesus tradition: if God was in him during his life time, he must have been in him from the beginning of time and will be in him to the end of time.

[22] Not only on the basis of their anthropological assumptions but also on the basis of empirical evidence economists have rejected the thesis that culture plays a role in the economic behaviour of subsistence peasants: "Farmers in economically underdeveloped countries respond significantly and substantially to economic incentives ... The burden of proof thus now lies with those who maintain that the supply behaviour of farmers in underdeveloped agriculture cannot be understood predominantly within the framework of traditional economic analysis" (Behrman quoted by Thirlwall 1994:90). The response of peasants to high maize prices in Zimbabwe seems to confirm this. On the other hand it has often been observed that traditionalists will stop producing when their income reaches a particular level, leading to the "backward-bending supply curve of labour" (Thirlwall 1994:110). It may be that these differences in response are due to differences in the level of acculturation to modern assumptions, values and norms.

[23] In Taiwan, for instance, ancestors are venerated with the hope that they will bless and prosper modern businesses.

[24] Many theologians (Gogarten, Tillich, Bonhoeffer, Ebeling, Pannenberg) have depicted this change of consciousness.

[25] Heilbroner 1985:136.

[26] Hans Blumenberg attributed the genesis of this mindset to medieval nominalism rather than faith in God (Northcott 1996:57ff).

[27] "One aspect of the culture of most past civilisations strikes everyone ... their sacred view of the world ... It exists to be cajoled, propitiated, rewarded, and thanked, not to be abused, invaded, violated, or ignored." In contrast, the Judeo-Christian tradition "bids man to seize and shape, appropriate and subdue nature for human purposes alone. Nature is therefore desacralized and objectified ..." And this includes labour as a "versatile machine" (Heilbroner 1985:133f.). One can, of course, trace this attitude back to Gen 1:28f, and this has often been done. But, as we shall see in chapter 11, this is a wrong interpretation. It is also the only biblical text which could be misunderstood in this way. It is modern utilitarianism, not Christianity, which led to the modern attitude towards nature.

[28] On the achievement motive see McClelland 1961; cf Seligson 1984:53ff.

[29] Fukuyama 1992:135ff; 141ff.

[30] Fukuyama 1992:304ff.

[31] Cf Northcott 1996:70ff.

[32] Nürnberger 1998:28ff.

[33] Cf Duchrow 1994:223.

[34] For an example see Wolff 1974.

[35] Riddell 1970.

[36] Cf Dumont, Rodney, Frank, Mende, Rhodes, and many others.

[37] The following is adapted from Nürnberger 1988:154ff.

[38] Cf the description of "retreatism" by Hagen 1962; cf also Mende 1973:181ff.

[39] "Societies which have sought to resist this unification ... have managed to fight rearguard actions that have lasted only for a generation or two." Fukuyama 1992:126.

[40] Hagen interprets this phenomenon psychologically as identification with the aggressor

(1967:93). Cf Berger, Berger & Kellner 1974:126ff.
[41] A good example was the peasants' rebellion in central Europe in the 16th Century.
[42] "The same individuals have within themselves both modernizing and fatalistic values, and at different times, in different circumstances, one or the other come forward." Morgenthau in Arkhurst 1970:38. Taylor 1965 superbly described the dualism between homestead and school in Africa.
[43] Cf Anderson & Vorster in Arkhurst 1970:317f.
[44] Hagen 1967:207ff, especially the analysis of concrete cases 411ff.
[45] Cf Berger et al 1974.
[46] For an analysis of the political circumstances see my essay "Democracy in Africa - the raped tradition". In: Nürnberger, K (ed) 1991: *A democratic vision for South Africa*. Pietermaritzburg: Encounter Publications pp 304-318.
[47] Examples are not hard to find, especially in Africa: Nkrumah, Bocassa, Obote, Mobutu, Amin, Banda, Jonathan, as well as many military rulers (Andreski 1968; Klitgaard 1988 and 1991).
[48] H M McFerson: Focus on democracy and development in Africa. *Journal of Peace Research* 29/1992:241.
[49] During the struggle against apartheid in South Africa, for instance, there was a tendency among urban youths to reject all kinds of authority, including those of their elders, parents and teachers. They considered the latter to be uneducated, ill-informed, timid and subservient. See Leatt et al 1986:105ff. Cf Hagen 1967:411ff; Lauterbach 1974:163ff; Berger et al 1974:119ff.
[50] See Fukuyama 1992:235f for Islamic fundamentalism and pp 237f for the African renaissance in the US as responses to failure to maintain dignity.
[51] Berger et al 1974:118f; contributions in Arkhurst 1970:111f, 119ff, 320ff; Andreski 1968:165ff.
[52] In Indonesia 11.7% of rural female Batak use contraceptives as against 78% of urban male Chinese (Tan & Soeradji 1986:75).
[53] There is a direct correlation between large families and poverty. According to the Bureau of Market Research (1985:21) less than 25% of Black families with 2 and 3 children were in poverty and over 50% of those with 8 and more children in 1985. The other side of the coin is that in urban centres families become smaller. In Pretoria, for instance, the percentage of children under 10 has dropped from 30,9% to 17% between 1970 and 1985.
[54] Bedjaoui 1979:45.
[55] Other early examples are the Turks and the Russians. More recently Koreans, Chinese and others have followed suit.
[56] Cf Fukukyama 1992:3ff.
[57] Milbank 1993 offers a postmodern Christian critique of modernity.
[58] I owe this insight to Friedrich Gogarten, *Der Mensch zwischen Gott und Welt*. Stuttgart: Vorwerk 1956, 3rd ed, 139ff.
[59] For missionary attitudes to the non-Western world, see Bosch 1992:291ff.
[60] "Biblical repentance involves more than a hasty tear and a weekly prayer of confession. Biblical repentance involves conversion. It involves a whole new lifestyle. The One who stands ready to forgive us for our sinful involvement in terrible economic injustice offers us his grace to begin living a radically new lifestyle of identification with the poor and oppressed." (Sider 1977:144).
[61] Christians, in their freedom, have the ability to combine "responsibility and authority with humility and consideration" (Sleeman 1976:172).
[62] Cf Wogaman 1986:40ff.
[63] Cf Sider 1977:167.
[64] Russell 1994:81ff.
[65] Cf Duchrow 1994:294.
[66] It is important to note that Paul's usage of the terms flesh and spirit is not in line with the Hellenistic dualism between matter and spirit. Flesh means human reality as a whole as inauthentic, spirit as authentic.
[67] Durning 1992; Goudzwaard & de Lange 1994.

 *The dependency syndrome*

What is the task of this chapter[1]

In the last chapter we dealt with cultural aspects of power discrepancies between centre and periphery. In this chapter we turn to their social-psychological dimension. Both liberal and Marxist anthropologists have observed that disadvantaged communities tend to be fettered by a personality type which is based on a strong sense of authority - which is, in turn, correlated with psychological dependency. A dependency syndrome can undermine personal responsibility, initiative and decision making. As such it can present a formidable obstacle to development. It may also result in the vulnerability of a community to domination and exploitation, thus leading to the "development of underdevelopment".[2]

There are a number of variations of the dependency syndrome. It is present in Third World and First World contexts; in families, churches, corporations and polities.[3] To understand the phenomenon, we shall discuss (in section I) two classic analyses: Mannoni's exploration of the "psychology of colonisation" in colonial Madagascar[4], and Paulo Freire's programme of a "pedagogy of the oppressed" in North-Eastern Brazil[5].

These two studies complement each other in many ways. First, they originated in two different Third World contexts. The Brazilian rural proletariat cannot be compared with Malagasy subsistence peasants; the Brazilian latifundia owners cannot be compared with the French colonial elite.

227

Second, the two authors operate from opposite sets of assumptions. Mannoni is a bourgeois intellectual, Freire a Marxist activist. Mannoni departs from a pragmatic economic approach, Freire from revolutionary involvement. Mannoni is a cultural anthropologist, Freire an educationist. The differences and similarities between these two approaches will give us an impression of the complexity and the depth of the dependency syndrome.

In section II we shall argue that, under favourable conditions, the local Christian community has the potential of overcoming dominance and dependency and to generate a new mindset characterised by emancipation, empowerment, and communal responsibility. A potential is not necessarily realised. The Christian faith has often taken on repressive forms;[6] it has often legitimised oppressive regimes, for instance in South Africa;[7] it has often interpreted Christian humility, selfless service and devotion as attitudes of dependence, subservience and acquiescence.

But the potential is there and it is located in the core assumptions of this faith: through Christ we are supposed to participate in God's creative authority, redemptive concern and comprehensive vision. Here and there this potential has been realised, especially in Latin American base communities, ecumenical bodies and Christian activist groups the world over. To analyse the evidence, however, and to explain why it is usually not realised, is beyond the scope of this chapter.

As in previous chapters we sketched ideal types which do not occur in social reality in pure form. The idea is, again, to construct models which can be modified and applied to various circumstances. Third World elites will, in many cases, not acknowledge even the existence of the dependency syndrome. The emphasis of their social analysis lies on agency rather than passivity - which is a healthy sign that the dependency syndrome is being overcome at least in these quarters.

Section I: Two analyses of the dependency syndrome

Mannoni: dependency and inferiority

Mannoni's psychology of colonisation is concerned with the traditional type of dependency. Many traditional societies are marked by patriarchal and hierarchical structures. The stream of authentic life is perceived to be flowing along the male lineage from the deceased to the living and on to the not yet born. Beliefs, values and norms are retrieved from the tradition and passed on to the following generation by the most senior members of the community, such as chiefs, clan heads, fathers, and women of particular status. The elderly are in command; the young have to listen and obey. Because the lines of authority extend beyond death to the ancestors, nobody ever comes of age.[8]

To highlight the character of the traditional mindset, Mannoni juxtaposes two mutually exclusive collective personality structures, the colonised Malagasy versus the colonising French. The first is characterised by a sense of dependency which is caused by a hierarchical social structure, the second is characterised by a sense of inferiority, which is due to the existence of competition within a liberal economy.

The typical Malagasy is embedded in a net of social relationships, within which vertical relationships are pervasive and decisive. According to Mannoni, the reasons for the stability of such a system over centuries can be found in the fact that they develop a dependent collective personality structure with deep psychological roots. The underlying motive is the need for security. Security is found in submitting to a strict, but benevolent and reliable authority, that is, to a kind of father-mother constellation with comprehensive dimensions.

Individuals belong to their communities so unquestionably that there is no room for doubt concerning their right of existence. Within such psychological surroundings feelings of guilt and inferiority do not arise. One is free from the agony of personal responsibility. The authority of one's parents, the traditions of the community, the decisions of the elders, and the directions given by the diviners in the name of the ancestors create no inner conflict.

Mannoni believes that the basic structure of dependency is transferable. When a greater power or a higher authority appears, it is not difficult to yield to the new master. The transfer of authority only requires some familiarity with, and adaption to, the values, norms and social conventions that rule within the new realm of power. This type of transfer occurred in the early phases of colonialism.

The power of the French colonialists in Madagascar was unbelievably great. They came from the other end of the sea; their weapons seemed to be invincible; they abolished slavery and enforced a new social order; they overthrew the sacred order of elders and ancestors without adverse effects to themselves. Only a higher power can commit sacrilege without coming to grief. The Malagasy therefore expected the French to take over the role of superiors, to provide protection, security, welfare and moral direction. In return, they would receive unconditional loyalty. It could be argued that, because traditional societies experience life as an integrated whole, conversion, or the transfer of loyalty, means taking over the entire cultural package of the new masters. In the French power sphere one had to think, speak and behave like the French.

Such a transfer of loyalty does not change the dependent personality structure; it only changes the parameters of life. The old personality structure, Mannoni argues, remained intact: neither the ancestor cult nor the old customs were given up; they were merely superseded by a new authority. Even Malagasy who became Christians continued to serve the old gods in private without sensing any clash of loyalties.

The psychological advantages of being embedded in such a system are considerable, especially when one's childhood does not prepare one for a conflictual existence. Moreover, according to Mannoni, dependent personalities are, within themselves, psychologically healthy. This is apparent in their easygoing approach to life and complex-free sexual relations. Problems start to arise only when the authority or power, to which they have yielded, becomes suspect. As long as this does not happen, a dependent personality does not yearn for freedom. In fact, in a free society it would be like a fish out of water.

Mannoni observes that it is possible for such personalities to learn and execute the most complicated procedures with great precision and perseverance, even if it were a routine that made no sense at all. But it is almost impossible for them to develop new procedures on their own initiatives. The reason is that the meaning of the act lies not so much in the goal one wants to achieve, but in the feeling of security conveyed by unconditional obedience and the stability of the routine. The Malagasy's degree of intelligence is not lower than that of the European, but it is loyalty rather than initiative that has determined its development.

These observations are important for understanding the problems encountered in development projects and partnerships. According to Malagasy thinking, the beneficiary of a project can assist the provider to set up and run the project, but it would be inappropriate to develop initiatives or responsibilities of one's own. One cannot usurp a role and a status that properly belongs to the superior. Should the providers leave, after having set up the project, this could only mean that the providers have abandoned their work. To act in the spirit of the relationship, the beneficiaries would then also have to give up the project.

Alternatively, the beneficiaries might look after the project with great dedication and loyalty during a time of absence of the providers, without ever dreaming of making any changes to the layout of the project or to its operation. One cannot interfere in something one does not own. It has long been discovered that owning a project is the key to its successful operation. But the very concept of ownership is problematic for a dependent personality type.

We hardly ever realise what mental conditions have to be met before the concept of "help towards self-help" can mean anything at all in such a context!

According to Mannoni, the opposite of dependency is manifest in the Western collective personality structure. The infantile need for dependency, which continues unhindered in Malagasy culture, is repressed in Western child-rearing at an early age. In Africa the child is carried on the back of the mother, in closest touch with every movement and emotion of the latter. It simply develops into an extension of the mother's personality. In the West the child is placed into a pram. It is on its own, facing the mother, surrounded by technological artifacts. Emancipation and individual responsibility are thus programmed into the psyche already during infancy.[9]

As the child grows up, it becomes what it makes of itself. Repeated failures and adult demands cause permanent feelings of guilt and inferiority. These agonies drive the young individual to ever greater effort, thus to higher achievements, without changing his or her basic assumptions about life. The result is the aggressive self-assertion of the Western competitive mentality on the one hand, and its permanent latent misanthropy on the other. Under constant pressure, the Western individual dreams of a world without strife, a world without demands, a desert island which can be populated with obedient creatures of one's own creation.

Robinson Crusoe finds Friday; Prospero needs Ariel and Caliban. The attitude toward life during the colonial era is expressed in these characters of literary fiction. Mannoni believes that colonial Europeans had chosen life in the colonies precisely in order to assert their repressed egos. Their idea of the nature of "primitive man" was already fixed in their collective psyche before they had ever met any native. It was a product of their own psyche. Arriving in the colony they encountered a social reality that seemed to match perfectly with their subconscious preoccupations.

Just as the dependent personality of the Malagasy yearned for the presence of strong leaders, so the submissiveness of the Malagasy has been a balm for the souls of the colonials. They were hurting from the competitive battle, plagued by feelings of failure, longing for some acknowledgement. It seemed as if the two personality types found each other at once. But in this misunderstanding on both sides lay the seeds of the tragedy of the future.

The interlocking of two personality structures comes into being instantaneously whenever a European appears within a tribal context, even if the tribe concerned is still independent. The only condition is that the Europeans be regarded by the natives as rich, powerful, clever, endowed with special abilities, or merely immune against local powers, and that they are deriving a feeling of superiority from this. The saying goes that a European is either killed or made a king.

The link between these two personality types was fatal to the progress of the colonies. Colonial Europeans lost the Western dynamic by indulging in a dream world of feudal lordship; the Malagasy clung to an unworthy figure of authority and degenerated together with it. Going beyond Mannoni's analysis, one may point out the fact that independence did not automatically alter the situation. As our analyses in chapter 8 have shown, the westernised elites were often themselves traumatised by the pressures of acculturation and the competitive spirit. In fact, many of these leaders fell into the temptation of ruling with enhanced authoritarianism.[10]

For the vicious circle to be broken, Mannoni believes, two things must happen. The European colonials must be thrown back into the competitive situation from which they had escaped. However, since they instinctively sense that they would no longer be able to compete, they oppose this with all their might. On the other hand, the dependent parties can only be liberated if the umbilical cords are cut and they are forced to assume self-responsibility. However, their collective personality structure had been laid down in earliest childhood, and was thus perpetuated from generation to generation. So they will always tend to fall back into the nearest available situation of dependency to escape the agony of inferiority caused by competition. If this alternative was not available, they would be uprooted, fall into anomie and sink into drug dependency and crime.

Freire: pedagogy of the oppressed

As stated above, Paulo Freire deals with the feudal type of dependency as found in north-western Brazil. He too describes two opposite collective personality structures and the interaction between them. In line with his Marxist assumptions, however, he calls the oppressor and the oppressed. The oppressor is characterised by sadistic love. He wants to own, occupy and control. In his voracious striving for power he changes human beings into things to be utilised for selfish purposes. He organises the social and intelligible world in such a way as to cement the existing power structure. The oppressing elite is united within itself, but it divides "the people" (a derogatory term among the elites) in order to gain tighter control over them.

More important is what happens on the level of collective consciousness. The myths of the oppressors lay down the present structures of social reality and these are internalised by the oppressed. Through constant indoctrination, the existing power structure is reproduced within the psyche of the oppressed. This happens with such severity that the mere thought of rebelling against the

Chapter 9 - The dependency syndrome | 233

existing order does not even occur to them. The education system of the ruling party reduces its subjects to mere receivers of educational material which is laid down, layer by layer, into seemingly empty vessels. Freire calls this a Bankers' Education. Individual experiences and powers of decision on the part of the oppressed are deprecated and, where possible, eliminated altogether. The education provided by the oppressors teaches them not to think. Their consciousness is "flooded" by the "cultural invasion" of the elite.

Even if they want to help the oppressed, the oppressors try to keep the situation under control and withhold responsibility from their charges. The arrogance, paternalism and superiority displayed by experts, and the technical nature of the power they offer, causes the recipients of "development aid" and "modernisation" to degenerate into passive spectators. They are not subjects but objects of the projects.

The power-centredness of the oppressors even distorts their own view of reality. In fact, they believe the exact opposite of reality to be true. It is not they who are violent; it is the oppressed, at least when they rebel against their oppression. It is not oppression which blocks the way of the poor towards welfare and freedom, but their laziness and foolishness. And when, due to a revolution, the oppressors are forced to give up their selfishness, they do not experience that as liberation but as oppression.

On the other side are the oppressed. Their consciousness is remote controlled, flooded with alien material, thus alienated. They experience being oppressed as a normal condition. Usually their self-confidence has been shattered to such a degree that they are passive recipients of orders. They develop no initiatives. In fact, the oppressors fascinate them. If the oppressed develop any ambition at all, they copy the style of the oppressors. The oppressors have become the reference group of the oppressed: it is they who determine what ought to be.

A strange love-hatred makes the oppressors untouchable in both a moral and a political sense. Instead of fighting the oppressors, the oppressed vent their aggressions on their families or their fellow workers. The causes of their misery are attributed to superpersonal powers, to fate, or to God. Their psychological life is characterised by "necrophilia", self-contempt and masochism.

When revolutionary leaders appear on the scene, the oppressed are wary of, and hostile towards, their potential liberators. To make them understand their oppressive situation and recognise the true intentions of the liberators is an uphill battle. They can be liberated only if the idealised image of the oppressor is demonised and if the deterministic interpretation of the oppressive situation is demythologised.

This the oppressed cannot do on their own; they need the help of outside catalysts. The reason is that they are, internally and externally, so totally under control, and have sunk so deeply into impotence, misery and fatalism, that they cannot liberate themselves. They cannot even imagine or desire their freedom. A new vision can only come from outside their oppressive situation. The revolutionary leaders originate from the class of the oppressors. But out of love they have abandoned their class loyalties and moved into solidarity with the oppressed. This double condition makes them potential liberators.

It is obvious that Marxist ideas about the "avantgarde" determine Freire's views in this regard. But Freire's is a critical reception of the Marxist paradigm. If, because of historic conditions, the liberators fail to win the trust of the oppressed, they have to fight on two fronts. As a result they may become bitter. Fearing and mistrusting the oppressed, and not believing in the capabilities of the people, they try to create the revolution without the people and for the people, instead of with the people. They also use the methods of the oppressors, namely propaganda, cultural invasion, force and terror. Instead of a liberating radicalism, they create fanatical sectarianism - which shows the same symptoms, whether it comes from the right or from the left. Education is postponed until after the revolution, as though it could then still be achieved without brutal force!

To counter this aberration, Freire demands that there must be a liberating dialogue with the people from the outset; dialogue is the inner precondition of successful revolutionary action. But even if the people accept the revolutionary leaders, the latter are in danger of reverting to oppressive patterns of behaviour. They have internalised the ruler mentality of their class of origin. They are constantly tempted to take over the initiative. They know everything better. They lack the patience necessary for the abilities of the oppressed to unfold. Leadership thus ends up in messianism.

Freire develops his educational method against the background of this danger. The people must consider the revolution to be necessary; they must be committed to it; they must take responsibility for its implementation. The people must think for themselves and act for themselves. Revolutionary educators should not provide a new doctrine, which would again flood the consciousness of the oppressed; they should facilitate an analysis of the situation by the oppressed themselves. They should not impart the contents of their own consciousness, but, together with the people, bring into the open, and reflect upon, the ideas that are already present in the consciousness of the people. In this process, the people gain their own insights and develop their own motivation. In short, revolutionary pedagogues function merely as catalysts, not as teachers.

For this process to happen, Freire developed a practical educational method. First, "generative themes" are gathered. These are themes that cause discomfort and immediately result in discussion. Generative themes always call up "borderline situations", that is, situations that limit one's sphere of action. They pose the challenge of "borderline actions", that is, actions that lift the boundaries. These themes are then coded, conceptualised or presented in visual images. The people then decode these images by finding them in their own world of experience. An interdisciplinary team explores them further and shapes them into educational material. This kind of material is true to life. It is taken up by the people with lively interest. Of course, it aims at liberation.

While the revolutionary leaders should not dominate, therefore, they have their own contribution to make. They do not simply adapt to the worldview of the people. The people can have narrow horizons, short-lived interests, a superficial understanding of their situation, and superstitious ideas about the causes of their misery. Therefore the revolutionary pedagogues have a duty to educate. Does this in fact again mean doctrine, flooding of consciousness, cultural invasion? Freire does not think so. Their task is to start with the people's own interpretation of reality, present it as a problem, and then transcend it, that is, lift it above its original contents towards deeper analyses and more comprehensive goals. Freire calls this interplay between the people and their leaders "cultural synthesis".

A comparison between Mannoni and Freire

In Mannoni's work we are faced with the dynamic interaction between two types of mindset. One belongs to an originally autonomous Third World culture which, though immersed in a process of acculturation, is still relatively intact. The other belongs to modern, liberal, European imperialism.

In Freire's case there is the static relationship between two feudal types of mindset. The first belongs to a population of land workers that, many generations ago, have been uprooted, enslaved and thrown into a cultural melting pot. The other belongs to feudal elites of mediaeval extraction. The differences between the two situations are considerable, and it is obvious that the two observers had to come to different conclusions.

But the scientific premises of the two authors are also opposites of each other. Mannoni is a cultural anthropologist, who places his observations into a psychoanalytical frame of reference. Freud, Jung and Adler are his mentors. He subscribes to the ideals of a value-free science. He thinks that the intrusion of political objectives into empirical research is fatal - not only for science itself, but also for politics. Science has to produce undistorted facts;

politics has to take decisions on the basis of these facts. The scientist cannot relieve the politician from making decisions. Science can only serve politics if it remains at a critical distance.

Freire, on the other hand, is a lawyer, a historian, and an educator. Most of all, he is a revolutionary leader. He cites Marx, Lenin, Fanon and Fromm. For him, insight is possible only on the basis of commitment. Reflection without action is verbalism, just as action without reflection is actionism. According to him, a supposedly value-free science can only lead to those static myths with which the oppressors fetter their victims spiritually. It is a method of domination. Insight occurs either without the voluntary participation of the people in their liberation, and then it is elitist and stabilising, or with the people in the process of their liberation, and then it is revolutionary. There is no third alternative.[11]

Evaluation

Whether the respective analyses of either of the two authors are acceptable is open to debate. Mannoni has certainly not understood the liberation movement. Freire has certainly not understood the impact of cultural and religious traditions on development. Both Mannoni and Freire see their efforts as pioneering work in need of correction or refinement. They each have to be read in that light. But even if we could trust their analyses, a sound diagnosis does not necessarily lead to an effective cure. A more differentiated analysis (Mannoni) may hamper committed engagement, while revolutionary commitment (Freire) may lead to a blinkered understanding of social reality.

The careful observations of Mannoni indeed leave one with a feeling of impotence and futility. Conventional solutions, like granting the franchise, do not solve the psychological problem; in fact, they make it more complicated, no matter how desirable such solutions may be in their own terms. Mannoni's advice to begin with traditional village councils and gradually build greater independence from the bottom upwards seems to emphasise the importance of the grassroots, as Freire would demand, but it is not clear how this would lead the people out of dependency. Although the analysis is profound, the solutions offered are not convincing. Despite the deep and engaged empathy of Mannoni, his descriptive method does not create sufficient motivation for liberative action. One begins to sympathise with Freire's opinion that a detached scientific approach only serves to stabilise the status quo.

In Freire's work the winds of change are blowing. His method breathes the assurance of a goal, and the experience of success. While Mannoni is wondering how a dependent personality can be restructured, Freire has gone ahead and developed an "educational" (rather a psychotherapeutic) method with which he is actively restructuring human consciousness.

This method simply presupposes that being human, as opposed to being animal, is characterised by the will to change the world. Passivity, receptivity and dependency are symptoms of dehumanisation. He who receives orders, and has to sell his labour, is a slave - no matter what his cultural background is - and a slave is not an authentic human being! Whoever cannot develop initiatives without a superior has become an object of the will of others. It is as simple as that.

Obviously traditional cultures would not agree with this verdict. As Mannoni has shown, dependency can be a cultural phenomenon. Most cultures in human history have considered vertical relationships between people to be normal, healthy and just. A hierarchical order was normally not seen as a problem. On the contrary, it yielded emotional security and social stability. It provided a fixed framework within which the boundaries between righteousness and sin, justice and injustice, blessing and curse, had been laid down, depending on whether or not the lord or the servant fulfilled their obligations.

While injustice always prompted disaffection, justice was defined, most of the time, in hierarchical terms. Considering the vast sweeps of human history, the preoccupation with emancipation and liberation is a recent phenomenon. Incidentally, Western societies were also characterised by hierarchical social structures, at least until the Enlightenment and with powerful leftovers reaching into the 20th century. In Victorian times serving a great master fulfilled one with great satisfaction and pride. In the Germany of the thirties, following a great leader in absolute dedication was a high ideal - an ideal abysmally abused by Hitler.

So Freire's belief that liberty is a constitutive element of human nature must be seen, not as a timeless anthropological fact, but as a culture-specific assumption. For Freire it is so self-evident that he never questions its validity. All true convictions seek the verification of what ought to be in a mythological past, an other-worldly realm, or an eschatological future. True to the Marxist conviction, Freire's vision is a "real utopia", an anticipated future, a future which has to be achieved. Leaders and followers must actively bring about a fully humanised human being in the revolutionary process. Their old existence must be left behind.

This is not easy. According to Freire, freedom demands psychological strengths which many people cannot muster. Cultural patterns may have been so deeply internalised and entrenched in social structures, that they will prove resistant to the onslaught of an emancipatory mentality. These obstructions to human self-fulfilment simply have to be dismantled. There is an incredible missionary fervour in Freire's approach, obviously based on the absolute metaphysical certainties of the post-enlightenment West. No true Marxist is ashamed of being a missionary!

Yet in his own terms, Freire does not seem to be radical enough. One of his basic assumptions is that people must not become objects. They have to understand the truth of their situation and develop the will to change it. Is this also true for the oppressor? While Freire is indeed concerned about the liberation of the oppressors, he does not develop a pedagogy of the oppressor, nor does he volunteer to become a catalyst for the reconstruction of the collective consciousness of the dominant class.

This would seem to be the prerequisite for the true and comprehensive liberation of the society as a whole. True liberation will hardly be achieved if the process of conscientisation transforms the idealised image of the oppressor in the minds of the oppressed into a demonised caricature. Such a demonisation may generate, in the oppressed, the will to fight and liquidate the oppressive class, but not the mindset necessary to reach a society based on human rights and equal dignity. It is not only the catalysts who can revert to the methods of the oppressors, but the conscientised oppressed as well. Countless examples could be cited for this phenomenon.

The goal of conscientisation should not be the demonisation of a social class, therefore, but its humanisation. The insight must dawn that there are no angels nor devils in the world; there are only humans - and all humans are prone to anti-social behaviour. It is the horizontalisation of vertical relationships in the collective consciousness of both groups which should pave the way for structural transformation.

Before we leave the topic we need to remind ourselves that the two interpretations do not exhaust the subject. Consider the possibility that Freire's assessment of the dependency syndrome also applied to traditional cultures: the veneration of ancestors could be a subconscious ploy of fathers, chiefs, diviners and elderly people to ensure that their authority and power would be respected. Consider, on the other hand, the possibility that the technical superiority of the foreign catalysts questioned the power and authority of the traditional leadership structure. To "demythologise" the foreigners' superiority, and to assert its own position, the latter would secretly undermine the credibility and legitimacy of a project during its installation and make sure that it deteriorated or collapsed when the providers had departed. In other words, the "inefficiency" of traditional communities in running modern installations could be due to a secret power struggle between the traditional and the expatriate leadership. Such a power struggle could also ensue when the inevitable rivalry between traditional and Westernised leaders emerged.

On the side of the oppressed, the work of Scott suggests that the phenomena associated with the dependency syndrome may be symptomatic of an entirely rational choice for the "lesser evil" under conditions of severe repression.[12] Repression in turn could be accepted and internalised because no alternative presented itself. This could be applicable not only in feudal, but also in traditional cultures. This would bring Mannoni's analysis in line with that of Freire. Apart from referring to different situations, therefore, these interpretations may complement rather than exclude each other.

Severe repression can systematically induce psychological dependency. Torture and brain washing techniques in oppressive regimes, such as the Third Reich, the apartheid regime, or Maoism in China, have deliberately cracked formerly autonomous personalities and transformed them into dependent personalities.[13]

Revision: What are the most important differences and agreements between the analyses of Mannoni and Freire?

Application: On a visit, you observe an almost "natural" sense of superiority and dependency in the relation between a white South African farmer and his team of workers - seemingly taken for granted by both parties. How would you interpret this pheno~menon?

Critique: (a) "Do you really believe in this Marxist nonsense, even after the collapse of communism the world over?" (b) "Mannoni's psychological argument shows that he is in the camp of the bourgeoisie. Not a new mindset, but the revolutionary liquidation of the class structure will liberate the oppressed."

Section II: Horizontalising vertical relationships

I have argued on many occasions that the Christian faith should have the effect of horizontalising vertical relationships at the level of collective consciousness, though this potential is not necessarily realised.[14] Faith in the Creator reminds people that they have brought nothing into the world and will take nothing out. Faith in the Redeemer reminds people that "Christ, being rich, became poor, so that we may be enriched by his poverty" (2 Cor 8:9). Faith in the Spirit reminds them that they all belong to the lowest level possible, that of condemned sinners, and have all been elevated to the highest possible status, that of "sons and daughters of God".

The equality of dignity between all human beings is, therefore, an essential Christian assumption. Once the oppressors have internalised that mindset, they should be able to encounter the oppressed at their own level without condescension or arrogance. Once the oppressed have internalised that mindset they should be able to encounter the oppressors at the same level of dignity without submissiveness or hatred. The task of the Christian catalyst is simply to spell out these consequences of the gospel and draw them out into congregational praxis. As Bosch writes: "In their being converted to God, rich and poor are converted toward each other."[15]

In terms of congregational praxis, the two groups must be led into confrontation with each other under the canopy of mutual acceptance. This must happen quite deliberately.[16] The Christian community is potentially a particularly well suited training ground for creating a new liberated mindset because of its basic assumptions.[17] In Christ God has accepted the unacceptable into his fellowship, suffering the fact that they are unacceptable. Those who have been accepted into God's fellowship unconditionally have been accepted into each other's fellowship unconditionally. This implies that the agony of mutual unacceptability has to be suffered.[18]

God has accepted the unacceptable not to condone, but to overcome what makes them unacceptable. The canopy of mutual acceptance makes it possible for honest confrontations to take place. In this way deficient norm systems, unacceptable patterns of behaviour and detrimental structural realities are exposed, questioned and superseded by more wholesome alternatives. These new patterns are worked out together. The insights gained in congregational praxis are then taken back into the struggles of society at the particular stations of life of the participants.

The prerequisites of this process are that the Christian community is willing to be an open community, which strives to be representative of the population, and that they do not hesitate to undergo the painful process sketched above. A cozy ingroup consisting of like-minded, culturally homogeneous and socially compatible people, which withdraws into its ghetto to avoid unpleasant confrontations and transformations, will not do.

These theological insights form a potent recipe for healing the social disease of dominance and dependency. The goal of achieving freedom, equal dignity and mutual responsibility is clearly stated. Yet neither subordinate groups nor elites are rejected because of their respective positions in society. Liberal contempt for the underachievers, feudal contempt for the plebs, and underdog contempt for the oppressors all make way for mutual acceptance, which is a suffering acceptance of the unacceptable.

Acceptance does not mean that the untenable social situation is condoned. The goal is to overcome it. But neither party is confronted with a demand they cannot fulfil, namely to change their social position or their social conditioning overnight. Instead they are taken into a dynamic process in which the power of the one party is utilised to empower the other party. Once again, the entire programme presupposes that relevant sections of society are represented in the church so that the church mirrors the conflicts of society.

Application to the wider society

In spite of its Christian origins, I believe this paradigm to be accessible to all thinking and well-intentioned people.[19] It is also applicable to the wider society, even if the latter happens to be pluralistic. I do not hesitate, therefore, to draw out parallells between what happened in the church and what happened, for instance, in the South African society after the collapse of apartheid.

As recent developments in South Africa have shown, mutual commitment to achieve justice can defuse an explosive situation and lead to peace. Mutual acceptance and mutual commitment to justice is built upon the realisation of mutual interdependence. Without doubt the South African model, whether ultimately successful or not, will be studied carefully in a world of increasing social tension. The progress in insight, which the model presents, is that neither the liquidation of the oppressor, nor the subjugation of the oppressed can yield acceptable and lasting solutions. Both parties must cooperate in finding a way out of current dilemmas and to rebuild society on more acceptable foundations.

It is important that the impact of power discrepancies and contradictory interests between the two parties are not denied, but exposed and placed on the agenda.[20] Concerning the first, the aim should be to harness the power of the elite for structural change rather than to destroy it. We shall argue below, for instance, that affirmative action should not be designed to marginalise the elite. Likewise, the growing energy of the underdog should be channelled into reconstruction rather than revolutionary destruction. Why should valuable potentials on both sides be wasted or channelled into mutual annihilation rather than utilised in complementary action! But this presupposes that there is a common value system, at least a commitment to each other's wellbeing.

And that again presupposes that those who suffer want have a greater claim to being heard than the affluent.[21]

A confrontational pedagogy of the oppressor

Freire belongs to the Marxist school of thought which believed that it was only the poor and oppressed who could achieve liberation. One had to be committed to the "preferential option for the poor". Usually that was sufficient reason for leaving the oppressors to their own devices and concentrating one's energies on the empowerment of the oppressed. But historical evidence suggests that this assumption must be questioned. Whoever has power to determine society must be held responsible for the use of that power. If the elite changes direction, the impact of that change can be so much greater. This means that one has to develop a "pedagogy of the oppressors" which antithetically corresponds to the "pedagogy of the oppressed".

To act as a catalyst in the direction of the oppressor was certainly not fashionable among activists in Latin America and South Africa. It was also not very pleasant. When dealing with the conscientisation of the poor, educators speak to people on a lower social level than their own, while among the elite they normally encounter people on a higher social level than their own. It is much easier to work downwards than upwards, to accept the respect and gratitude of the poor, the uneducated and powerless, than contend with the disdain and indignation that one encounters when working among the rich, the educated and the powerful.

Is there a subconscious motivation to act as benefactor, enjoy the status of a superior and act out one's Prospero complex? This would be the attitude of Mannoni's colonist in a new form. Conscientising the elite does not question the "option for the poor". It merely extends the field of operation to the much more difficult but equally important opposite dimension. One's ability to persevere with conscientisation in the upward direction can, therefore, serve as a test to see whether one has begun to work at the problem of the power trap in one's own existence.

Even more important is the realisation, gleaned from the theology of confrontation and borne out by the South African experience, that the solution to the problem lies in the agonising interaction between the two partners rather than by the action of any one of them (or a catalyst for that matter) in isolation. Going at it alone, and trying to impose one group's will on the other, just does not work. The "generative themes" which could serve as a starting point for the discovery of one's own problematic position in the system must be formulated not by one's own group, nor by the catalyst, but by one's respective adversary.

Privileges that are taken for granted by the elite, for instance, can be formulated positively as claims of the underdog. We remember that it was with Western ideals that the Westernised elites of the Third World embarrassed their colonial masters on the level of ideology[22]. If democratic rights were valid in the motherland, they had to be valid in the colony.

In economic terms, the prospects that the poor will reach the inflated living standards of the rich are slim[23]. By raising these demands, therefore, the privileges of the rich themselves are questioned. The mechanisms by which these privileges limit the possibilities of the underdog can be analysed and lifted into the consciousness of the elite. Moreover, the fact that we cannot all attain the living standards of the elite because of ecological constraints, forces us all to look for sustainable alternatives to the present economic system. We shall pursue this argument in Part III.

A pedagogy of the catalyst

The functions of the catalysts have to change accordingly. They no longer move from the elite to the underdog, but mediate between the two. They act as facilitators of acceptance and confrontation in both directions. A whole new set of skills is required for this.

Catalysts must also develop a different understanding of themselves. According to Freire, the catalyst turns into the revolutionary leader. This is a good Marxist principle. On the other hand he says that "the people" must come to their own insights, formulate their own goals and act of their own accord, otherwise nothing is gained. This contradicts the Marxist idea, at least in its crude form. If one takes this thought seriously, one is lead to a conclusion at which Freire does not arrive.

For him, the final goal is a happy marriage between the leader and the people, in which the leader leads and the people are being led. This is also the case when the leader claims to do nothing but act according to the people's wishes. Practice has shown how deceptive such a claim can be. Mannoni's analysis explains why the people identify so readily with strong leaders. The latter relieve them from the burden of having to make their own decisions and shoulder the responsibility for their actions. When this occurs, the bonds of dependency are not broken, and the people do not come of age. In Freire's work this paradox does not seem to be resolved.

A more radical formulation of Freire's approach would be that no leader should come from outside the ranks of "the people". If the people have sunk into fatalism and inner impotence, there has to be a catalyst for change. But the motive of this catalyst must be to arouse the initiative of the people themselves and to disappear forthwith. If catalysts want to take over the leadership, their motives are suspect. The result may very well be that the people do not achieve the stage of taking responsibility for themselves. What does it help if one state of dependency is replaced by another![24] The identification of an obedient people with a talented leader is the most subtle and the most popular instrument of a successful dictatorship. Too often revolutionary leaders have turned into life presidents. And the Hitler regime, among others, has shown that this problem is not confined to the Third World. This trap can only be avoided if the catalyst renounces any claim to leadership right from the outset. This is even more important for catalysts of the confrontational model developed above.

But can leaders who emerge from among the people not also become tyrants? Without doubt they can![25] They are human beings, not angels or devils. No human being is immune against the seduction of power. But this merely says that the principle of being a catalyst must also hold good for leaders who come from the people. Conscientisation can only be considered successful if the leaders act with the same motivation as the educators, namely to lead the people to self-responsibility. This would imply that the latter transfer leadership only temporarily, on the basis of a democratic mandate and with full accountability. This is how democracy takes root in a formerly dependent culture.

But what happens if catalysts are already in charge of the situation, whether by virtue of inherited patterns of authority, or by virtue of superior technological sophistication, economic resources, or political power? To make the ascension of a new leadership possible, they have to dismantle their own positions of power and embark on a journey downwards. Impossible? The South African example has shown that it is not. Another example, that we want to pursue below, is provided by the experience of the missionary.

The missionary example

Let us explore the facets of this model by analysing the change from expatriate to indigenous leadership when African churches began to gain their independence.[26] How can a "young church", as they were called at the time, assume control over, and responsibility for, its own life under the conditions analysed by Mannoni? The sudden imposition of structural equality between superiors and subordinates is of no use if the mental constellation between the two parties remains intact. Mannoni and Freire both saw this, though each in his own way.

It is necessary for the status of superiors to be demythologised in both their own eyes and in the eyes of their subordinates. It is also necessary for the subordinates to be empowered not only technically but also spiritually (or psychologically) to take up their new responsibilities. The reorganisation of collective consciousness on both these levels is a process in time and involves a redefinition of the entire system of meaning, its system of values and norms, as well as a redefinition of roles and statuses.

While the first generation of missionaries were unchallenged superiors,[27] the last generation became catalysts of empowerment. The difference between these missionaries and the Marxist avant-garde was that the former did not aspire to lead the indigenous church, but provided the space for the church to develop its own leadership potential. There were indigenous Christians who had acquired an approximation of the status of the expatriates, but they were not readily accepted at that level yet by the rank and file. To facilitate the process of horizontalisation, those at the top had to descend. This descent by some of the leaders was a pioneering act which challenged other leaders to descend as well. It also presented a challenge to the subordinates to ascend.

The theological insight that seems to be crucial for an understanding of this process was first formulated by the Apostle Paul: "You know the grace of our Lord Jesus Christ who, being rich, became poor for us, so that through his poverty we might become rich" (2 Cor. 8:9). We have quoted this pivotal text before. The context of this verse implies that, by being enriched, the beneficiaries are enabled to participate in the process of becoming poor to enrich others. The goal of the exercise is a sort of dynamic equilibrium (Greek: *isotes*).[28] In modern terms we could speak of the horizontalisation of vertical relationships.

In a rather subtle way the act of descending sets the criteria for a legitimate ascent of the new leaders. Paul shows in 2 Cor 8 that, for the sake of the community, the ascenders have to participate in the descent once their ascent has been completed. I am not sure that this goal has been achieved in the case I am referring to. What often happened was that, once in power, the new leadership reverted to feudal attitudes, and the rank and file acquiesced. The process of descent and ascent can then easily be misunderstood as a mere power struggle between expatriate and indigenous elites. At a deeper level, authoritarian attitudes often reflect an unresolved sense of insecurity. We analysed the reasons for this phenomenon in section III of chapter 8. One has to assert oneself because one does not yet possess a kind of "natural" or self-evident authority. Obviously the same danger lurks in the new South African political constellation.

However that may be, ascent and descent are processes in time. In the early stage the catalysts are above the subordinates. If they win the trust of the subordinates, the dependency of the latter will be transferred to the catalysts. This stage cannot be avoided. The rank and file is still in dependency and would not understand the "humility" of descending leaders, because the dependency syndrome needs powerful superiors. As Mannoni tells us, they would misunderstand humility as weakness and feel orphaned, turn away disappointed, perhaps even disgusted. Weak leaders are not respected, indeed not wanted, by followers under the best of circumstances.

Those who have the ambition to ascend also need a strong and critical partner whom they can trust. But for that to happen the latter must affirm their ambition and actively work for their ascent. Yet the relationship is full of ambiguity. At first the status and experience of the old leaders seem to present insurmountable obstacles to the advance of aspirant leaders. To allow the potential for the subordinates to ascend, responsibilities must increasingly be transferred to indigenous leaders. Sooner or later the cutoff point is reached where, structurally, former leaders become subordinates, and former subordinates become leaders.

This is the most critical point in the process, because the balance of power now definitely tips in favour of the ascending party. But its acquisition of power is still uncertain and vulnerable, especially since the performance of the new leaders is being compared with the performance of the old by both parties, as well as by the rank and file of the subordinates. To stabilise their status, the ascending party will, consciously or subconsciously, seek to destroy the authority, the status and the reputation of the descending party. For this they might fall back on the secret resources of their own cultural heritage, which are not necessarily at the command of the expatriate party and which are experienced as deviousness by the latter.[29] There is some affinity here to Freire's destruction of the image of the oppressor in the mind of the oppressed.[30]

Psychologically all this is important, but difficult to endure for the descenders. Emotionally the latter cannot understand why the ascending party is responding to their caring concern with so much animosity. Moreover, the descending party is passing the psychological threshold where power has to be forfeited in real terms. Such a step is coupled with anxiety. One feels humiliated, rejected and useless. One longs for gratitude and acknowledgement. One's gifts and the investment of one's life seem to go to waste.

Moreover, the descenders cannot help but believe that they could have done the job better than the ascending party, but have to refrain from interfering. The tender plant of responsibility may not be disturbed; it must be nurtured and protected. Advice from the descending party is rejected or given out as the wisdom of the ascending party. Initiatives emerging from the descending party are boycotted or taken over as being the ascending party's own achievements. The descending party may become bitter, lonely and frustrated - corresponding with the extent to which they had invested themselves in their former subordinates.

The frustration on the side of the descending party may be matched by secret feelings of guilt on the side of the ascending party, and these feelings may be suppressed by an even more self-assertive attitude. After the cutoff point, the ascending party may experience a short high. Now the time to prove oneself has arrived. But the ascender cannot get rid of the inner authority of the descender. Again, anxiety turns into aggression.

The integration of the descender

At this stage the descender will be tempted to leave the field. Many of the most sensitive and perceptive missionaries heeded the call "Missionary go home!" This would be in line with Mannoni's recipe. Do parents not have to leave their children alone at some stage, no matter what these will do with their newly found freedom? What about the "creative withdrawal" with which Jesus catapulted his disciples into maturity and responsibility?[31]

While this may be so, one can ask whether the process can come to a healthy conclusion if the descender leaves the scene too abruptly. For traditional communities loyalty is of immense importance. One is expected to stick it out with one's group, even when this implies suffering and deprivation. The descender now truly becomes one of the rank and file, in spite of his/her previous position. It must become clear that the process of descending, or a simple rotation in leadership, does not mean demotion. Can the principal of a seminary again become a lecturer after his/her term of office has expired? If not, the principle has not been understood.

There is another immensely practical consideration. Leaving the field is only possible for expatriates. South African whites, for instance, have no other place to go. And even if they found another place, this would be a loss to the country. The approach mentioned above does not foresee a departure at all but integration on a new basis.

This means that the creative disengagement of the descending party should liberate, but not abandon the ascending party. In my opinion, Mannoni needs to be corrected on this point. He knows that the departure of the former authority is traumatic for the dependent personality, but he believes that the rupture is necessary.[32] My view is that the presence of the descending party is of vital importance for the process of emancipation to reach its goal.

The descender must learn to adopt a role which is not threatening to the ascender. The ascender must learn to adopt the role of mandated leadership. Only when the process of ascent has been completed and the status of an established leader has been achieved, will the ascending party be able to think of its own descent. As long as one is struggling for status and identity in the face of a potentially devastating competitor, one does not have the psychological strength to descend. This, however, is the condition for the descending and the ascending parties ultimately to relate to each other on the same status level.

So the descenders may not withdraw in the last minute in order to remain distant superiors. Only then will they understand what it means to belong to the weaker party. Only then can they show that it is possible for a superior to be a subordinate without psychological disintegration. Only in this way can they demonstrate to their former subordinates that their descent is really genuine and not just a manoeuvre. And only in this way can the subsequent descent become a respectable option for the ascending party.[33]

For this to happen, however, there must be an underlying certainty of mutual faithfulness in spite of all hurting and all anxiety. Once again, loyalty is fundamental for the functioning of a common venture. The basis is, again, the unconditional acceptance of both parties by God in Christ which translates into the hidden but determinative unconditional acceptance of each other by the two parties concerned. The ascending party has to be certain that the descending party will not abandon them but remain available. The descending party must be certain that the ascending party does not actually want to get rid of them. Both parties must know that they are accepted and supported by the community as a whole.

This loyalty is sustained by faith. All parties must be able to fall back on a power that infinitely transcends their own capacity of accepting the unacceptable, namely the power of God's unconditional acceptance in Christ. Under God, the positions of the two parties become relative. They are both miserable sinners; they are both sons and daughters of God. The atmosphere of acceptance is created when the proclamation of God's benevolence incarnates itself into the life processes of a community of unconditionally accepted individuals. Ascent and descent do not happen in a vacuum, but in the context of a community which draws on the suffering acceptance of God[34].

Mutual acceptance has an indispensable structural dimension. The descending party can only remain loyal if it is not denied a place in the system. Affirmative action policies in the new South Africa, for instance, have led many white professionals at the peak of their careers to exchanging their jobs for retrenchment or retirement packages. Among young white professionals the perception is rife that they have no future in the country. As a result, many of them emigrate. This is a great loss to the country as a whole. It should be clear that our model does not deny the importance of affirmative action as such. It is, in fact, an affirmative action procedure. The point is, rather, that affirmative action must not lead to the marginalisation of the descender, otherwise it is counterproductive.

The perception of not having a future after their term of office has expired has also led many democratically elected leaders in Africa to entrench themselves and turn into dictators. It is a fundamental and widespread problem that needs a solution if the model we espouse is to work to the benefit, and not the detriment, of the community.

The aim of the exercise is equality of dignity. Only when the descending party has really reached the bottom, and when the ascending party has really arrived at the top, can they again ascend and descend to meet at the same level. Then both have covered the whole way. They know what it means to be human, in weakness and in strength. Their respective identities and their right to exist are no longer linked to their social status or their achievements, but to the ultimate origin of their existence as human beings.

Before we leave the topic, we have to mention that horizontalisation and affirmative action exercises come with certain trade-offs. The outgoing incumbent can normally be expected to have greater experience, wider contacts, accumulated confidence among partners, subordinates and superiors, perhaps even a better training than the newcomer. This is part of the verticality analysed above.

One would expect, therefore, at least a temporary drop in efficiency and productivity for the organisation concerned. To a certain extent this also holds true for the normal turnover of staff due to retirement or resignation. In the latter case, however, such losses are usually compensated by the greater drive and energy of younger staff and their more up to date knowledge. This cannot be presupposed in horizontalisation and affirmative action exercises. However, without allowing the newcomers to get to the levers of power, they are unlikely ever to learn the ropes. Some sacrifices for the organisation are inescapable. It is an art to mobilise the potentials of all members and, at the same time, to balance out the trade-offs as far as possible.

Revision: Explain the principles of horizontalisation which can be derived from fundamental Christian beliefs.
Application: Analyse the hidden mechanisms operative in the affirmative action policies of the South African government against the background of the analysis above.

> *Critique: (a) "The idea that we are all equal in status is a utopian dream which has led to chaos, inefficiency and misery wherever it was applied. A viable society simply cannot do without structures of authority and subordination." (b) "Your pious pussy-footing is only designed to undermine the militancy of the oppressed; no half-hearted compromisers will ever break the iron grip of the ruling class." Comment.*

Let us summarise

In this chapter we dealt with the social-psychological dimensions of power discrepancies. To understand the phenomenon of dominance and dependency we compared, in Section I, two classical analyses, that of Mannoni in colonial Madagascar, and that of Freire in feudal Brazil. *Mannoni* confronted the dependency syndrome in African traditionalism with the inferiority syndrome generated by the competitive spirit of modernity. In the colonial encounter these two personality types clinched to form a progress-inhibiting neo-feudal situation. The colonials eagerly accepted the role of superiors which the natives readily ascribed to them. Only a break of this "happy marriage" could lead to the liberation of both.

Freire confronts the ruthlessly exploitative mentality of the oppressor with the submissiveness of the oppressed. The tragedy is that the brainwashed victims of domination have come to idealise their masters. Only the intervention of catalysts, coming from outside the situation of oppression (thus from the ranks of the oppressors), can lead to an awareness of their true situation and the determination to throw off their yoke. Freire developed an ingenious educational programme to demystify the oppressor and empower the oppressed.

Our *evaluation* found that Mannoni's recipe is not likely to lead to determined praxis, while Freire's recipe is not likely to lead to genuine liberation. In Section II we asked, therefore, whether the fundamentals of the *Christian faith* could not offer some further clues. The core assumption of this faith is that we are drawn into God's unconditional acceptance of the unacceptable in Christ. However, the idea is not to condone, but to confront and overcome the unacceptable condition. If it does not withdraw into a homogeneous and protected ghetto, the Christian community should be able form an ideal training ground for this principle - which could then be carried into the wider society.

However, the dissolution of the dominance-dependency syndrome is a process in time. An analysis of the relation between the last generation of expatriate leaders and the first generation of indigenous leaders in an *African church* at the time of independence gave us clues concerning the painful descent of the former leaders and the equally precarious ascent of the incoming leaders. Only when the descender has reached the bottom and the ascender has reached the top can equality of dignity be established. The psychological traumata caused by this process need the constant assurance of their unconditional acceptance by the community as a whole and the integration of the descender into the structures of the community.

Notes

[1] First published as "The dependency syndrome in marginalised cultures and the liberative potential of the Christian community". *Missionalia* vol 24/1996, pp 148-170. The original research was done for Nürnberger 1987, chapter 8. The financial support of the University of Natal Research Fund is gratefully acknowledged.
[2] This is the slogan of the "dependency school" which came to prominence in Latin America during the 1970s.
[3] Block analyses it in the modern corporation (1993:147ff).
[4] Mannoni 1956.
[5] Freire 1970.
[6] This has been the rule rather than the exception, whether one thinks of the Constantinian church, the mediaeval church, the Reformation and Counter-Reformation, the Puritans or modern forms of piety. Alves (1985) argues that the very doctrine of his own denomination, the Presbyterian church in Brazil, is repressive.
[7] At first sight the biblical witness seems to confirm authoritarian assumptions in many places. But a careful historical analysis reveals a powerful emancipatory thrust based on the core assumptions of this faith. For detail see Nürnberger 1991c.
[8] For an analysis of such a mindset see Nürnberger 1975.
[9] The "object relations" theories of the psychoanalytic tradition, drawing on Freud, insist that the earliest years of infancy are crucial for personality development. In an interesting and related discussion of the work of the feminist Nancy Chodorow, Brian Morris explains that "For all infants the first few years of life are preoccupied with issues of separation and individuation, and this essentially involves the breaking of the primary bond with the mother. The way in which a person . . . comes to develop a secure and individuated sense of self is crucially dependent . . . on the nature of the early mother-child relationship. Important . . . is the keeping of a balance between dependency and autonomy. Fear of dependency could lead to the development of overrigid ego boundaries, while fears of autonomy can lead to a regression towards infantile forms of dependency." (1994:173f, citing Chodorow 1978 and J Grimshaw 1986).
[10] For a deeper analysis see Nürnberger 1991a and Nürnberger 1988:156ff.
[11] It is interesting to refer to Balcomb's "Third Way Theology" (1993) in this regard.
[12] "Thus the peasantry, in the interest of safety and success, has historically preferred to disguise its resistance. If it were a question of control over land, they would prefer squatting to a defiant land invasion; ... if it were a question of rights to the products of the land, they would prefer poaching or pilfering to direct appropriation ... It is for this reason that the official transcript of relations between the dominant and the subordinate is filled with formulas of subservience, euphemisms, and uncontested claims to status and legitimacy ... The goal of slaves and other subordinate groups, as they conduct their ideological and material resistance, is precisely to escape detection (Scott 1990:193).

[13] For detailed studies see Heinz Wiesbrock: *Die politische und gesellschaftliche Rolle der Angst*. Mannheim: Europäische Verlagsanstalt, 1967.
[14] For instance, Nürnberger 1983:53ff, 1988:301ff.
[15] 1992:104.
[16] This would be my interpretation of the church as a "site of struggle".
[17] For the following see Nürnberger 1983 and Nürnberger 1991.
[18] This insight is superbly formulated in Eph. 2:8-10 and 14-16. Suspending his law as condition of acceptance, God accepts us into his fellowship as a gift of grace, not because of our moral achievement. As a consequence the law is also suspended as a condition of acceptance between us and others. Unconditional mutual acceptance again implies that we suffer each other's unacceptability as God suffers ours.
[19] See an interesting parallel to the argument below in Management Science: Block 1993:41-51.
[20] Practical ways of implementing this "descent" of the powerful and an "ascent" of the weak are described by Sleeman, e.g. "aid as an end in itself" and "positive discrimination in trade and payments" (Sleeman 1976:107ff).
[21] A point stressed by liberation theologians, but see also Wogaman 1986:24ff, Duchrow, Küng, and many other "Western" theologians.
[22] See, for example, Jean Comoroff 1985; Comoroff and Comoroff 1985.
[23] For example, see Korten 1990, Asian NGO Coalition, et al 1993, Meadows 1977, Meadows et al 1992.
[24] The effects of the failure to turn leadership over to indigenous administrators of development projects is well documented. Yet despite the recognition of the need for people to manage their own economies, we still see whole "armies of expatriates" working to create new institutions for development and manage these themselves (Berg and Whitaker, 1986).
[25] Von Schelling presents an interesting case study of this double struggle for hegemony in Brazil (1992:248-276).
[26] What follows is based on my own experience. As a former missionary (though not expatriate), who belonged to the "last generation" of missionaries holding leadership positions in an African church, I had the opportunity of living through the transition from expatriate to indigenous leadership.
[27] For their own self-understanding as superiors, see Bosch 1992:291ff.
[28] 2 Cor 8:13f.
[29] Anthropologist James Scott identifies the cultural patterns of resistance among dominated peasant societies, and shows how rumour is a particularly effective tool to be used against colonials (1990:144-8)
[30] In my view this is the source of the widespread disparagement of the missionary movement by the leadership and academic elite in Africa at present. Obviously the missionaries have made grave mistakes but this fact alone does not account for the emotional hostility often expressed by the latter.
[31] I owe this idea to a lecture given by the German theologian E Jüngel.
[32] The most devastating critique of those missionaries who took the call "Missionary go home!" seriously was "You have thrown us away."
[33] So far it has been very difficult for deans and bishops to become priests again in the case I am referring to.
[34] The Base Christian Communities in Latin America and the Small Christian Communities in East Africa have been cited as examples of redemptive community in just such instances as described in this paper. See, for example, Boff 1977; Cook 1985; Kiriswa 1982; Konings 1981; and Okeyo 1983. Philpott (1993) has produced a classical South African case study and Cochrane (1994) gains interesting insights from the same material; Draper (1993) discusses the topic from the study of the Third Quest material concerning the historical Jesus and his community.

Interests and ideology

What is the task of this chapter?

The last two chapters provided us with evidence of the role played by patterns of collective consciousness in socio-economic processes. In chapter 6 we have seen that mindsets are partly based on convictions, partly on interests, partly on popular conventions. Because of their immense impact on social reality, we close this series of chapters with an analysis of interests and their ideological justifications.

In Section I we expose *self-interest* as a parasitic form of consciousness which utilises both fact and faith to legitimate itself. We also indicate why we believe economics and the economic dimension of life to be natural candidates for de-ideologisation. In section II we shall argue that the *Christian faith* is both vulnerable to ideology and capable of playing a catalytic role in the process of dismantling ideological fortifications.

Section I: The phenomenon of ideological legitimation

The primary cause of social change may be the experience of immanent needs. Hunger leads one's feet inexorably in the direction of food. But humans are not simply driven by instincts. They need to make sense of their world. Therefore needs, such as the need for food, rest, security, beauty, and sexual gratification, never operate in a "naked" form; they are always inter-

253

preted, prioritised and legitimated in terms of a set of beliefs, values and norms. On the basis of this system of meaning, right of existence and authority are granted or withheld.

So the *system of meaning* is fundamental. Social structures are informed, underpinned and legitimised by mental structures. Where this is not the case, cognitive dissonance leads to all kinds of inefficiencies, conflicts and pressures in the system. There needs to be a measure of consensus for any social system to tick. By implication any deliberate attempt to transform a social structure must be based on a change of collective consciousness - creating precisely this kind of cognitive dissonance and, with it, pressures for change.[1]

Of course, people do not only strive for the fulfilment of immanent needs such as food and sexual gratification, but also for power and glory. But the need for status is a form of the need for *acceptability* (or right of existence); the need for power is a form of the need for *authority*. Moreover, honour is often linked with sexual and economic claims. So it is transcendent needs, meaning, right of existence and authority, arising from immanent needs which determine human behaviour.

A system of meaning does not fall ready-made from heaven. Meaning emerges in response to needs. That is also true for the traditions found in the biblical witness. Abraham needed male progeny; Israel in Egypt needed its freedom; Israel in the desert needed a land to settle; Israel raided by the Philistines needed the central authority of the king; the lepers and cripples in the gospels needed healing. Whole systems of meaning can cluster around a particular set of immanent needs and their transcendent implications. These interpreted, prioritised and legitimated needs are called *interests*.

The irrational justification of the pursuit of personal interests with clever arguments is called *rationalisation* in psychology. The covert or overt legitimation or concealment of the pursuit of collective self-interests at the expense of the interests of others is called *ideology* in radical sociology. Note that this is the negative usage of the term ideology; there is also a positive and a neutral usage, which, to avoid confusion, we do not utilise in this book. Ideology, as we shall use the term, selects certain facts, interprets them in its favour, and integrates them into a seemingly coherent set of arguments.[1] Once an ideology has been dogmatised, this dogma generates its own dynamic and exert its own power.[2]

Ideology, in this negative sense of the term, is one of the most potent obstacles to change. It can also be the most potent source of change in the wrong direction. Typical characteristics of ideology are: *selective perception* (only particular facets of reality are recognised),[3] *scapegoating* (blame is shifted to opponents or complete outsiders), *bias* (a lack of objective judgment), *self-isolation* and *censorship* (the group may not be exposed to a different version of the truth because this may reveal the flaws of its own), *stereotyping* (people and groups are placed into "boxes" with vastly generalised, often mythological characteristics, such as "blacks", "communists", etc.), *deprecation* (opponents are belittled or demonised), *fabrication* ("facts" concerning motives, actions, or histories of opponents are invented), and *fanaticism* (both the absolute good and the absolute evil have been located and one is obliged to join the battle against the latter).

Ideology has devastating consequences in society:

(a) The *truth* is concealed both to the ideologised themselves and to the victims of their propaganda. Ideology largely remains unconscious.

(b) *Communication* is impeded. A discussion between ideologised groups is a dialogue between the deaf.

(c) *Solutions* are obstructed. Ideologised groups lose their capacity to acknowledge and rectify their own mistakes and shift the blame for every calamity to their opponents. In this way each group fails to do what is in its own power to do and the situation stagnates.

(d) *Conflict potential* builds up, which may ultimately erupt and cause havoc.

These negative consequences are particularly grave in a stratified society. The elite is the elite because it commands power; the underdog population is in its underdog position because it lacks sufficient power to counter-balance the power of the elite. The elite follows policies which enhance its interests. It also commands mass media and educational facilities to propagate its interpretation of social reality. So its ideology is the dominant ideology in society. In fact, it regards itself as the only legitimate "embodiment of the spirit and the mission of the society that in fact it dominates."[4]

Because the underdog population covets the position of the elite, it accepts the latter as its *reference group*. A reference group defines acceptable behaviour. The relation between elite and underdog population groups is, therefore, one of dominance and dependency. This does not exclude the development of ideology among the lower classes. Underdog ideologies typically either explain current subservience or justify the struggle for liberation and social advancement. They also shift the blame elsewhere to explain why they do not fulfil accepted norms.[5] Where the underdog population is sufficiently conscientised to offer its own interpretation and to demand that relationships be horizontalised, the contradictory interests between elite and underdog clash and serious conflict may result.

Some research findings

Apartheid South Africa, my own context, offered particularly fertile grounds for the emergence and growth of selective perception and ideological argument. In my empirical studies on the relation between ideologised convictions and structure related interests, I found a clear correlation between lower incomes and socialist leanings on the one hand, higher incomes and free enterprise leanings on the other. 94% of white politicians feared the "communist threat", against 24% of black elites. The same discrepancy existed between those who believed that blacks have "good reasons to take up arms against the Government" and those who did not.[6]

White politicians were clearly subject to self-deceptions: 78% believed that white leaders could depend on the loyalty of blacks in the case of war against South Africa, while only 24% of black elites did.[7] Such delusions flourish under conditions of self-isolation. A similar phenomenon was caused at the other end of the spectrum by a lack of information: workers earning less than R 200 per month believed that 85% of post-tax company profits were spent on management bonuses and salary increases, while only 1% went to worker bonuses and salary increases. Nothing was attributed to shareholders by these workers; apparently they did not know of the existence of shareholders. It was also clear that the stance of blacks with higher salaries approximated that of whites and the stance of average whites with lower salaries approximated that of average blacks.[8]

The ideologisation of economic life

These empirical studies highlight the impact of social location on interests and the impact of interests on perceptions and convictions. This is a widespread phenomenon with political implications. People report and support economic theories and policies when they are in line with what they want to believe, not when they make economic sense.[9] In chapters 8 and 9 of this book we found evidence of the subtle power of ideology in the encounter between traditionalism and modernity. To me and many others it seems evident that liberalism is the legitimating ideology of an economic elite, though not all liberals will agree with this verdict. In chapter 11 we pursue this issue further.

But traditional assumptions also fulfil a legitimating function without being conscious of the fact. The traditionalist system of meaning, with its tight set of values, norms and role allocations, legitimates the authority and the privileges of elders, males and diviners. The concept of sorcery is used most effectively to crush dissent and provide scapegoats. The paralysing effect of traditionalism is largely due to these ideological functions of the underlying mindset.

In a situation of class conflict, groups interpret social reality according to their location in the social power structure. The poor and powerless tend to see themselves as victims of the avarice and the oppression of the rich; the rich and powerful tend to see themselves as beneficiaries of their own ingenuity and diligence.[10] Affluent people tend to perceive the power of the economic centre as a life-giving sun which sends the light of knowledge and the warmth of technological development into the darkest corners of the globe. In contrast, spokespersons for the poor tend to perceive the economic centre as a greedy and cruel octopus which pushes its tentacles into all corners of the periphery to suck its life-blood into its insatiable digestive system.[11] In the North-South dialogue these opposing perceptions are epitomised by two classical book titles, "The white man's burden" by Rudyard Kipling, and "How Europe underdeveloped Africa", by Walter Rodney.

As we have seen, a lack of information can play a substantial role in the creation of faulty perceptions. But biased judgments cannot be attributed simply to the ignorance or naiveness of the proverbial man and woman in the street. In fact, additional information tends to reinforce rather than dismantle prejudice. The reason is that "facts" do not speak for themselves; they need interpretation. Interpretations are determined by the need to justify interests.[12] In post-apartheid South Africa, social ills such as escalating crime or rising unemployment is attributed by whites to black incompetence and by blacks to the heritage of apartheid.

The mass media spread such doctored "information" according to the interests of those who control it. Where media are controlled by authoritarian regimes, they will report and support what is in line with the dominant ideology. But a "free press" alone will not do the trick. In Western countries, private entrepreneurs want to make a profit; so they will report and support what will sell to the public, which is either sensation or legitimation of collective interests.[13]

The ideologisation of economics

If interest rather than information is the main spring of skewed perceptions, it should surprise nobody to find the effects of selective perception and ideological bias even in well researched and tightly argued academic treatises.[14] *Sociology* was marked by the conflict between liberal and radical schools during the heyday of ideological conflict, particularly on the European continent.[15] *Economics,* with its long standing claim to empirical objectivity, came up with contradictory approaches to the same phenomena, each offering their own sets of analyses and recipes, each underpinned with flawless statistical evidence and mathematical precision, yet arriving at totally different conclusions.[16]

A crude summary of these positions must suffice.[17] *Liberals* emphasise development and growth. They believe that income discrepancies create the necessary incentives. A free market ostensibly balances out the relative weights of different contributions and benefits to the national and global economy. Humanity is perceived to be on a common journey towards "the creation of a universal consumer culture based on liberal economic principles".[18] If some lag behind, they simply have to pick up their socks. If some are ahead of others, they may be moved to assist those left behind, but surely the most advanced should not be blamed for their excellence!

For *radicals,* in contrast, the global class struggle provides the basic frame of reference. The rich and powerful have oppressed, exploited and crippled the poor and powerless during colonial times. Neo-colonial hegemony continues to hold them in the bondage of dependency. The aim of the "struggle" is equality, the means resistance and revolution. The poor must shake off their yoke; the rich must release their victims, restore their wealth and allow them to develop. The least they can do is to establish fair trade relations. Surely it is not the poor who are to blame![19]

Referring to the model of causation developed in chapter 5 we can say that liberals emphasise the positive factors inherent in the centres and the negative factors inherent in the periphery, while discounting the effects of the asymmetrical interaction between them. Radicals emphasise asymmetrical interaction, while discounting the factors inherent in centres and peripheries respectively. The same is true for Third World dependency theorists.[20] One-sided perceptions even extend to the diagnosis of prejudice itself. For Marxists, liberal economics is nothing but the ideological justification of capitalist profit maximisation.[21] For liberals, Marxist economics distorts reality to legitimate a politically oppressive and economically unworkable system.[22]

Traditionally, both schools have overlooked or discounted the environmental effects, both of the population explosion and of industrial growth. In the capitalist West, preoccupation with economic growth crowded out ecological concerns; in the socialist East, pollution and resource depletion were proclaimed to be typically Western phenomena, due to capitalist social irresponsibility. Southerners reacted with particular anger against ecological prescriptions emanating from the North, arguing that industrial nations had become rich by depleting and polluting nature, and that it was now their turn to exploit and pollute.[23] If Northern countries were eager to preserve rain forests, it was said, they would have to pay for the luxury, for instance by cancelling international debts.[24]

Biased interpretations of the relation between poverty and pollution can take many forms, from very subtle to very crude. At a popular level the rich blame the poor for "breeding like rabbits" and call for greater seriousness in population policies. The poor declare the "population explosion" to be a myth invented by the rich to avoid the challenge of a just distribution of resources, or more cynically, a scare designed to reduce the numerical strength of the poor. They particularly refer to the fact that the average child in industrial centres was likely to consume at least ten times the amount of resources and energy used by the average child in the Third World during its life time.

The last example shows that many of these arguments are not necessarily wrong; it is the one-sided emphasis and escapist purpose which makes them ideological. We could add endless examples. The point is that prejudice is not restricted to uneducated citizens and unscrupulous politicians, but pervades all levels of society, including academics and political leaders. Some analysts of ideology have argued that, because academics have fixed salaries and security of status, they can afford unbiased knowledge, attain objectivity and draw up disinterested agendas. Their task should be to enlighten and de-ideologise the public.[25] While there may be some truth in this argument, its validity is limited. In fact, academics may be a particularly potent source of "false consciousness" because the disinformation emanating from their selective perception and skewed interpretation is endowed with the academic status of established fact and reasoned argument. Experts who turn against the tide are largely ignored; those who toe the line enjoy more credibility than they deserve. Krugman's analysis of the conflict between Keynesianism and monetarism reads like a litany of the impact of selective perceptions and collective interests, not only on the business cycle, but also on the underlying theories and the policies to counteract it. His examples also show that it is not possible to distinguish between serious economists and political chancers as neatly as he wants to make us believe.[26]

Is there a cure?

Indeed, ideology can be overcome! We have seen that ideology utilises elements of fact and faith to legitimate the pursuit of collective interests. More devious regimes manipulate them deliberately to mislead the public. They also impose censorship to forestal any critical and constructive debate. But even apart from such designs, ideological distortions of fact and faith remain largely subconscious.

They can be exposed and dismantled in two ways. One possibility is that the *situation* changes so drastically that a particular perception is no longer capable of interpreting actual experience. It may even become counterproductive in terms of the interests of the perceivers themselves. Examples are the sudden collapse of Nazi ideology in 1945, Soviet ideology in 1989, and apartheid ideology in 1994. It is amazing how staunch communists could turn, almost overnight, into enterprising capitalist entrepreneurs. This shows that self-interest had been their prime motivation all along, while conviction was only used as a legitimating device. At the moment free enterprise liberalism is in a dominant position, but ecological constraints may soon impose new perceptions and attitudes onto a hitherto unwilling populace.

The other way is to lead contrasting perceptions of reality into *confrontation* with each other. Unconscious prejudice is exposed fairly quickly when two parties on opposite sides of the ideological spectrum meet and analyse controversial issues together. Both their versions cannot be true. The ensuing struggle for the truth can lead both parties to question their own assumptions and seek for a more appropriate picture of reality. If the motives on both sides are genuine, the struggle for the truth may lead to a more balanced and less simplistic picture. The same can happen when the insights of various academic disciplines are confronted with each other. That is what we try to do in this book. But the psychological and social pressures of collective interests should not be underestimated. People must *want* to discover the truth, otherwise their encounter will end up in a shouting match.

The common quest for truth is both painful and rewarding. It liberates and motivates. In the end all parties gain.[27] However, there are three prerequisites for the success of such an encounter: (a) the parties must accept each other unconditionally; (b) they must operate on the same level of dignity, even if they do not command the same social power, and (c) they must be committed to absolute honesty. Each party must concede that the other may have a point, and acknowledge that their own interests may distort their perceptions. *Acceptance* makes honest confrontation possible; it allows errors and wrongs to be exposed without fear of rejection or condemnation; it creates a redemptive attitude towards the opponent. Acknowledgement of the *same level of dignity* horizontalises vertical relationships. *Honesty* excludes the use of ideological arguments to gain political or propagandistic leverage.

Unfortunately negotiations are often characterised not only by irrationalities, but also by power struggles and calculated deviousness. We shall argue in the next section that it is the particular task of the Christian church to provide the space for honest confrontations.

> *Revision:* Describe the root causes and the consequences of collective self-justification.
> *Application:* Analyse some of the contradictory statements made by employers, unions and political parties on opposite sides of the spectrum in your country, highlighting their particular ideological slants and their underlying interests.
> *Critique:* (a) "Do you honestly believe that the attempts of the oppressed and exploited to articulate their valid demands lie on the same level as the blatant lies concocted by the capitalists and imperialists? It is you who is ideologically biased!" (b) "Recent history has demonstrated to the whole world that there is no alternative to the capitalist free market system when it comes to economic performance. The poor countries must cease complaining and begin pulling their weight." With which of these statements do you sympathise? Could this be due to a bias of your own?

Section II: Faith and ideological distortions

Confrontation under the canopy of acceptance

We now explore the role that a religious community can play in facilitating this liberating and empowering confrontation. Again it is the Christian faith which we single out for the purpose. One cannot speak about convictions in general; convictions are always specific. We are Marxists or liberals, Buddhists or Christians, Muslims or Hindus. The question whether convictions have the power to dismantle ideological distortions can, therefore, not be answered in general; we have to explore the potential of a particular conviction to address the root causes of the ideological phenomenon. What follows is based on insights gained during the apartheid struggle in South Africa, when such confrontations took place on a substantial scale.[28]

The core of the Christian message is that God, the source of reality, is for us and with us and not against us. Its particular content is the driving force which makes the Christian message enter all cultures, classes and social situations. This is the *centrifugal* dynamic of the Christian faith. In each of these contexts it relates to the specific needs of the people concerned and, for better or for worse, gets mixed up with their hopes and fears, their resentments and interests, their perceptions and interpretations of reality. In other words, the message is syncretised and ideologised in divergent directions. As a result, Christians can be found all over the social and ideological spectrum and display a wide variety of social stances.

But God's redemptive love also calls these different strands back into the one and only Christian community. This is the *centripetal* dynamic of the Christian faith. The first effect of such an encounter is that relationships are horizontalised. All members know that they have received their lives naked when they were born, and that they will hand them back naked when they die. All members are exposed as miserable sinners who have lost their right of existence, thus their status before God. All members have been forgiven and granted the highest status possible for humans, namely the status of "sons and daughters of God". This is an ancient royal title!

When these assumptions work themselves through concrete relationships, they dismantle arrogant and condescending attitudes of the elites on the one hand and submissive and resentful attitudes of the underdogs on the other. In spite of their vast differences in income and power they begin to operate on the same level of dignity. But this does not happen automatically. Normally the church is fragmented in relatively homogeneous and self-perpetuating social entities, composed of like-minded people who strive to avoid all conflict. This means that the first dynamic has stalled, while the second is still outstanding. To become effective, the church must resume its mission and give expression to its unity. Both missionary and ecumenical encounters must be organised, even institutionalised.

Wherever this happens the church becomes an "impossible community". On the one hand the horizon of the "ingroup" widens to cover the former outgroup. Groups that are "natural" outgroups to each other, all become part of the "ingroup of God". Their interests, perceptions and interpretations of reality have to be accommodated. Primary group loyalties and social pressures to conform are relativised.

On the other hand, the norms with which the respective groups measure each other, do not simply evaporate. They continue to function as criteria which define acceptable thoughts, attitudes and patterns of behaviour. This function is called the "law" in theology, as opposed to the gospel of grace. The fact that the constituent groups are mutually unacceptable becomes painfully clear. Each group tends to believe that its own norms are valid and binding, thus part of the law of God. But not all the criteria of acceptability of all the social groups can be identical with the law of God. So the confrontation between different sets of criteria makes the struggle for the truth imperative.

When the encounter between such ideologised groups goes beyond the niceties of tea parties, the confrontation can become so disconcerting that the new community breaks apart. That is the reason for the existence of isolated and homogeneous Christian communities in the first place. But God's redemptive acceptance of the unacceptable, shared by the believers, does not allow anybody to be excluded or written off. Confrontations have to take place under the canopy of acceptance. This has two consequences.

On the one hand, differences have to be tolerated - and that implies sacrifice and suffering. On the other hand, the pain of dissension induces the parties to struggle for the truth together.[29] They begin to analyse the causes of the conflict. As a result, the perceptions and interpretations of each group are challenged, corrected or augmented by those of the other. They also struggle for the true content of the Word of God. By going back to the sources, ideological distortions of the truth are challenged and filtered out.[30]

This can lead to a more balanced picture of social reality and a more appropriate theology. The relative legitimacy of opposing views becomes clear. Though different perceptions will not collapse into uniform patterns, compatible diagnoses can emerge and complementary tasks and policies can emerge. In short, the stalemate is broken and a way forward opens up.

Feedback into society

All this happens within the Christian community between individuals in personal encounters. But individuals represent groups situated in social contexts. When individuals return to their respective social contexts, confrontations under the umbrella of acceptance have to happen all over again. Having returned to their social contexts, they have to contend with the power of primary group loyalty. People do not easily step out of line, even when their observations, their sound reason, or their faith tells them that the perceptions and assessments their group are wrong.

People who have gone through an ecumenical experience, therefore, are torn between the expectations of their own social groups and the new vision they have acquired in the cross-cultural encounter. This tension is healthy and fruitful. They will no longer feel comfortable with the half-truths current in their home groups, nor with the demonisation of the respective outgroups. They will feel the urge to generate a more inclusive vision, transcend divisive myths, question problematic truth claims, and establish points of contact.

All this will not make them popular. In severe cases, they may have to sacrifice their careers, their personal safety, their family harmony. "Traitors" and "heretics" are never far from ostracism and persecution. It is with their very bodies that they build bridges across the chasms of society. There is no way of avoiding this conflict if the insights of the ecumenical encounter are to benefit the society. Without such confrontations - relatively mild in my case - I would not have arrived at the models spelt out in this book.

Of course, confrontations do not only happen within a particular faith community but also between faith communities. Secular humanism, for instance, currently challenges traditional Judaism, Christianity, Islam and African traditional religion to rethink their stances on the role of women in society and their function in development. This challenge, I believe, presents the Christian faith with an opportunity to rediscover and reinstate its own fundamental insights concerning the equal dignity of all human beings.[31]

Revision: Describe how a consistent Christian faith could counter-act ideological legitimations and distortions.

Application: Take two newspapers or journals which represent opposite points of view. Try to discover the relation between collective interest and the interpretation of events in each case.

Critique: (a) "Christianity has been one of the main sources of ideological distortions throughout its history. Crusades, imperial conquests, racist oppression, capitalist exploitation, the use of nuclear weapons, the cold war, even genocide, have been justi~fied by Christian church. It really takes some self-deception, cynicism and arrogance to portray the Christian faith as a cure for this scourge." (b) "The pursuit of collective interests is as unavoidable and normal for a healthy society as the interpretation of social reality in terms of these interests." How would you react to these two statements?

Let us summarise

We need new a new vision and new directions. Visions and directions are provided by facts, convictions and interests. This chapter concentrated on *interests* and their *ideological legitimation*. In section I, we saw that the pursuit of collective self-interest has to be justified against the set of convictions prevalent in society. This leads to distortions both of the facts and the convictions concerned. Each group tends to interpret reality in its favour, project the blame for social calamities on its opponents and neglect its own responsibility. These ideological legitimations of collective interests can effectively block insight and subvert the political will to change an untenable situation. We have mentioned examples from economic life, especially differences in perception between elites and lower classes.

Ideological cover-ups and distortions are largely subconscious. They are revealed when opposing groups confront each other under the canopy of mutual acceptance, honesty, and concern for the truth. In section II we argued that *the Christian community* could play a catalytic role in dismantling ideological distortions. The missionary dynamic of the biblical faith leads to its entry into all kinds of social context. The result is that it becomes ideologised in contradictory directions. The ecumenical dynamic leads these syncretistic forms of faith into confrontation with each other under the umbrella of the unconditional acceptance of God. In the ensuing struggle for the truth, ideological distortions are exposed and replaced with common analyses and complementary commitments.

Notes

[1] For examples of the operation of ideology in social reality, see Nürnberger 1988 Part II.
[2] Cf Storkey 1986:195ff.
[3] See Korten 1990:54f.
[4] Heilbroner 1985:130.
[5] The economist Axelle Kabou from Cameroon is said to have exposed such attitudes in Africa in her book *Et si l'Afrique refusait le developpment*. Unfortunately the book is not accessible to me.
[6] Nürnberger 1988:207-213; 243.
[7] Nürnberger 1988:244f.
[8] Nürnberger 1988:251f.
[9] "A major part of monetarism's appeal was that it seemed to confirm the conservative prejudice that government activism is always a bad thing ... (conservative thinkers) were unconsciously moved to overlook its flaws because it fitted their political philosophy" (Krugman 1994:52).
[10] The rich tend to "invent plausible justifications" for their wealth. This "rationalization" often comes in form of ideology (Sider 1977:49).
[11] According to the South "the environmental crisis results from poverty in the South, wastefulness in the North, external debts and destructive trade relations." According to the North it results from "over-population, poor management and lack of technology." The Southern therapy is "sustainable growth in the South; debt relief; fair trade relations; modest consumption in the North" and the therapy "birth control in the South; environmental impact assessment in development projects; transfer of technology and management skills; debt relief with green conditionality" (Von Weizsäcker 1994:167).
[12] Habermas speaks of "insight-guiding interests" (erkenntnisleitendes Interesse). In: Connerton 1976:330ff, 363ff.
[13] Ehrlich & Ehrlich give examples (1996:189ff).
[14] For a collection of myths in "scientific" work see Ehrlich & Ehrlich 1996.
[15] The Frankfurt School, also called "critical theory" (Horkheimer, Adorno, Marcuse, Habermas) dominated the German scene for decades, while American main line sociology ignored radical sociology.
[16] For a comparison of the two main schools see Hunt & Sherman 1981.
[17] For detail see Nürnberger 1998.

[18] Fukuyama 1992:108.
[19] For an instructive dialogue between these two schools see K Boulding: Twelve friendly quarrels with Johan Galtung. *Journal of Peace Research* XIV/1977 75ff and J Galtung: Only one quarrel with Kenneth Boulding. *Journal of Peace Research*,24/1987. A comparison in text book form is offered by Hunt & Sherman 1981.
[20] For a discussion of these schools see Kubalkova & Cruickshank 1981; Caporaso 1987:1ff; 22f; 28ff; 92ff; Seligson 1984:105ff.
[21] For an example of a Western analysis of capitalist ideology, and economics as an expression of it, see Heilbroner 1985 and 1988. Critics include Tawney, Hirschman, Galbraith, Wallerstein, Joan Robinson, the Frankfurt School, the Dependency School, liberation theologians such as Hinkelammert, and others.
[22] Examples are Bauer, Friedman, Rostow, von Hayek, Gross, Berger, Novak.
[23] Third World representatives argued in this fashion from Stockholm 1972 to Rio 1992.
[24] C Schütze gives a survey of "ecological blackmail" since Stockholm 1972 in *Der Überblick* 30/1994, 41-43.
[25] This Mannheim's stance (1972); see Larrain 1979:101ff. Ehrlich & Ehrlich speak of "anti-science" (1996:11ff) and assume that true scientists are agreed (1996:233ff). This assumption is naive.
[26] 1994:6ff.
[27] Go Brundtland, chairlady of the World Commission on Environment and Development: "We joined the Commission with different views and perspectives, different values and beliefs, and very different experiences and insights. After these three years of working together, travelling, listening, and discussing, we present a unanimous report" (1987).
[28] For instance, within the "main line" churches, the SA Council of Churches, the National Initiative for Reconciliation, and various other ecumenical bodies. The "Evangelical Academies" in post-war Germany provided an excellent forum where explosive emotions could be mediated in a population traumatised by ideological abuse, economic misery and political humiliation.
[29] The church means "watching over one another in love". This makes "loving defiance" possible. Churches may not be "comfortable clubs of conformity" (Sider 1977:163).
[30] This is at variance with the view that the church could form an alternative society which could confront the world (cf Duchrow 1994:228). The church is part of the world and needs to be confronted as much as the world.
[31] That gender equality belongs to the fundamental implications of the biblical faith may not be immediately apparent, but it is borne out by history. The Babylonian myth of creation (*Enuma elish*) maintained that humans were created from the blood of a rebellious god to be slaves of the gods, implying that they were by nature rebellious and needed to be subdued. Jewish theologians responded with the statement that humans, both men and women, were created in the "image of God", an ancient royal title (Gen 1:26-27). According to Paul God's unconditional acceptance in Christ implies that there is "neither Jew nor Greek, slave nor free, male nor female" in Christ, that is, in authentic humanity (Gal 3:23-29). In both cases this fundamental insight was buried under the weight of conventions, even in the Pauline school. 1 Cor 11:3ff and 14:34ff cites the older creation story (Gen 2:18ff) which legitimates female subordination. In 1 Cor 12:13 the equality between men and women is dropped from the triad. In Eph 2 only the equality between Jews and Gentiles is retained, while women and slaves are taught subservience (Eph 5:22ff and 6:5). The worst case of ideological legitimation of female subordination is found in 1 Tim 2:11ff. It contradicts both the original (Gen 2) and Paul's teaching on salvation by grace accepted in faith.

PART III

Towards a new economic paradigm

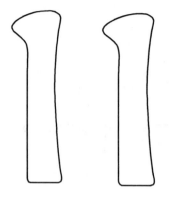

The pending paradigm shift in economics

What is the task of this chapter?

As stated in chapter 7, the picture which emerges from our analyses is rather dismal. It seems that humankind is moving inexorably towards disaster. An awareness of the dangers hardly exists. Among the masses of the population the daily struggle for survival and prosperity is characterised by narrow horizons, short-term decisions and the expectation of immediate satisfaction. Powerful elites have entrenched themselves in major institutions and pursue their collective interests regardless of the consequences. Governments want to stay in power and hardly think beyond the next election.[1] Many experts share these truncated perceptions and motivations. Those who do not, seem to be voices in the wilderness. When confronted with the facts, one gets the feeling of powerlessness and hopelessness.

We have said in chapter 7 that this impression is wrong and must be overcome.[2] Even a brief glance at past history reveals that humankind does not move through time in predefined directions. There are constant surprises, revolutionary upheavals and dramatic changes.[3] The present trend cannot continue for much longer without creating havoc among humankind on a gigantic scale. Change is inevitable. Some say that, for better or for worse, the next century has to become the "century of ecology".[4] Others say that "the twenty-first century would either be religious or not at all."[5] These two statements complement each other. Ecology and conviction are intimately connected. What is needed is a massive change of collective mindsets and a determination to make a difference.

Great movements in history often begin with small initiatives. For a considerable length of time they proceed slowly and seemingly void of significance. But slowly they gather momentum. At some stage they may gain a critical mass and cause cataclysmic change. Such movements can go in both directions - towards catastrophe or towards optimal wellbeing. The relevant questions are, therefore, whether we understand what is happening; whether we are prepared for change; whether we shall be victims of fate, or take control of reality and ease in the necessary transitions.

Those who have access to information and skills bear a grave responsibility. Before academics can enlighten the broader public, however, they have to find their own directions. The same kind of responsibility rests upon representatives of religious convictions. In chapter 7 we challenged a pivotal *religion,* namely the biblical faith, to get its house in order. Here we challenge a pivotal *social science,* namely economics, to rethink its fundamentals.

Sciences make use of models or paradigms to explain reality. Paradigms are abstractions of entities and relationships that are deemed to be decisive from the vast pool of phenomena. These abstractions form a system of knowledge; the rest forms the background. This background remains largely unknown, underemphasised and underutilised.[6] As research unfolds and findings accumulate, current paradigms may no longer cover the expanding body of knowledge. Then they must adapt to the situation or become obsolete. A paradigm may have to be dropped altogether in favour of another.

In Part III we argue that a major paradigm shift is overdue in economics, whether as an academic discipline, a set of public policies, or a pattern of behaviour. Economics is not alone in bearing responsibility for the material survival and prosperity of humankind, but it is the leading academic discipline in this field. If economics does not change its orientation, it deprives the entire movement of its intellectual backbone. It may also lend ideological legitimation to detrimental trends in the economy (national and global) and in economic policy making.

This chapter deals with the *broad outlines* of a potential transformation of economic thought. Section I reflects on paradigm shifts in the social sciences and argues that a paradigm shift in economics is overdue. Section II spells out a few requirements for a new economic paradigm. Chapter 12 suggests a change in the fundamental *points of departure* of economics and its repercussions. Chapter 13 confronts the economic-ecological problem with concepts gleaned from the *natural sciences,* namely entropy, evolution and acceleration. The application of insights gained in one discipline on the material of another is a potent method of challenging entrenched assumptions.

Section I: A pending paradigm shift in economics?

The relativity of knowledge

In chapter 10 we dealt with the ideologisation of economics. But there is a phenomenon which is both more respectable and more subtle than blatant ideology. We call this phenomenon *paradigm shift*. To take paradigm shifts seriously, one does not have to accept radical postmodernist relativity, sometimes associated with the name of Thomas Kuhn, which believes that there is no reality out there at all.[7] This would be ridiculous. Just try to walk through a wall and see how far you get! The point is, rather, that all human knowledge is limited, selective and perspectival. The same reality can be perceived in a great number of ways.[8] It is precisely empirical research, directed at establishing "pure facts", which has to be subjected to scrutiny.

There was a time when divine revelation, as found in sacred Scriptures, and promulgated by a religious hierarchy, was believed to be absolutely true. Since the Enlightenment, Western academics have believed that the application of strict empirical and rational criteria in science is bound to lead to objective facts. They had the vision of a truth which was no longer under dispute, not because it was imposed by unquestionable authorities, but because it could be shown to be true by observation and reason.

This vision has made an indispensable contribution to human emancipation and insight. We cannot want to undo these achievements. In our times, however, both theology and science have lost their innocence.[9] The awareness of limited horizons, ambiguity and relativity has grown even in exact sciences such as physics.[10] We know today, or we should know, that sciences are dealing, not with reality as such, but with theories or paradigms or models which do not depict, but abstract from, reality.

A theory is a model imposed on reality by our consciousness.[11] An integrated set of assumptions, observations and conclusions gives us a reasonably consistent picture of how reality might operate. Its function is to reduce the complexity of reality, make it more accessible for our understanding and more manageable for our intervention. This is done by selecting and combining aspects deemed to be decisive from the vast pool of phenomena into a system and relegating the rest to the background. All systems have an environment which they do not cover.

Because paradigms select and combine facts, they are based on judgment and cannot be considered value-free. Surely scientific paradigms are not arbitrary inventions of individuals with a lively imagination. Most scientists try to be as objective as possible, to account for their assumptions and procedures, to subject their inferences to empirical substantiation, and to remain open for correction. But paradigms do contain, however implicit and rudimentary, a conception of what reality is *supposed* to be. Therefore paradigm construction is always based on some kind of conviction. Moreover, paradigms have a tendency to underpin what human beings *want* reality to be.[12] So interests also play a role. There is no disinterested science.

In so far as they have both a *normative* and a *legitimating* component, paradigms give direction to personal actions and incarnate themselves in social structures. But they are historically relative. They emerge, grow and decline. There are always plausible alternatives around. Eventually any paradigm makes way for another paradigm.[13] But many economists, emphasising empirical evidence and mathematical rigour, find it difficult to accept that they are dealing not with pure facts, but with replaceable theories.[14]

Why outdated paradigms persist

If we had a purely academic perception of science, we would expect that knowledge was continuously adapted to changed circumstances. Reality shows, however, that the fate of paradigms is not due exclusively to the advance of research and the internal logic of the science concerned. "Irrational" factors such as personal convictions, peer group loyalty, social pressures, political sensitivities, changing collective interests, the location of the scientist in the social structure, the ongoing flow of what is deemed to be acceptable thought and behaviour by the dominant layers of society, even mathematical and logical aesthetics, may become more determinative than internal developments within a body of scientific insight.[15]

Dominant paradigms do not easily give way, therefore, even when their validity has been questioned beyond repair, or when they no longer cover actual experience.[16] People uphold and defend their certainties when they are attacked. This is particularly true for the social sciences.[17] There are essentially two reasons for this:

(a) Paradigms are part of a greater *system of meaning* which dispenses acceptance, belonging and authority subject to its own set of values, norms and goals. The identities, roles and statuses of scientists and non-scientists are defined in terms of this "sacred canopy" (Peter Berger). In the age of modernity, science has assumed the role of a religion for millions of people.[18] When foundations are threatened people rally to their defence.

(b) A dominant system of meaning is *incarnate in social institutions.* It legitimates entrenched interests. Paradigms are linked to careers, promotions, budgets, apparatus, data bases, peer groups, and communication networks. Once a system of meaning is discredited, the entire social edifice may disintegrate.

Though understandable, the consequences of such a reticence can be quite alarming. Throughout history dominant systems of meaning have functioned like mediaeval fortresses with impenetrable walls, defences and prisons. They have frozen the flow of life, impeded communication, closed the minds of people, and led to hatred, bloodshed and misery. Enough reason to become humble in one's certainties and alert to emerging alternatives which may be more capable of serving the human enterprise! The flow of history cannot be stopped; we either leave the past behind and anticipate the future, or we become the victims of time.

The deficiencies of main line economics

The deficiencies of main line economics have been spelt out again and again by economists themselves.[19] The greatest weakness of the dominant liberal approach is that it is not designed to depict the system as a whole. Nor are its tools capable of handling problems of this magnitude and complexity. By and large, classical economic theory concentrates on *significant but small segments* of the economic problem, for instance the allocation of resources and products through the market mechanism, international trade, and monetary policy. Even these segments are analysed subject to abstract theoretical assumptions which do not take account of the multi-dimensional nature of economic reality.

Non-economic factors, whether physical, psychological, social, cultural, religious, or ethical, are largely considered to be economically irrelevant. "Rational" behaviour is defined in individualistic and utilitarian terms, with some tokenism in the direction of communal and environmental concerns. Economics abstracts from *time* (acceleration of economic processes), *space* (economic geography) and *power* (concentration of economic potential).[20] *Marginalisation,* which is a fundamental economic issue, is not dealt with by the liberal school. *Ecological* questions, such as the limits of the resource base and the disposal of waste, are largely ignored or discounted.[21]

All this simply will not do. Of course, in the vast field of economics one always finds niches where these issues are addressed. As an economist friend of mine has pointed out, every critique and every suggestion found in this chapter will have been addressed in some way by somebody in some corner of the discipline. Moreover, Western economic schools differ considerably in the extent to which these issues are overlooked. Our critique is probably most applicable to "supply side" economics; slightly less to Friedman and the Chicago School; considerably less to Neo-Keynesian economics;[22] least to subdisciplines specialising on some of these issues, such as welfare economics, resource economics, environmental economics, consumer economics, economic geography, even economic psychology; and perhaps not at all to "institutional economics".[23]

When we speak of "main line" economics, therefore, this is a vast generalisation. However, for our purposes the generalisation is fully justified. The questions are (a) whether the concerns mentioned are considered to constitute fundamental parts of the training of the vast numbers of economists churned out by universities each year and absorbed in the private and the public sectors, (b) whether they determine the research of leading academic institutions to any appreciable extent, (c) whether they are at the top of political agendas, and (d) whether they are heeded by the most powerful national and multinational corporations. The answer is, in all these cases, negative.

The present crisis of economic paradigms

Classical economics has tried to achieve the kind of certainty which the natural sciences seemed to have reached in the 19th century.[24] This certainty was meant to replace the deceptive certainties of revelation. Yet the pattern was very similar. Reminiscent of the religious orthodoxies of the 16th century, economics has laid a foundation of axiomatic "truths" on which an extensive system of deductions is built. Research follows along the lines of the system, just as scriptural material is fitted into the system of a theological orthodoxy.[25]

This is most apparent when one views economic schools as "sects", each of which claim absolute truth for itself and fights the heresy or the fallacy of the other. For one and a half centuries economic thought was torn apart between two competing paradigms, represented best by their most extreme forms, capitalist liberalism and Marxist socialism. Neither was able to accord the other one any respectability whatsoever. A decade ago both seemed to be alive and well, but which one was true? Probably neither.

The *liberal* school emphasises growth, development, free market and free trade. It celebrated its triumph in the economic policies of Ronald Reagan and Margaret Thatcher. For this school the world is unlimited and for the taking. Those who have less simply have to catch up with those who have more, without impeding the progress of the latter. The great concern of the *socialist* school is equity and the alleviation of suffering. It has been knocked out of action for now by the collapse of Marxism in the East and the apparent failure of socialist policies the world over. Even Keynes is not very respectable at present.[26] But the socialist agenda has not been dealt with and the frustrations and agonies of masses of impoverished people accumulate.

A third school has slowly been gaining ground which centres around *ecological* concerns, but it has not yet managed to get its act together. Some people merely feel sorry for endangered species, others see the demise of humankind as a whole looming over the horizon. Some link equity issues with ecological concerns, others merely want to repair the damage caused by capitalist development.

There are valid insight and untenable assumptions in all three schools.[27] While the feasibility and credibility of the socialist option has been plunged into crisis,[28] the claims of capitalist liberalism seem to have been vindicated. Its theory radiates confidence; it is heavily entrenched in the global bastions of political, military and economic power; it is still poised to conquer the world. Many liberal observers believe that a watershed of global significance has been reached with the demise of socialism. Those who do not manage to jump onto the bandwagon of liberalism seem to be doomed to backwardness and misery. An enthusiastic author has spoken of the "end of history".[29]

But socialism had emerged as a reaction to the failures of capitalism in the first place. In spite of all its adaptations, the dominant system has not been able to overcome the four basic and interlinked economic problems mentioned above: (a) the grinding poverty and growing marginalisation of billions of people, (b) escalating discrepancies in economic potency and life chances between rich and poor societies, (c) deepening destruction and pollution of the natural world, and (d) growing conflict over scarce resources at a time when armaments have become forbiddingly destructive. In our times these problems begin to acquire proportions which threaten to undermine not only the global economic system, but the very survival of humankind.

Economists tend to ignore, rationalise away, or play down the failures of capitalist liberalism. Where one has to cover up, there must be something wrong. As has been alleged again and again not only by Marxists, but also by liberal economists, economics functions as the legitimating ideology for capitalist interests.[30] Uncritical acceptance of profit and utility maximisation as normal and normative economic behaviour,[31] unquestioned faith in the possibilities of linear and unlimited growth,[32] undeterred confidence in the wisdom of the market mechanism, and the merits of free trade have come to guide the thoughts, attitudes and behaviour not only of business people, but also of many politicians and academics. Questions concerning the sustainability and equity of the system are swept away by a disarming kind of optimism. Even where the issues of poverty, ecological destruction and armed conflict are specifically addressed, it is not necessarily realised that they are due to the very nature of the underlying paradigm, rather than to a few regrettable but avoidable imperfections of the system.

Symptoms of a paradigm shift

For those who care to look at the total picture, the direction in which the global economy seems to be moving is thoroughly disconcerting.[33] There are strong indications that a paradigm shift may be in the offing.[34] Here we do not refer to schools within capitalist liberalism, which have challenged each other for decades, but to a possible change of the basic parameters of the current system.[35] As mentioned above, some observers believe that it will be forced upon us at the beginning of the next century by the deteriorating situation.[36] Attempts to come to grips with current economic trends abound, but an authoritative new paradigm has not jelled as yet.[37]

There is a chasm between those who demand economic justice and prosperity and those whose main concern is the environment. Some concentrate on endangered species, others on the gap in the ozone layer or global warming. Some demonstrate against nuclear tests, others offer ambulance services in war-torn and disintegrating societies. Some design intermediate technologies,[38] others attack technological growth as such.[39] Some locate the problem in the failure of social institutions, others in a short-sighted and selfish value system. Some identify inadequate communication systems as the key ailment,[40] others point to obsolete cultural traditions. Some call for a reform of international trade and radical debt relief, others believe that the financial system as such is the main cause of current impasses.[41] Some believe that economic growth must be married to ecological responsibility, others call for a "fundamental ecological transformation of our socio-cultural systems, collective actions, and lifestyles".[42] There is also the ongoing debate between the Keynesian, the neo-classical and other approaches.[43]

Attempts to come to grips with particular aspects of the problem are valuable in themselves because they add to the general ferment. But most authors focus on their favourite fields in isolation from the entire context. As a result they often contradict each other. Particular thrusts are no substitute for an understanding of the syndrome as a whole and a comprehensive vision for the future. This is a tall order. Some have suggested that it may be premature to even think of a new paradigm.[44] Certainly our task is not to provide final answers but to make tentative contributions to the emergence of something more compelling than what we have at present. In fact, there are no final solutions as long as history continues to unfold. But the challenge to become more alert and more responsible is not out of place at a time when symptoms of a pervasive disease can no longer be overlooked.

The task of the other social sciences

Economists geared to mathematical rigour and empirical evidence are likely to resent my "rhetorical twaddle".[45] I can only respond by saying: if the cap fits, wear it. If not, improve on my effort. I am an outsider; I do not consider myself to be infallible; my task is to facilitate an interdisciplinary self-searching exercise. The issue is too important to take things personally. It is also too important to leave to experts operating in a professionally constrained field. We need to relinquish our seemingly invulnerable professional fortresses and begin to listen to each other's observations and concerns.

Other human and social scientists are perhaps positioned best to challenge economists with alternative analyses, critiques and proposals. It is one of the more profound insights of theology that truth - both in the form of challenges and in the form of valid insight - has to come from outside established mindsets, whether individual or collective.[46] This book is meant to do precisely that. In the end a new paradigm can only emerge through cross-fertilisation and cooperative interdisciplinary effort.[47]

Revision: Which arguments can be advanced for the claim that a major paradigm shift in economics is in the offing?

Application: Go through the financial pages of a newspaper or the topics discussed in an economics journal. Can you detect some of the phenomena referred to above?

Critique: (a) "The proof of the pudding is in the eating. Free enterprise and the free market have worked wherever they were applied. What you present here is irrelevant and misleading ideology, not a reflection of reality." (b) "You may be right, but you will not change the course of history as a lone individual. Cries in the wilderness have no effect whatsoever." How would you react to these statements?

Section II: Requirements of a new economic paradigm

A new vision

Positivism and pragmatism scoff at visions. After all the ideological delusions humankind has gone through during the 20th Century there is a strong case to be made for factuality and pragmatism. But factuality and pragmatism are no excuse for a lack of direction. To be pragmatic means to employ feasible means to achieve practical aims. But aims must be spelt out; their prerequisites and consequences must be accounted for.[48] This is what a vision is all about.[49] A vision is not a blueprint. It is an anticipation of the future as it *ought* to be.[50] This general sense of direction can be translated into particular directives for personal action and public policy.[51] But visions as such do not materialise; they are like horizons which retreat as we move forward, ever luring us on to new vistas.

The boldest vision so far has been the Marxist vision of a "classless society". This vision is rooted in the apocalyptic version of the Jewish-Christian heritage.[52] Apart from the flawed assumptions of the Marxist metaphysics, which caused its practical failure, this vision is deficient in one fundamental respect: it ignores the limits of the planet and believes in the possibility of indefinite material progress and universal affluence.

Despite its alleged "pragmatism" liberal economics also has a vision, expressed best by the so-called American dream: "from rags to riches". This is an almost unforgivably shallow and narrow vision. It concentrates on material wealth at the expense of more fundamental concerns of humanity. It concentrates on the individual and ignores communal and social interests.[53] It is unable to deal with critical issues such as mass unemployment and marginalisation. It concentrates on immediate satisfaction and ignores the demands of long term sustainability.[54] The beauty and wealth of nature and the ecological damage caused by the modern craze are beyond its horizons.

The search for a new paradigm will have to begin with a different vision of what we believe humankind as a whole should achieve and could achieve. The economics of the future should be built on

the vision of the comprehensive wellbeing of all human beings, present and future, within the context of the comprehensive wellbeing of their entire social and natural environments.

The vision of comprehensive wellbeing does not overlook the fact that all choices imply trade-offs. It refers to an optimal situation, rather than to a paradise in which no sacrifices will be necessary. The adjective "comprehensive" indicates that we have to target any *specific deficiency* in wellbeing in any dimension of reality. The idea is not to achieve everything at once, but to follow the direction indicated by the vision step by step as possibilities unfold. Vision and pragmatism are not alternatives; they complement each other. We do not need to hide the fact that this vision is derived from the Jewish-Christian heritage. This has been spelt out in chapter 7. Here it suffices to say that there is an alternative to the present impasse.[55]

A vision is often smothered by the fatalistic enumeration of seemingly insurmountable obstacles. Such fatalism is not necessarily disinterested, particularly not in the economic sphere, and one should not be overly impressed by the skeptics. People who have a stake in the status quo, whether ideally or materially, will not want to give a new vision a chance. They will also put on the brakes in practical terms. The realisation of a vision demands the collective will to persevere, the willingness to make sacrifices, and the deliberate development of competence in the direction of the vision. It is also wrong, in principle, to confuse the question of what ought to be with the question how it could be achieved. What ought to be must be formulated in clear and uncompromising terms, otherwise it cannot give direction.

The economic problem - economic goals

Because the definition of a problem follows one's vision, it must be in dispute between the old and the new paradigm. The old economic paradigm sees the economic problem in scarcity and the necessity to make choices according to what it calls "opportunity cost". This means that when making a choice, costs and benefits of one option have to be confronted with the costs and benefits of an alternative option.

This is a useful approach if it operates within inclusive and long term horizons. But it is also a very narrow definition of the economic problem. It can be argued that in situations of absolute affluence there is no scarcity, except for the limits of time and energy to enjoy everything that could be enjoyed. In absolute poverty there is no choice, except to starve. The new economic paradigm should define the economic problem in terms of needs and means. In the next chapter we shall try to spell out the consequences of this shift in fundamentals.

As a result of its problem formulation, the old paradigm has an extremely narrow and distorted concept of what the economic enterprise should try to achieve. According to Heilbroner the capitalist system is driven by the motive of amassing capital as a means of amassing wealth, which is motivated by a craving for prestige and power.[56] Power is, in the first instance, the power of the managerial elite.[57] In terms of consumption it concentrates on material wants (not needs) of individuals without regard to other needs, the interests of others and the environment.

Jevons says: "To satisfy our wants to the utmost with the least effort - to procure the greatest amount of what is desirable at the expense of the least that is undesirable - in other words to maximise pleasure, is the problem of economics."[58] Economists should be too proud to agree to such a debasement of what could be a noble and indispensable discipline.[59] The new economic paradigm should define the goals of the economic enterprise as

**the balanced satisfaction
of all valid human needs
in all dimensions of life
for all humans, present and future,
within a healthy natural context.**

This formulation has far-reaching repercussions. Mainline economics concentrates on economic growth.[60] The goal of developing countries is to catch up with the industrialised nations as far as they possibly can, while the goal of industrialised nations is to grow even further. Paradoxically, the industrialised nations must grow so that developing countries can grow, because growth in the periphery depends on the markets of the centres, not on their own. Peripheries also depend on the capital provided by the centres.

On the basis of current ecological insight, however, it is impossible for the vast majority of humankind to reach current standards of living in the industrialised nations. The new economic paradigm will have to come to terms with this fact. More importantly, it has to reject the goal itself as inappropriate and undesirable. Growth at all costs does not represent an increase in wellbeing. The priorities on the new economic agenda include sustainability; sufficiency; equity; concern for the weak and vulnerable, and balance in the satisfaction of needs. We shall spell them out in chapter 15.

The "liberation" of nature and human cooperation

Current economic behaviour, economic policy and economics as a science are all still rooted in the ancient culture of isolated tribes who were dispersed in a vast and hostile environment. In that type of situation nature had to be subdued and exploited if the tribe was to survive and prosper. Living space and hunting grounds had to be defended against enemies, rivals and intruders. Whether these were plants, animals or other humans made no difference. That is the background against which the biblical injunction to "fill the earth and subdue it" made sense (Gn 1:28).

The situation has changed fundamentally since then. Far from being the superior force to be conquered, nature has become the abused and tortured slave of an increasingly prolific and ruthless humanity. Human societies have become more and more integrated and interdependent. Armed conflict has become devastating. The most advanced weaponry has become so formidable that attack or defence on that level would amount to collective suicide. Apart from having become a scourge to nature, humankind has become its own worst enemy. The most dangerous threat to our survival and prosperity no longer emanates from human enmity or natural forces, but from the escalating entropy of the entire biological system which humans have set in motion.

The new economic paradigm has to take the new situation into consideration: not capital but nature is the most indispensable, the most precious, and the most vulnerable endowment entrusted to humankind. It needs to be preserved, given space to breathe and pampered more than any other asset on earth. Coming generations will have to live from this highly vulnerable asset as much as we do. Other human beings need to be accepted as partners, not enemies. They have a right to survival and prosperity, as much as we have, and we need to cooperate with them to achieve humankind's collective goals.

A new cosmology

Orientation towards a comprehensive vision will inevitably change our scientific approach. Economics must acquire horizons which reflect the current state of human insight. In *traditional* societies, even in mediaeval Europe, reality had stable foundations, narrow horizons, short time spans and limited possibilities. But it had transcendent escape roots from space, time and powerlessness to heaven, eternity and divine omnipotence. Moreover, there was a strong sense of accountability to a higher authority. We may ridicule these assumptions as "superstitious", but it is an evolutionary fact that humans can and must transcend their immediate experiences. The only question is how they do it.[61]

During the *Enlightenment* reality became unified, secular and subject to natural laws. The progressive discovery of natural laws promised infinite power over nature. Space and time became relative. The stability of the mediaeval world dissolved into the infinite space of the universe and the endless flow of history. Under the impact of *Newtonian* refinements reality was increasingly seen in terms of static material which can be broken down into isolated building blocks which again can be put together in new combinations. Reality was seen as a machine which could be manipulated and reconstructed at will. Ever more sophisticated artificial constructs were seen to represent progress and the creation of wealth. We lost sight of the fact that reality is an evolving system in which relationships are more fundamental than components and where wholes are embedded in larger wholes.[62] Any artificial construct leads to painful destruction elsewhere in the system.[63]

The effects of this change of world view were far-reaching and ambiguous. On the one hand, the prospects for progress seemed to become unlimited. Nothing seemed to be impossible. And these possibilities evoked an irresistible fascination with science and technology. In fact, science became the dominant ideology. According to Heilbroner ideology is "an explanatory view of the world" which, "despite its vaunted 'positivist' approach and its shunning of truth by revelation ... fills a social requirement indistinguishable from religion."[64] This is particularly apparent in the case of economics with its definite metaphysical assumptions.

On the other hand, modernising humanity lost its orientation.[65] Deprived of their transcendent foundations and stable metaphysics, Westerners sought refuge in the subjective experience of the individual - whether this experience took the form of *observation* (empiricism), *reason* (rationalism), *intuition* (mysticism), or *devotion* (pietism). Philosophy began to disintegrate. Theology lost its plausibility. Accountability to an overarching authority made way for human autonomy. Soon this autonomy was privatised.

Beginning with empiricism and culminating in positivism, *morality* became suspect in scientific circles. The ideal of a value-free science emerged.[66] The utilitarian assertion that "whatever served the individual served society" provided a "blanket moral exemption" for personal acquisitiveness. It ruled out of bounds the old canons of virtue and justice - "canons that were always founded on a scrutiny of motives and an 'external' assessment of social results" - by replacing them "with a system that declares these canons to be arbitrary and therefore null and void."[67] An objective reality out there, to be investigated by science and manipulated by technology, and the subjective longing of individuals for personal and immediate satisfaction were all that mattered in the vast and empty expanses of reality.[68]

But such a world view has become obsolete. It has also become counterproductive. In terms of human motivations, concentration on individual emotional experience belongs to the Enlightenment era, not to the incoming 21st Century. Yet the appeal of the multi-billion marketing industry finds it convenient to celebrate subjective happiness, irrespective of the consequences for one's body, for communal relationships, for society, for nature. With that it radicalises and cheapens the trend to reduce reality to the sphere of individual experience. Moreover, this experience has become a tradable commodity, whether it is the thrills of subversive literature, sport or sex. Heilbroner believes that the total commercialisation of life is "perhaps the most self-destructive process of modern capitalist civilisation".[69]

In terms of *scientific methodology* economics has simply fallen behind. Physics, its great example, has long gone beyond the mechanical worldview of the 19th century and the ideal of an objective and value-free science. It operates with the concepts of quantum theory, relativity, entropy and evolution. In contrast to irrelevant idealism and materialist reductionism "the new physics introduces the concept of the world as a unified and inseparable whole."[70] We shall come back to that in chapter 13.

Moreover, leading physicists have attacked positivism as irresponsible and misleading.[71] Sciences have rediscovered the limitations and the relativity of all human insight. The theory of science has shown that science is built on assumptions which can be questioned.[72] The clean Cartesian split between subject and object has become problematic. We know that the results of observation depend on the disposition of the observer.[73] Religious and ideological determinants must be acknowledged.[74]

Concerning morality, sciences have been confronted with the dangers of nuclear and gene technology and called for scientific accountability. Moreover, they discovered the essentially cooperative structure of life. "An increasing body of evidence procured from a variety of sciences demonstrates the revitalizing and ennobling power of unselfish love."[75] Blind individual profit and utility maximisation at the expense of the rest of reality is most certainly out of step with what happens in the rest of nature:

> Detailed study of ecosystems over the past decades has shown quite clearly that most relationships between living organisms are essentially cooperative ones, characterised by coexistence and interdependence, and symbiotic in various degrees. Although there is competition, it usually takes place within a wider context of cooperation, so that the larger system is kept in balance. Even predator-prey relationships, that are destructive for the immediate prey, are generally beneficent for both species. This insight is in sharp contrast to the views of the Social Darwinists, who saw life exclusively in terms of competition, struggle, and destruction.[76] ... An organism that thinks only in terms of its own survival will invariably destroy its environment, and, as we are learning from bitter experience, will thus destroy itself.[77]

Interdisciplinary accountability

Compared with these developments in the natural sciences, mainline economics suffers from cultural lag. "Many economic theorists appear to be unaware that the scientific philosophy of modern physics has changed radically ..."[78] The reasons seem to lie in the general inability of scientists to transcend their specialisations, to transcend the constraints of models entrenched in the conventions of their disciplines, and the peculiar tendency of economics to isolate itself.[79]

The limitations imposed on human insight by *specialisation* are hazardous.[80] The current body of scientific knowledge is growing to proportions which no human being can hope to overlook or handle. While the fields studied by experts become minute, the body of total knowledge grows in immensity. But the isolation of ever smaller particles of reality from their overall contexts can yield highly misleading results and the lack of coordination can lead to extremely dangerous practical consequences. Because the problems we are facing are of a new order of magnitude, we need a new approach altogether. Specialised studies must take the entire contexts of their objects of study into consideration.[81] Because of the increasing complexity of the scientific body of knowledge, this can only be done in the context of a competent multidisciplinary team.[82]

Concerning *isolation,* classical economics has often adopted an attitude of self-sufficiency. The contributions of geography, psychology, sociology, anthropology and other human sciences were widely believed to be irrelevant to the functioning of an economic system.[83] Kenneth Boulding, a celebrated economist, speaks of "economics imperialism",[84] that is, the attempt to explain all social phenomena in economic terms and, by implication, declare the other social sciences as incompetent.[85] The facts are that many development projects have foundered because of a lack of consultation between experts, field workers of different persuasions, and the people on the ground who are supposed to benefit and who are most affected.

In the attempt to come to grips with the wider dimensions of reality, the physical sciences constitute one extreme of the academic spectrum; the other is represented by anthropology, philosophy, religious studies and ethics. These disciplines are indispensable because without the "encapsulation" of competition by civic values the economy will become "barbaric".[86] The most questionable assumption of economics is its particular view of human nature, the famous or rather infamous *homo oeconomicus* (= economic human being), which we shall soon discuss.[87]

The point to be made here is that economics cannot take over the task of disciplines which focus on cultures, convictions, values and emotions.[88] The particular field of economics is to provide tools for cost-benefit analyses and rational choice in all sorts of contexts and on all sorts of levels, not to pronounce judgment on the validity of a particular anthropology. By assuming that it was competent to define human nature, especially human motivations, and by implying that selfish human attitudes and actions are normal, thus legitimate and acceptable, economics has dabbled in fields of competence other than its own. This cannot always be avoided, and it is in order as long as one is open for dialogue and correction emanating from other disciplines.[89] If not, one may spread more myths than facts - myths which, in the case of economics, have become self-fulfilling prophesies.

In short, a new economic paradigm will have to build openness to interdisciplinary confrontation, consultation and coordination into its very foundations. It must expose itself to the process called "unification" in the theory of science: to be valid, different theories must be shown to be compatible with each other when linked in an appropriate way.[90]

The new discipline of *systems analysis* can provide the framework for such an inclusive approach.[91] The "hard facts" unearthed by the physical sciences, the malleable structures of society analysed by the social sciences and the soft data provided by anthropology and science of religion must be integrated with each other.[92] The theoretical dimension provided by the theory of science and the normative dimension provided by ethics must be combined. The impact of social structures on collective consciousness (materialism) and the impact of collective consciousness on social structures and processes (idealism) must be shown as part of a system of relationships and feedback loops.

Methodological accountability

Significantly, the discovery of relativity has not led the natural sciences to an endorsement of the relativity of human values and the suspension of human responsibility. As mentioned above, they rediscovered the importance of both methodological and moral accountability. Scientists have begun to reflect very deeply on their presuppositions. They have also begun to look at the human, social and ecological consequences of the applications of their findings.[93] In view of this rising awareness in the natural sciences, as well as the general insights of the theory of science, the new economics will have to shed the naive positivism of the old approach.[94]

Positivism simply takes for granted what it observes and discards everything else as irrational and unscientific.[95] By the adoption of such a stance one forsakes the ability to be self-critical.[96] In fact, there is no "pure" or "objective" science, least of all an objective social science like economics.[97] Convictions and interests enter the scientific process right at the beginning, where the basic assumptions are made. "Without vision there can be no analysis".[98] If assumptions are not accounted for, their potentially ideological character is not exposed. Positivism has proved to be highly vulnerable to ideology.[99]

Ideology, as we use the term, is a combination of selected facts with rational arguments to produce a system of meaning which justifies the pursuit of collective self-interest at the expense of others.[100] Heilbroner draws attention to the fact that ideology is the explanation the ruling elite gives to itself; it believes in this distortion of the truth because it needs legitimation. While underdog population groups also have legitimating ideologies, elites tend to take their own world view for granted. It never occurs to them that there might be an alternative view that could claim validity.[101]

In this sense the claim to objectivity is also dishonest. A pure science would have to be purely descriptive; in fact economics offers policy directives and these directives are valued by politicians. Obviously policy directives presuppose value decisions. And the values built into the assumptions of economics, notably those derived from "economic man" as profit and utility maximiser, are not only shallow, they are also designed to legitimate the pursuit of the self-interest of those who have the power to maximise their profit and utility.[102]

Towards an appropriate anthropology

Economics widely believes that the economy is driven by a universal compulsion of consumers to maximise utility and of producers to maximise profits. The interests of community, society and natural environment do not figure in the concept of the *homo oeconomicus* (economic human being). Such deliberate and calculated selfish behaviour is considered to be typical of the human being, thus normal. And what is normal conveniently becomes normative. A behaviour which deviates from this general trend is called "irrational". Preferences are taken as facts, not as cultural acquisitions which are relative, subject to critique and open to transformation.[103]

Modern humans may indeed want to be doped by the "drug" of an immediate satisfaction of ever growing wants. They may want to yield to any vulgar desire irrespective of personal, communal, social and ecological consequences. But to call human weakness and deception "rational behaviour" is total folly.[104] The endless quest for aggrandisement which, according to Heilbroner, lies at the roots of the drive to amass capital is "so patently without rationality, and so perilously liable to bring psychological discontent, that Adam Smith was forced to find its rationalisation in a delusion imposed upon us by the Deity."[105]

The consequences of the assumptions of liberal economics concerning human nature are far-reaching. They are also not true. Uninhibited greed has never been the only, let alone the only acceptable human motivation.[106] As we have argued in chapter 8, most cultures throughout human history have recognised that humans are social beings by nature and that selfishness is counterproductive for individual need fulfilment and dangerous for communal wellbeing. For this reason these cultures have developed mental barriers and institutionalised social controls. They have also channelled human motivations into positive directions. There is no reason why a culture of responsibility cannot emerge, persist and develop, unless it is deliberately and systematically dismantled.[107]

And this is precisely what happens in the liberal economy: a financially powerful advertising and entertainment industry is allowed to expose the human mind to an incessant bombardment of images aimed at spreading dissatisfaction with what one possesses, breaking down human inhibitions and creating new wants to push up sales. There is nothing natural or inborn in the success of this approach; it is all artificial and deliberate. Millions are invested in the attempt to restructure collective consciousness in the desired direction. It is one of the greatest projects of social engineering in history. A lengthy quotation from a work of the natural scientist Brian Swimme may be in order here:

> Advertisers in the corporate world are of course offered lucrative recompense, and, with that financial draw, our corporations attract humans from the highest strata of IQs. And our best artistic talent. And any sports hero or movie star they want to buy. Combining so much brain power and social status with sophisticated electronic graphics and the most penetrating psychological techniques, these teams of highly intelligent adults descend upon all of us, even upon children not yet in school, with the simple desire to create in us a dissatisfaction for our lives and a craving for yet another consumer product. It's hard to imagine any child having the capacities necessary to survive such a lopsided contest, especially when it is carried out ten thousand times a year, with no cultural condom capable of blocking out the consumerism virus ... Put it all together and you can see why it's no great mystery that consumerism has become the dominant world faith of every continent of the planet today.[108]

The critique, rather than the enhancement, of selfishness has become the most formidable task in modern times.[109] In a liberal society, where everything goes, we only can counteract this self-interested brain-washing of masses of people in the interest of big business by appealing to authentic human motivations. It is simply not true that material acquisitiveness is the prime motive of human behaviour. It is need which motivates people. Under modern conditions genuine material needs can be satisfied with relative ease. Beyond that humans are motivated by transcendent needs, the most important of which are the need for meaning, the need for the certainty of one's right of existence and, derived from that, the authority to act. We analysed the structure of these needs in chapter 6.

The need for one's right of existence, for instance, manifests itself in phenomena such as social acceptability, belonging and status. This need has deep psychological roots, some of which we have discussed in chapters 8 and

9.[110] Here it suffices to say that if the norms of one's social environment dictate that one should have a large progeny, then this becomes the prime occupation. If the norms dictate that one should drive a Mercedes, then it is the Mercedes that provides the motivation. So it is the manipulation of social norms which lies behind the advertising industry. Right of existence can equally depend on one's certainty that the investment of one's life is meaningful. There are countless people who work for NGOs, the caring professions, charities, household chores, or missionary endeavours for next to no remuneration.

Of course, many economists have long understood that human motivation is more complex than presupposed by the idea of the "economic man". Yet the assumptions connected with pleasure, utility and profit maximisation are not questioned or tested by main stream economics. Due to what economist Joan Robinson called the "self-sealing" capacity of economics, which is, without doubt, due to pandering to powerful collective interests, mainline economics has not taken note of the protests even of prominent economists themselves.[111] Surely the new economic paradigm must be built on a different kind of anthropology.

Moral accountability

Critique of assumptions is one side of the coin; the other is critique of goals and consequences. As the natural sciences now insist, the justified demand for the freedom of science must be linked firmly to the demand for scientific responsibility. That is the realm of ethics. Ethics transcends our insight into *what is* towards a perception of *what ought to be*.[112] What ought to be again rests upon a system of meaning offered by convictions and world views. Theology and philosophy try to fathom this "depth dimension" of human experience.[113] Chapter 7 of this book is a case in point.

Why should it be necessary to transcend experienced reality? The life sciences have shown convincingly that humans have evolved beyond the stage of animals. Animals are guided by reasonably stable and reliable instincts and can afford to take the given world for granted. Humans do not have such crutches. For better or for worse, humans have to explore, interpret and transform the world. In other words, human life is determined not only by nature, but also by culture. Once we interfere in given reality the outcome can be either advantageous or detrimental. In the words of a distinguished economist, "as science develops, it no longer merely investigates the world; it creates the world which it is investigating."[114] Once you become creative you must know what you want to create. More fundamentally, you must know what you are *entitled to create*. Animals have no such problems.

The problem has become infinitely more serious in modern times because science has provided tools for technological manipulations unheard-of in previous history. Just think of nuclear power, gene manipulation, virtual reality, psycho-drugs, ocean-going factory ships with dragnets, and so on. Because of the immense powers unleashed by scientific discoveries and technological developments, natural scientists, usually associated with the highest degree of objectivity, have recently begun to emphasise ethical accountability in no uncertain terms. It is high time that economics follows suit.

Universal horizons and the dignity of the individual

In the cultural evolution of humankind, horizons have widened from clans to tribes, from tribes to nations, and from nations to humanity as a whole. This evolution of universal awareness has been documented profoundly as early as the Hebrew Scriptures.[115] In modern times the general trend has been greatly reinforced by modern systems of information, communication and interaction. Through radio, television and the internet we are in touch with catastrophes occurring in the remotest corners of the earth. We are also informed about different cultures or newest technological developments. Never before in history has our common humanity had a chance to become so real to so many people at the same time. This evolutionary achievement must not be lost under any circumstances because our future depends on it.

Mass communication has also made it possible to understand and appreciate nature. The present task is to create an awareness that the destiny of humanity is inextricably embedded within the context of nature, not somewhere above nature. This implies that the horizons of the new paradigm should be inclusive. It should place individual phenomena, such as the firm or the nation state, into their global social and natural contexts. It should also take heed of the large sweeps of history and recognise that we may just be passing the decisive threshold where nature can no longer absorb the impact of industrial and population growth.

Our private behaviour and our public policy must shift from short term, partial, and haphazard goals to long term, inclusive and reflected goals. Nothing less than the comprehensive wellbeing of the whole human being, and of all human beings, in the context of the comprehensive wellbeing of their entire social and natural environments will do as a vision for the future. It is this vision which has been emerging hesitantly for a few decades now and which needs to be developed forcefully.[116]

As the evolution of the Hebrew religion also shows, the change from ethnic particularism to inclusive universalism implies individualisation. If all people count, every person counts. Every individual is immediate to God.

The New Testament has powerfully reinforced this insight of post-exilic Judaism, and the Reformation has asserted it against the hegemonic claims of the institutional church. The dignity of the individual has found its secular expression in the Declaration of Human Rights. Again this is an achievement of cultural evolution which we cannot afford to lose. It means that every single individual has the same right to survive and prosper as any other individual, present or future. This principle has the practical corollary that development must be initiated, accomplished and controlled by decentralised grass roots communities rather than by "big government" or "big business".[117]

However, the dignity of the individual must not be confused with individualism.[118] Dignity has nothing in common with selfishness. The trend towards egocentricity emerged only during the so-called Enlightenment.[119] It was subsequently legitimated and exploited by liberal economic theory. The dignity and freedom of the individual was interpreted as the right to profit and utility maximisation. It was overlooked that growing private ownership of resources by some implies increasingly restricted access to resources for others.

Still aware of the values of social responsibility, early liberal economists rationalised away the social consequences of individual selfishness by means of the theory of the "invisible hand", which ostensibly transforms individual and collective avarice into social prosperity.[120] This may be true as far as the growth of Gross National Product per capita is concerned, but Gross National Product is not what collective prosperity is all about. What happens in fact is that those with higher economic potential outcompete economically weaker sections of the society in a self-propelling spiral. In the next chapter we show how this can lead to the marginalisation and impoverishment of large parts of the population. The "invisible hand" has proved to be an invisible foot which tramples on the weak and vulnerable.

Heilbroner suggests that both the craving for power and prestige, thus for domination, and the capacity for submissiveness are situated in the residues of our experiences on the way from infant dependency to adult emancipation.[121] Should he be right, this craving is as infantile as the drive to smoke cigarettes as a substitute for the mother's breast. As we mature, we grow out of such "instincts". If advertisers appeal to these instincts they make no contribution to human emancipation, as they constantly suggest, but reinforce human immaturity.

The point to be made is that, in terms of the cultural evolution of humankind, the ideological legitimation of human craving for immediate satisfaction of every want and desire, however shallow and fleeting, is a *retrogressive* step. The emerging awareness of the whole, this precious cultural achievement, is being lost - whether in terms of the whole spectrum of human needs, the dignity of others, or the health of the natural environment. In fact, it is deliberately being squandered for the sake of short term and private gain. In contrast with the humanist democratic tradition, on which it feeds, the liberal economic paradigm has no antenna for the dignity of the human being, which is the dignity of all human beings. A new economic paradigm has to retrieve this tradition and make up for lost ground.

Revision: Can you summarise the main thrusts of the proposal for a new paradigm made above in contrast with the current paradigm in economics?
Application: Spell out the consequences which the new approach would have for economic behaviour in your country, particularly in connection with poverty, social tensions and ecological stability.
Critique: (a) "What you clamour for is nothing new. It has repeatedly been tried and it has always failed. Why do you continue to spread dreams which do nothing but cause immediate resentment and eventual disillusionment?" (b) "You are a privileged academic who can afford to develop grand ideals of equity and sustainability in the tranquillity of your office. People on the ground have to struggle for their next lunches, the schooling of their children, or a balance sheet acceptable to their shareholders." (c) "Comprehensive wellbeing is a utopian dream which belongs to the world of fantasy. In the real world you have trade-offs and entropy. We cannot all be in paradise without costs." Comment.

Let us summarise

Human sciences are, like all intellectual endeavours, subject to *paradigm shifts*. Paradigms have to adapt to changing situations and the accumulation of knowledge, or they become obsolete. Because paradigms entrench themselves in collective mental and social structures, define identities and legitimate collective interests, a change of paradigm tends to cause upheavals, not only in the academic discipline concerned, but in broader society as well. The obsolete does not yield without resistance, the new does not appear without its birth pains.

In this chapter we argued that a major paradigm shift is called for in *economics*, whether as a social science, a set of principles in public policy, or a pattern of private thought and behaviour. Capitalist liberalism seems to have been vindicated by the dramatic collapse of its arch-rival, Marxism-Leninism, in the former Soviet Union and its satellites. But its own failures begin to assume proportions which can no longer be overlooked with impunity. A great variety of attempts to find alternative approaches have emerged without showing any consistency as yet.

This observation led us to the attempt of formulating the outlines of a new economic paradigm. Its main ingredients include the following:

(a) Economic pragmatism should be guided by a *vision*. This vision should project the *comprehensive wellbeing* of all human beings, present and future, in the context of the comprehensive wellbeing of their entire social and natural contexts.

(b) The economic problem should, therefore, be defined as the *balanced satisfaction* of all *valid human needs* in the context of a healthy social and natural environment.

(c) The new paradigm should express the potential of overcoming the conflictual and dominating spirit of our tribal ancestry, give *nature* its due and entrench human *cooperation*.

(d) Universal horizons reinforce the *dignity of the individual* and constrain individualist selfishness. The current concentration on the immediate satisfaction of material desires represents a retrogressive step in cultural evolution. The distorted concept of the human being as a profit and utility maximiser must make way for the dignity of a responsible, communal and caring being.

(e) Economics should be integrated into an *interdisciplinary team*. It should take the insights of the social and natural sciences into consideration. It should heed the demand of the theory of science to account for its basic assumptions, as well as the ethical demand to account for the consequences of its applications.

Notes

[1] Laszlo 1977.
[2] Cf Küng 1997:167ff.
[3] Examples are the collapse of the Hitler empire, of Marxism-Leninism, of apartheid in South Africa.
[4] Von Weizsäcker 1994.
[5] Boff 1995:80, quoting Malraux.
[6] " ... as anyone who has done economic theory knows, the style of our models strongly determines their content - issues that are awkward to address are generally speaking not addressed" (Krugman 1993:x).
[7] Cf the "obituary" to Thomas Kuhn found in *The Economist* (EU), July 13, 1996, 97.

[8] Cf Storkey 1986:67ff.
[9] For the debate on the relativity of scientific theory see Gerrard 1989:1ff.
[10] For a discussion of this discovery in physics see Rohrlich 1987:3ff.
[11] Rohrlich 1987:114ff.
[12] "But it has also been the age of the policy entrepreneur: the economist who tells politicians what they want to hear. The way that a small group of 'supply siders', preaching a doctrine that even conservative economists regarded as nonsense, is one of the wonders of our age" (Krugman 1994:xiv). However, the "faultline" is not just between "serious economic thinking and economic patent medicine", as Krugman wants us to believe. Radical ideology critique of main line liberal economics goes to the heart of the discipline itself and not all of that is simply malicious. Cf Heller, Walter H: "What's right with economics" *American Economic Review* 65/1975 1-26.
[13] Kuhn 1970. Cf also Duchrow 1994:133.
[14] On the reaction of mainline economics to the theory of science and methodological questioning see J S Hart 1997: Can philosophy of science be helpful to economics? *The South African Journal of Economics,* Vol 65:4 1997, pp 510ff.
[15] Krugman puts it humorously but succinctly: "I have to live with these people for the next thirty years" (1993:8).
[16] Stackhouse refers to a Harvard study called the "American Business Creed" which shows that "most business leaders hold an outmoded ideological view of their sphere of expertise that does not correspond with what they do every day or with the kinds of social interactions that actually dominates their lives" (1987:124).
[17] "No science will agree to junk its tried and true axioms, even when they become trying and untrue, until a new and more powerful set of axioms is available" (Shlomo Maital 1982, quoted by Daly & Cobb 1989:8).
[18] A critical analysis of this phenomenon is offered by Jürgen Habermas in *Technik und Wissenschaft als Ideologie.* Frankfurt/M: Suhrkamp, 1973.
[19] Examples are Schumpeter, Keynes, Galbraith (1958); Boulding (1970); Schumacher (1974); Heilbroner (1985 and 1988); Holland (1987), Davidson & Davidson (1988); Lester Thurow (1983); Lutz & Lux (1979); Karl Polanyi; Henderson (1978); Naqvi (1993); Gerrard (1989), Etzioni (1988), David Smith as economic geographer (1979, 1994), Daly (since 1973) Krugman (1993), and many others.
[20] It is noteworthy that it is Marxist and Christians who should have pointed out the importance of power in the economy! E.g. Meeks 1989:58ff.
[21] It is revealing to see that the "seven basic trends" concerning economic growth enumerated by Samuelson & Nordhaus (1989:861f) do not include the growth of economic discrepancies between rich and poor, the growth of unemployment, the growth of misery and population numbers in peripheral groups, the growth of the destructiveness of armed conflict, the growth of the waste problem, the growth of pollution ... and so we can continue! One cannot help but believe that economists have blinkers on. Krugman says that " ... one can imagine an unequalizing spiral in which the world endogenously becomes differentiated into rich and poor nations" (Krugman 1993:94) and admits that all this has been known to leading scholars in the past, for instance, Alfred Marshall (1920), Allyn Young (1928), Gunnar Myrdal (1957), Alfred Hirschman (1958), Allen Pred (1966), Nicholas Kaldor (1972). "Yet while the ideas may be familiar, they have never become part of the mainstream of economic analysis" (Krugman 1993:98).
[22] For a discussion of these schools see Krugman 1994.
[23] Quoted by Duchrow 1987:158ff.
[24] For a good description see Knight 1986.
[25] The metaphysical character of economics has often been exposed by prominent economists. Economics tried to equal the mechanistic certainties of 19th century physics (Mirowski 19989). Even today "many economists regard nineteenth century classical physics as the ideal model of scientific methodology ..." (Drakopoulos 1994:333. See 333-339). According to Johns, classical economics has simply fallen in love with its abstract logical structure and has lost its awareness of the real world out there (Johns 1985:142f).

Chapter 11 - The pending paradigm shift in economics | 293

[26] There is, however, a "new Keynsian economics" (Mankiw & Romer 1991).
[27] For a comparison and a discussion of capitalism and socialism see the companion volume, Nürnberger 1998. For the theological challenges posed by these three schools see Nürnberger 1987:233ff.
[28] There are other examples where social systems and their legitimating ideologies have disintegrated, for instance, African Socialism in Tanzania, apartheid in South Africa, fascism in Central Europe, or - looking further back - feudalism in Europe.
[29] Fukuyama 1993.
[30] It was upheld by leading economists until very recently: Friedman (in Hausman: *The philosophy of economics*, 1983) and Lipsey (*Positive economics*, 1984). Myrdal 1953 was ahead of his time.
[31] See McKenzie 1985:17, 21.
[32] Critically analysed by, amongst others, Mishan 1993; Daly & Cobb 1989, Hoogendijk 1993, Lutz & Lux 1979.
[33] Daly & Cobb say that they would not want to disparage the lifelong efforts of many sincere economists, but "at a deep level of our being we find it hard to suppress the cry of anguish, the scream of horror - the wild words required to express wild realities. We human beings are being led to a *dead* end - all too literally. We are living by an ideology of death and accordingly we are destroying our own humanity and killing the planet (1989:21).
[34] In his "steady-state economics" (1973, 1991) Daly already foresaw a paradigm shift in economics. "The social and economic anomalies (economics) can no longer address ... are now painfully visible to everyone", but economics fails to address these issues and the population is becoming restive (Capra 1983: 199). "One aspect of the contemporary situation is an increasing awareness that the human race, in pursuit of material gain by exploitation of nature, is racing towards destruction of the planet and itself" (Council of the Club of Rome 1991:xviii). Cf also Meeks 1989:15ff.
[35] See Froyen 1996, part II for a good introduction to the history of modern economic schools: mercantilist, classical, Keynsian, monetarist, neo-classical, neo-Keynsian.
[36] Daly & Cobb 1989:21; von Weizsäcker 1990:9ff; 1994:5ff.
[37] Greens, for instance, are split over technology: new technologies seem to have less impact on the environment and give room for decentralisation. The economic imbalances caused are not considered, because greens tend to be middle class and love technology. Those who are critical keep themselves busy with a subsistence economy of their own. There is no alternative ideology or utopia that could unite and channel different forms of opposition (Junne in Caporaso 1987:78f).
[38] Schumacher and his Intermediate Technology Group.
[39] Mishan 1993.
[40] For instance Klitgaard 1991.
[41] For instance Hoogendijk 1993.
[42] This is the "deep ecology movement" (Drengston & Inoue 1995:xix).
[43] E.g. Hoover 1988; Skousen 1990.
[44] At this stage of the debate, Capra says, we should be formulating "a network of interlocking concepts and models and, at the same time, developing the corresponding social organizations" (1982:285).
[45] An economist friend of some standing feared that this would be the reaction of economists when reading this manuscript. If he were kind enough to take my book seriously, Krugman would probably put me into the category of a "crank" (1994:90).
[46] Luther spoke of the Word of God being a *verbum externum*, a word coming from outside one's own hangups and desires. In my theology I have developed this idea for social contexts. See chapter 10, section II.
[47] Note the remarkable insights gained by Krugman (1993) when he rediscovered the virtually forgotten contribution which geography could make to economics.
[48] Cf Villa-Vicencio 1993:231ff and his "theological checklist" pp 238ff.
[49] Cf Meadows, Meadows & Randers 1992:224f.
[50] As spelt out in Ernst Bloch's famous "Principle of Hope", a vision is not a fairy tale but a "real utopia", that is, the anticipation of a situation which is not yet attainable but

which may become real.
[51] See Cobb 1992:31ff, Villa-Vicencio 1992:238ff and many others for examples.
[52] See Nürnberger, K 1987: The eschatology of Marxism. *Missionalia* 15/1987 105-109 and chapter 4 of Nürnberger:1998.
[53] Lutz & Lux give a vivid example: according to the Pareto optimum there is a net increase in welfare if a rich landlord gets richer and his poor tenant stays at the same level because nobody loses. In fact, the further increase in the discrepancy between the two incomes is likely to imply considerable losses in psychological and social welfare because the tenant will become bitter and the relationship suffers (1979:94).
[54] "The bounded vision in modern economics ... has contributed towards millions of individuals living in social and economic poverty as the result of prolonged periods of mass unemployment" (Gerrard 1989:11).
[55] Milbank (1993) offers a theological critique of the modern intellectual enterprise, including liberalism and Marxism. Built on the twin pillars of imperialism and personal property, it is characterised by self-aggrandisement, conflict and violence. The Christian alternative is the integration of the diverse in a social body through forgiveness and charity.
[56] Heilbroner 1985:33ff.
[57] Management Science begins to see the fallacy of this. Cf Block 1993:34ff et passim.
[58] Quoted by Lutz & Lux 1979:45.
[59] Apart from the Marxists, humanist economists critical of this stance have been, for instance, John Ruskin (Unto this last, New York, Wiley, 1888), John Hobson (Economics and ethics, London: Heath 1929), Thorstein Veblen (The place of science in modern civilization and other essays, New York, Caprice, 1915/1969), E F Schumacher (Small is beautiful, New York: Harper & Row, 1973), A Etzioni (1988), also the economic geographer David Smith (1979, 1994). See Lutz & Lux 1979:38-53.
[60] Cf Duchrow (1994:220).
[61] Cf Kieffer 1979:12ff.
[62] Capra 1992:69ff. See also McFague's reflection on an organic vs a mechanistic cosmology, where all bodies are "reflections of God" (1993:91ff, 134ff).
[63] See Capra for this important aspect. "An economy is a continually changing and evolving system, dependent on the changing ecological and social systems in which it is embedded. To understand it we need a conceptual framework that is also capable of change and continual adaptation to new situations" (Capra 1992:196).
[64] Heilbroner 1985:135.
[65] The interpretation of the Enlightenment as a truncation of Jewish prophesy by Daly and Cobb (1989:384ff) is insufficient for an explanation of the ensuing individualism and secularism.
[66] Cf Storkey 1986:69.
[67] Heilbroner 1985:115.
[68] Cf Campbell 1987, who traces modern hedonism back to emotional aspects of Puritanism, reinforced by Sentimentalism and Romanticism, producing consumer fantasies after immediate needs were met, to which production then responds.
[69] Heilbroner 1985:117.
[70] Boff 1995:39.
[71] Heisenberg (The physicists conception of nature, 1962:76-78) and Bohr (Atomic physics and human knowledge, 1958).
[72] Leading figures are Popper, Kuhn and Feyerabend.
[73] Drakopoulos 1994:342.
[74] "Science is, and always has been, based on a judicious mixture of empiricism and faith. These things can degenerate easily into gritty factuality, dogmatism or credulity" (Knight 1986:70).
[75] Kieffer 1979:35.
[76] Capra 1983:302.
[77] Capra 1983:313.
[78] Drakopoulos 1994:333.
[79] For a discussion of this phenomenon see Daly & Cobb 1989:34; Capra 1983:195ff.

Chapter 11 - The pending paradigm shift in economics | 295

There are notable exceptions, for instance Heilbroner (1985:13ff), who delved deeply into the nature of social formations, Skitovsky (1976), who worked on the psychological background of the Western economy, Frank (1988) who highlights principles and emotions in economics, Hirschmann (1977), and others. The proverbial arrogance of economics towards sociology, anthropology, social psychology and other social sciences, was rooted in the ambition of being an exact science similar to physics (Daly & Cobb 1989:26ff). There are observers who claim that neo-orthodox economics took over the mathematical model of mechanical physics in the 19th Century and got stuck in that paradigm even when physics moved on to wider vistas (Mirowski 1988). Mathematical sophistication grants prestige within the discipline, rather than enlightenment concerning what happens in the real world (Daly & Cobb 1989:31). See the particularly devastating critique of the mathematical fetish by Wassily Leontieff, a Nobel laureate in economics (ibid p.32). Logical deductions from questionable axioms often simply override empirical research. "According to the traditional view, economists proceed by deductive reasoning from assumptions that are self-evident, on the basis of either introspection, or common knowledge about the economic system. The truth of the premises is assumed to guarantee the truth of the conclusions. Under these circumstances there is no need for hypothesis-testing" (Hillinger 1992:55). Daly & Cobb give mind boggling examples of this observation (1989:35-41).

[80] Daly & Cobb 1989:32ff.
[81] Hurst 1996 provides a good impression of the multi-faceted character of inequality from a sociological point of view.
[82] Capra speaks of "the beginning of a fundamental change of world view in science and society, a change of paradigms as radical as the Copernican Revolution ... The more we study the major problems of our time, the more we come to realise that they cannot be understood in isolation. They are systemic problems - interconnected and interdependent ..." (In Ray & Rinzler 1993:231f).
[83] Krugman, for instance, speaks of the "tendency of international economists to turn a blind eye to the fact that countries both occupy and exist in space" which leads to "the exclusion of important issues and, above all, of important sources of evidence" (1993:2f). "Regional comparisons offer a huge, almost untapped source of evidence about how our economy really works ... Most important to my mind, however, is the support that the study of economic geography offers for a basic rethinking of economics" (Krugman 1993:98).
[84] Boulding 1970:131.
[85] This is surely not characteristic of all economists, but certainly of a significant proportion of the main stream. Development economists, for instance, normally add a chapter on "non-economic factors" (e.g. Kindleberger 1958). Some have investigated the psychological roots of economic motivations very deeply (e.g. Scitovsky 1976, Lauterbach 1974), etc.
[86] Davidson 1988 referring to Etzioni.
[87] See for instance Simons 1996:25-43, Daly & Cobb 1989:85ff and Lux & Lutz 1979:45ff for critical discussions.
[88] For emotions see Frank 1988; for convictions see Nürnberger 1988, part II.
[89] Lester Thurow says "Economics cannot do without simplifying assumptions, but ... the judgment has to come from empirical analysis (including those employed by historians, psychologists, sociologists and political scientists) of how the world is, not of how our economics textbooks tell us it ought to be." Daly & Cobb 1989:7.
[90] Rohrlich 1987:27.
[91] See Meadows et al 1992:3, Capra 1982:285ff, Checkland 1981; Mesarovic & Pestel 1974, Laszlo 1972. An overview and comparison of methods is found in Tudor & Tudor 1995.
[92] Checkland 1981.
[93] "Once the collective set of values and goals has been expressed and codified, it will constitute the framework of the society's perceptions, insights, and choices for innovation and social adaptation." When the value system changes, often in response to environmental challenges, new patterns will emerge (Capra 1983:196).

[94] Gerrard (1989:6ff) shows that, contrary to the beliefs of its followers, classical theory is not the only possible approach even in economics.

[95] Cf Gerrard 1989:12ff.

[96] "Any 'value-free' analysis of social phenomena is based on the tacit assumption of an existing value system that is implicit in the selection and interpretation of data." The protagonists of a value-free science "are not more but less scientific" (Capra 1983:196).

[97] The celebrated economist J Schumpeter said that "before embarking upon analytical work of any kind we must first single out the set of phenomena we wish to investigate, and acquire 'intuitively' a preliminary notion of how they hang together or, in other words, what appear from our standpoint to be fundamental properties. This should be obvious. If it is not, this is only owing to the fact that in practice we mostly do not start from a vision of our own but from the work of our predecessors or from the ideas that float in the public mind." (Quoted by Heilbroner 1988:165.)

[98] Heilbroner 1988:198 (quoting Schumpeter). Note that the term "vision" here means set of assumptions, not a future shape of reality as in our own considerations earlier on.

[99] "This worldwide environmental abuse is rooted in a set of ideas which have become the assumptions and axioms of modern human behavior ... (which) are saturated with emotions of pride, confidence, advantage, winning and losing ... It is a foundational worldview - or, more correctly, a faith. It offers a horizon of meaning. It grounds action. It provides an overarching social structure by which one lives and for which one is willing to die ..." (Joranson & Butigan 1984:3f).

[100] As Heilbroner formulates it, "lying on behalf of an idea or an interest" (1988:186).

[101] It perceives itself as the "embodiment of the spirit and mission of the society that in fact it dominates" (Heilbroner 1985:130).

[102] The ideological character of liberal economics has been analysed in some considerable detail by Heilbroner 1985:107 and 1988:185ff.

[103] Cf Boulding 1970:119.

[104] See McKenzie & Tullock 1985:17f for "rational behavior". For a critique see Daly & Cobb 1989:5ff. For the concept "rationality" in economics see Heap 1989. For the role of principles and emotions in behaviour see Frank 1988.

[105] Heilbroner 1985:54.

[106] Cf Küng 1997:210f.

[107] For a Christian approach see chapter 7.

[108] Swimme 1996:16ff.

[109] Sider argues that accumulation and compulsive consumption are "unbiblical, heretical, demonic", the "Big Lie of our secular, materialistic soicety" (1977:41).

[110] Brian Easley (1983), for instance, contrasts the natural female creativity with the male technical creativity. He concedes that the capitalist society needs wasteful throughput, but argues that the typically male craving for dominance, based on male envy of the womb, and male concentration on penis creativity, are the psychological roots of science - which can still lead us into nuclear destruction. This is another example of the right of existence syndrome. Cf also Easley 1973.

[111] Lutz & Lux 1979:51.

[112] Cf Moltmann (1984:147ff)

[113] For the moral dimension in economics see Davidson 1988.

[114] Boulding 1970:121.

[115] Witness the widening of horizons from Abraham's clan to the apocalyptic conception of the universe as a whole and of history as a whole. See chapter 7 for more detail.

[116] Von Weizsäcker sketches the development of the global vision from J F Kennedy onwards (1994:37ff).

[117] Cf Cobb 1992:48ff.

[118] See Cobb 1992:87ff for his anti-anthropocentrism.

[119] For the following cf Daly & Cobb 384ff.

[120] For an explanation see McKenzie & Tullock 1985:111ff.

[121] Heilbroner 1985:46ff.

A more fundamental approach

Market and Marginalisation

What is the task of this chapter?

In the last chapter we suggested that we need a new economic paradigm. In this chapter we shall try to lay the theoretical foundations for such a paradigm. All we do at this juncture, however, is to replace supply and demand, interacting on a free market, with *need and capacity* as the critical point of departure in economics. This small shift in the foundations (described in section I) is sufficient to link up the results of our analyses with economic theory.

In section II we shall utilise these insights to shed light on the link between market and marginalisation, that is, the mechanisms underlying the interaction between centres and peripheries mentioned in chapter 5. In section III we shall respond to objections arising from the success stories of the "young tigers" in South East Asia which managed to enter the club of industrialised nations.

Section I: Need and capacity as points of departure for economics

Let us begin with the most rudimentary form of the economy, namely hunting and gathering, as practised by our ancestors for countless millennia. There is a need for food, clothing and shelter, and there is the capacity to fulfil that need by taking what nature provides. Need and capacity are directly

297

related to each other. The situation is not much different in the next two stages of economic development, animal husbandry and subsistence agriculture. People consume what they produce and produce what they consume. But they extend their natural resource base, which makes them less vulnerable and more prosperous than in the former stage.

A *modern* economy, in contrast, is determined not by the relation between capacity and need, but by the relation between supply and demand on the "market place". At least, this is what economists tell us. The market mechanism forms the "first steps" in any conventional text book on economics. The fundamental problem, according to this approach, is not need, but scarcity. Scarcity is linked, again not to need, but to wants, and wants are linked to utility or pleasure.[1] What is not scarce in this sense is also not the object of economics.[2] We need air, but air is not scarce, so it does not present an economic problem. Does this mean that capacity and need are irrelevant? Surely not.[3] The link between the two pairs of concepts is money. Earnings prompt capacity to generate supply, and purchasing power turns need into demand (see diagram 12-1).

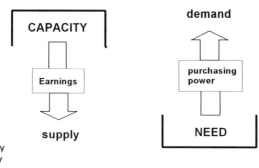

Diagram 12-1
The role of money in creating supply and demand

The diagram shows that need is identical with neither market demand nor consumption. The crucial ingredient for need to translate into market demand is economic power, in this case *purchasing power*. When discussing market demand, most economists simply take income for granted, but this practice does not do justice to economic reality. Where there is great need but no purchasing power, there is no market demand. Of course, where there is purchasing power and no need there is also no market demand.

Chapter 12 - A more fundamental approach | 299

This seems obvious. Why is need then not taken as a point of departure in economics? The argument that need cannot be distinguished from want and that wants are unlimited, thus not measurable in quantitative terms, is hardly valid.[4] Certainly there is a difference between the need of a homeless person on a winter night for a blanket and the want of a tycoon to breed race horses. Nor can the needs and wants of limited human beings be unlimited. It is also not impossible to develop statistical tools to measure these differences.[5]

On the side of production economic power is embedded in *productive capacity*. But productive capacity is identical with neither supply nor production. For productive capacity to translate into production and supply, it must be prompted by the prospect of making a profit, that is by potential earnings. Earnings again feed productive capacity. Producers will only offer their produce on the market if they get a reward which is greater than the cost of production. Nor should they, because then the economy would be running at a loss at that particular point.

It is only after having clarified these relationships that we can deal with the *market mechanism*.[6] It coordinates supply with demand and establishes the quantity and the price of the goods and services to be traded. Demand will be higher the lower the prices, while supply will be higher the higher the prices. Where the two meet, the market balances out and the deal is struck. The readiness to sell a given quantity at a given price then determines actual production, while the readiness to buy a given quantity at a given price determines actual consumption.

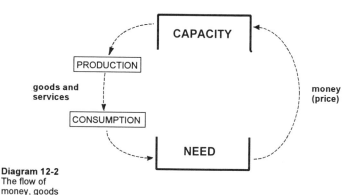

Diagram 12-2
The flow of money, goods and services

The result is that goods and services flow from productive capacity to the fulfilment of needs (that is from producer to consumer), while money flows in the opposite direction from purchasing power to earnings, which can again be translated into capacity (diagram 12-2). For the sake of simplicity we ignore the fact that there can be unsold stocks on the side of the supplier and unconsumed reserves on the side of the buyer. We also ignore distortions to the market such as monopolies and monopsonies.

Where does purchasing power come from? It could have been inherited, won in sweepstakes, or received from pension schemes. But ultimately it is derived from the earnings of some sort of production or another, even if this production only consisted of finding a diamond in the desert. Normally purchasing power is fed by rewards for initiative, expertise, management and labour (salaries and wages), on the one hand, and by dividends from investment on the other. Both are derived from production, thus from earnings.

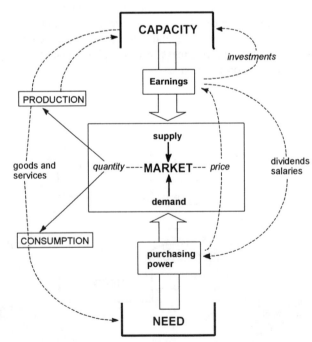

Diagram 12-3
Linkages between capacity and need

Some of the capacity is not utilised for the production of consumer goods, but for the production of *capital goods* and *infrastructure* (machines, roads, communication networks, and so on) which help to build up productive capacity further. Correspondingly some of the earnings are not paid out to the owners of the factors of production but reinvested in productive capacity. That is called institutional saving. The owners of purchasing power can also invest part of their assets in productive capacity, that is, in private saving.[7] When we put together all the elements discussed above we obtain the situation depicted in diagram 12-3.

In a market economy, therefore, money flows in at least three directions: (a) from purchasing power to earnings, when goods and services are bought on the market place, (b) from earnings to purchasing power, which enables need to form demand, and (c) from earnings to productive capacity, to build up that capacity. To simplify matters, we have ignored flows to and from the state in this model, which can be substantial.

Economic growth and recession

This model gives the impression of a balanced system where goods and services circulate endlessly. But in real life the economy is a dynamic flow from the resource base to waste (see chapter 2). Within this general flow the different streams discussed above can grow or decline. When productive capacity exceeds consumer demand and is used to build up productive capacity the sluice gates from the resource base to waste are opened further, and the increase in flow is called *economic growth*. On the other hand, when productive capacity is not built up but allowed to wear out and run down, the economy declines. This happens, for example, when the pressure of the need of the population is so great that it claims all earnings for consumption.

Economic growth thus takes place when the volume of production is greater than the volume of consumption and the surplus is used to build up productive capacity. The greater the proportion of production and earnings channelled into capacity building, the greater the rate of growth. A phase of rapid economic growth is called a *boom*. Economic growth leads to an increase in the overall prosperity of the system. At the same time it leads to an increase in pollution, resource depletion and waste.

The rationale of building up productive capacity is to make higher levels of production possible. But production yields no earnings unless it is channelled into consumption. To absorb increased production, consumption must grow accordingly, otherwise there will be surpluses. The production of surpluses which cannot be sold is an exercise in futility. Therefore the entire economic process hinges on the growth of consumption, thus on the growth of consumer demand, thus on the growth of purchasing power and the presence of unfulfilled need. Where demand declines, production declines, earnings decline and productive capacity changes into surplus capacity. This is called an economic *recession*.

But if it is the aim of the economy to boost economic prosperity, why should rising consumption ever be a problem? The answer is that consumption may not keep pace with capacity building for either of two reasons:

(a) There can be a lack of *purchasing power*. When more earnings are spent on capacity building than on purchasing power, the result is surplus capacity and the result is a recession. The situation becomes worse when resources end up neither in capacity building nor in purchasing power but get lost on the way through inefficiencies or leakages. A leakage occurs, for instance, when excessive profits are taken out of the economy by foreign investors, or when corrupt politicians transfer state revenues into private bank accounts.

(b) *Needs* as a package can decline because particular needs are being fulfilled. In fact, this should be the goal of the economic process. But when the market is saturated, whether in respect of certain consumer goods and services, particular sectors, or the economy as a whole, there is no growth in consumption. Then there is no growth of profitable production, thus surplus capacity is formed and again a recession ensues. This will not happen as long as production can be channelled into exports. But the international market can also be starved of purchasing power or become saturated. "Ultimately, demand growth must outpace productivity gains to sustain or raise employment levels, and the fabled forces of the market are not very evidently poised to accomplish this."[8]

Segmented economic growth

So far we have dealt with the relation between capacity and need, mediated through the flow of goods and services in one direction and money in the other direction. We now have to introduce two complications. The first is the segmentation of the population within a single economy into groups with varying levels of economic potency, the second the interaction between two economies operating on different levels of power.

The model depicted in diagram 12-3 may give the impression that it is the same group of people who have needs and possess productive capacity, who produce and consume, who spend their earnings on investment and on purchases. In fact, the population which participates in the system is highly segmented. Goods, services and money flow much stronger through the lives of some groups than through the lives of others. In other words, some are affluent, others are underprivileged. In previous chapters we have designated economically powerful groups as belonging to the centre and weaker groups to the periphery.

The reason for this phenomenon is that earnings are not allocated equally to all participants in the economic process, but distributed according to the contribution participants make to production. This again depends on the particular share they have in total capacity. Capacity consists of integrated factors of production, as mentioned in chapter 5, and these factors of production are owned by particular people. Some are capital owners, and receive dividends, others are not; professionals receive higher salaries than unskilled labourers; some operations are more efficient than others, and so on.

Distribution based not on need or equality but on different entitlements leads to the segmentation of the population into groups with varying economic potentials. Obviously those earning higher incomes have more purchasing power and their needs tend to be fulfilled. Those earning lower incomes have less purchasing power, thus greater unfulfilled needs. We have discussed these imbalances in chapter 3, section III. Inevitably the situation arises where those with the greatest unfulfilled needs have the smallest amount of purchasing power, while those with the greatest amount of purchasing power have the smallest amount of unfulfilled needs. This is true, at least, for the most basic needs of a healthy, communal survival. We shall soon come to higher levels of need.

The same segmentation happens on the supply side. Because their basic needs are largely fulfilled, those who receive higher proportions of total earnings can afford to channel more of their share into capacity building, whether in the form of entrepreneurial risk taking, technological advance, capital investment, education and training, health and nutrition, or whatever. This again increases their share of the earnings. Moreover, capacity building has a cumulative effect: technological development and productively invested capital both tend to snowball.

Within a country, capital competes with labour as a factor of production. Rapid technological advance implies that capital increasingly rewards higher skills but utilises less labour. Because of the growing sophistication of the process and the effective utilisation of non-human energy, those who have only their bodily energy to contribute, that is unskilled labourers, receive decreasing relative rewards and may become redundant altogether. Without jobs, they have no income, and without income, they have no purchasing power. So they become irrelevant both in terms of production (supply) and in terms of consumption (demand). We call this phenomenon *marginalisation*. In contrast, the elite controls the greater part of the purchasing power in the system. As a result, it could have the lowest level of unfulfilled need. We shall presently see that this is not necessarily the case.

The same picture repeats itself on the international scene. In recent times technology has advanced at breath-taking speed. Economies which have not entered the technological age, or which cannot keep pace, remain more and more behind.[9] Socialist economies in Eastern Europe, which are industrialised, have largely collapsed under the pressure of high-tech competition - so what chance is there for the poorest of the poor to compete?[10]

The result is that, as capacity grows, its ownership is increasingly concentrated in the hands of a financial elite. In a capitalist economy, capacity owners are able to hold the entire process to ransom.[11] In times of recession, the production process can be slowed down or terminated. If a particular location is not conducive to the optimum utilisation of capacity, segments of the process or the entire operation can be moved to greener pastures, irrespective of repercussions this can have for the less powerful contributors. This manoeuvre may lead to greater overall efficiency in the use of existing capacity for the economy, but unless countervailing processes are institutionalised, the utility of this efficiency will be concentrated on particular sections of the population at the expense of others.

Artificial need creation and luxury consumption

The processes mentioned above work cumulatively, and therefore accelerate, unless they are corrected by countervailing processes.[12] Let us consider the effect this has on the system as a whole by utilising the centre-periphery model which we have developed in chapter 3. There we have defined affluence and poverty in terms of the relation between economic potency and need. Where need exceeds potency (translated into income), there is a *poverty gap;* where potency exceeds need there is an *affluence gap* (see diagram 12-4, which bears repetition here). Both gaps tend to grow.

As stated above, purchasing power is concentrated in the centre population which has the lowest level of unfulfilled needs. When basic needs are fulfilled, the result could be a rising level of contentment. *Contentment based on sufficiency* should be the goal of the entire economic enterprise. But main line economics denies that there can be such a thing: human appetites are alleged to be limitless - and they are entitled to be limitless. We have argued in the last chapter that this contention is based on a distorted concept of human nature, and that the corresponding practice is a case of cultural degeneration. In many cultures contentment is not only considered to be a prime value, but it is actually achieved on much lower levels of consumption than those attained in industrial countries today.

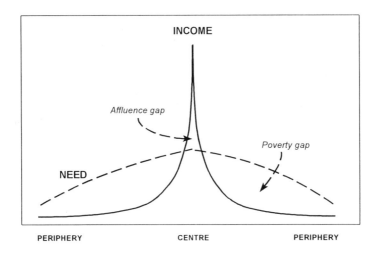

Diagram 12-4
Affluence and
poverty gaps

This is not just a moral issue. It is a *systemic* problem. In an industrial society contentment would imply that less and less purchasing power was utilised to buy consumer goods, demand would drop, production would have to decrease and the economy would go into a recession. If sufficiency was the goal of the exercise, this would hardly be a catastrophe. But in the modern economy it is growth which is the goal, not contentment. This is why contentment cannot be allowed to emerge in a modern economy.[13] To keep the economy going, contentment must be undermined, dissatisfaction must be induced, and consumption must be activated.

It can only be activated where the purchasing power is located. The periphery population has no purchasing power which could translate need into market demand. The productive capacity of the centre is directed, therefore, to the centre population where the purchasing power is located, not towards the periphery where the need is located.[14] The economy could actually stall in spite of unfulfilled need on one side of the economy and underutilised capacity on the other. This anomaly is caused simply by the unequal distribution of capacity and earnings.[15]

In the centre, needs have largely been fulfilled. If the centre wants to continue to produce and sell, it has to push up the need levels of the centre population.[16] This is done in various ways: advertising, aggressive marketing, planned obsolescence (goods are deliberately produced with a short life span), production of prestige articles, idealisation of luxurious life styles, and so on. Marketing has become a highly sophisticated profession and advertising has become a multibillion dollar enterprise attracting the best brains and equipment.[17] On the whole, the flow of goods can only be maintained if products are constantly thrown into the dustbin to make way for others.

One has to add that the state can also push up demand, for instance through military expenditure or space research. It can also raise the level of social services such as free university education. Many governments of affluent societies have caused these societies to live way beyond their means by allowing budget deficits to accumulate.[18]

Any dispassionate observer would have to concede that the current economic system is economically irrational. The productive capacities of the nation and the world are utilised proportionately more and more for the production of luxury consumer goods and services for a minority (or for other items not strictly necessary), while the basic needs of the majority of the population remain unfulfilled. Economically this is counterproductive because the purpose of the economy should be to allocate scarce resources to fulfil the most pressing human needs. The economy is simply grossly inefficient and suboptimal in terms of resource allocation.[19]

The process is also suboptimal in terms of the interests of the centre population itself. The satisfaction of artificially created needs may still serve to improve the quality of life of the centre population up to a certain extent. But human beings have only limited amounts of time, space and energy at their disposal. As the process continues, a situation is reached when goods and services become so plentiful, that they begin to crowd each other out. The result is that they remain largely underutilised or get wasted altogether. One cannot be an active businessman in Johannesburg, enjoy a yacht in Rio de Janeiro and a villa in the Alps all at the same time. Similarly, children of affluent parents tend to be swamped with sport outfits or mechanical toys which are never fully utilised or enjoyed. Here the economic law of declining marginal utility comes into play very forcefully.[20]

The creation of material needs also crowds out other needs. Families no longer go on a hike to enjoy God's lovely creation but speed through the landscape on super-highways. They no longer sing to the guitar around the fire place but stare at the television screen. Children no longer develop their creativity in making toys of their own, but play computer games. Communal and spiritual concerns are neglected because the rat race leaves no time. People overeat and develop bulging waste lines. The oversatisfaction of some needs at the expense of others leads to ill health and unhappiness. Spiritual emptiness leads to more material craving. Especially the youth expects to be on a constant "high". Where there could have been contentment and enjoyment there is a mad rush for ever more stimulation.

But the subjective consequences of imbalances in the satisfaction of centre needs are not the only problem. As the situation deteriorates, the misery of the periphery begins to spill over into the centre: drugs, diseases, terrorism, illegal migration, regional conflicts, and so on.[21] Worse than that, the quality of the environment in the centre deteriorates rapidly and precious resources are squandered.

This kind of imbalance in the economy should annoy pragmatists and antagonise ethically sensitive persons. But mere indignation will do nothing to change the "normal" operation of market forces. The engine which powers this process is *competition*. As Heilbroner put it, competition is "the inescapable exposure of each capitalist to the efforts of other [capitalists] to gain as much as possible of the public's purchasing power."[22] The "public" is, of course, the centre population. Due to the way the economy is organised, production has no choice but to concentrate on artificially created needs rather than respond to the vast unfulfilled needs of the poor, which cannot be translated into market demand and which have become irrelevant to the market mechanism, thus to the economy as such.

The real problem is that we have internalised the principles of the system to such an extent that we consider them to be normal and inevitable. Nothing could be further from the truth.

By the way: recession and inflation

This may be a convenient time to make a few points concerning the business cycle and the effects it has on the affluent and the poor. Whenever production outstrips consumption, producers are forced to reduce their output because they cannot sell what they produce. Workers are laid off, machines run below capacity, no further capital is invested. Workers who lose their jobs also lose their income, and thus their purchasing power. As a result, market demand is reduced further and so the economy spirals downwards. What economists call "surplus capacity" simply indicates that our factories would be able to produce more goods if there was a market for them.

It is a crazy situation: needs grow, while the capacities to satisfy those needs are progressively shut down. The capacities are not used because there is no market and there is no market because the people have no income. This vicious circle is called a *recession*. After a while, business confidence may be restored, capital owners begin to invest, manufacturers employ more people, who buy more goods and the economy again spirals upwards into a *boom*.

This is not the place to discuss the intricacies and possible solutions to the problems of boom and recession. What concerns us here is that a recession hits the poorer sections of the population hardest. They depend on work and wages for their livelihood. True, firms which have relied too heavily on credit facilities, who supplied luxury goods that are no longer wanted, or whose operations were inefficient, may go bankrupt. But great stakeholders in the stock exchange belong to affluent population groups and have reserves to fall back on. Employed people, in contrast, have only their labour to sell. Pensioners who have invested their life savings in unit trusts are also vulnerable. A recession may mean that the income of the lower population groups increasingly falls below the subsistence level and real suffering may be the result.

In a boom, market demand may begin to outstrip supply. Productive capacity may become insufficient to cater for what people want. Where demand is higher than supply, the market mechanism causes prices to rise. Rising prices are just another expression for money losing its value. When prices rise, people want a higher income and workers demand higher wages. This increases the cost of production which again inhibits the growth of production. Again this is a vicious circle, which is called inflation.

We have only described the basic mechanism. Supply and demand are complex entities and therefore inflation can be caused by many factors.[23] The point to be made here is that inflation reallocates assets within the population. The incomes of those who earn dynamic salaries (that is, salaries which are constantly adjusted to rising prices), or those who own productive assets such as factories, rise together with the inflation rate. These are the more privileged. In contrast, the purchasing power of the poorer sections of the population, whose salaries and pensions do not rise at the same rate as the inflation rate, or who live from their savings, declines. Inflation in fact causes a transfer of assets from the poor to the rich.

For a variety of reasons the rate of inflation is usually much higher in peripheral countries than in centre countries. That is one of the many reasons why the discrepancies in income between the rich and the poor are much more marked in the periphery than in the centre. It is also one of the reasons why the discrepancies between rich and poor countries rise.

The way forward

Our reflections so far have unearthed two fundamental and interlinked problems. The first is the imbalance in capacity ownership and control; the second is the imbalance in the amount of purchasing power at the disposal of various population groups. Obviously the latter is the result of the former. It is clear that the playing field of the market is not even. The theoretical solution to this problem would be to transfer productive capacity to the disadvantaged. In terms of the periphery one usually thinks in terms of a greater access to land, raw materials, capital, education and training, and the like.[24]

All this is not wrong, but it is only part of the story. As we have seen, the artificial bloating of consumption in the centre is fundamental to the problem. This is partly due to the "demonstration effect" of high consumption patterns of affluent reference groups, partly to aggressive marketing. If a solution is to be found, I believe we can no longer take it for granted that these two phenomena are "natural" and should be allowed to continue. The ascription of status to conspicuous consumption reveals a distorted value system. Once the more affluent sections of the population become aware of the havoc which their extravagance causes further down in the social system, they might want to become more modest in their spending and resist the ruthlessness with which advertising and marketing agencies undermine the sound sense of the population.

But the more basic problem is how the economy as such can be organised in a more rational way. We should have learnt from the collapse of Marxism-Leninism that the socialisation of the means of production is not the answer. What we need is a more balanced participation of the population in the "free enterprise" system and a reduction of throughput from resource base to waste.[25] For that to happen capacity owners and "saturated need" consumers must realise the senselessness of surplus capacity, artificially boosted luxury demand in the midst of misery, and wastage in the midst of ecological deterioration. We shall take up these issues in Part IV.

Revision: Try to reproduce the diagrams used in this section and explain the economic relationships and mechanisms which they represent.

Application: Imagine the type of economy prevalent in your country half a century ago, where expectations were much lower, fewer gadgets were produced, and less energy was consumed. Were the people, on average, more dissatisfied than they are now? Would it be a catastrophe if everybody, except the absolutely poor, returned to a 1950 standard of living?

> *Critique: (a) "Your argument aims at redistribution and thus leads straight into the trap of communism, a system which has created endless misery to countless people before it collapsed under the weight of its own misconceptions. To argue the way you do is quite irresponsible!" (b) "You cannot build economic theory on a concept which is as vague and undefinable as 'need'. There is no clear distinction between needs and wants, and wants are both subjective and unlimited." (c) "The rich take more than their share and the poor are at their mercy - that is the long and the short of it! Your complicated theories only serve to conceal the issue under a blanket of sophistication, which the poor will not understand and the rich will not buy." (d) "The history of economic systems during the past century has shown that the market system is by far the most efficient, productive and equitable. All interference in the operation of market forces can only lead to greater inequity and misery." How would you respond to these statements?*

Section II: The distortion of the peripheral economy

The revolution of rising expectations

In the last section we have discussed the trend towards luxury production and consumption in the centre. Although advertising and marketing is mainly directed at the centre market, it has profound effects on the need structure in the periphery. The goods manufactured by the centre are also tempting for the peripheral population: radios, cars, freezers, cool drinks, guns, tools, corrugated iron, barbed wire, cigarettes, and so on. Which peasant would not want to exchange a donkey cart for a four-wheel-drive!

But one cannot desire what one does not know exists. The awareness of available commodities is created through the *demonstration effect*. New products are seen in shop windows, on television screens and in the houses of neighbours. The effects of sophisticated advertising campaigns, designed to push up consumption in the centre, spill over into the periphery. Here they have a much greater psychological impact than in the centre because there is such an abyss of unfulfilled needs. People are made to feel that they are missing out on life, if they do not own a television set, drink a certain beverage, or wear the latest fashion. Things which people in traditional cultures never dreamt of gradually become "necessities" and people suffer want if they have to go without them.

So the needs of the peripheral population rise more rapidly than those of in centre. This is not only true for the Third World, but also for the Second World. East Europeans have been led into soaring expectations by Western life styles and adverts. This impatience has become powerful enough to topple communist governments, which the present author does not regret, but now they are saddled with massive economic paralysis and disillusionment. Similarly, black South Africans have been yearning for white living standards for decades and now want what they consider to be their share. That the economy, and thus the new government, cannot provide elitist life styles for the entire population, is not readily accepted by those who have seen the previous government secure such lifestyles for their white voters. One result is a soaring crime rate in both situations. Crime is simply the shortest way of getting what everybody seems to be entitled to.

The new need pattern is largely mediated by the peripheral elite. The peripheral elite consists of those who have the power to take all important decisions in a peripheral society. So they form the *reference group* for the peripheral population there. Their way of life is considered to be "smart" or "normal", that is, life as it is supposed to be. So the lower classes try to copy their behaviour and, especially, their patterns of consumption. This happens all over the world. But in poor societies it has much more far-reaching consequences than in wealthy societies.

Third World elites are, to a considerable extent, cultural and economic offshoots of the centres located in industrialised nations. As such they share the standard of living and the typical consumption patterns of Western societies. The peripheral elite has its own reference group, not in the rank and file of the centre population (the periphery in the centre), but in the centre elite (the centre in the centre). We have discussed this phenomenon in chapter 3, section IV. If you want to be "somebody", you must have an international airport, drive a Mercedes, drink whisky-soda - whether the peripheral society can afford such luxuries or not. So the need pattern transmitted by the peripheral elite to its poor environment is much higher than the need pattern of the average citizen in the centre.

As a result, the poor move away not only from traditional to modern patterns of consumption, but also to inflated patterns of consumption. The need curve (discussed in chapter 3, section III), rises rapidly in the periphery and approaches the level it has in the centre. This phenomenon has often been described as the "revolution of rising expectations". The spell of new products and their status enhancing capacity is so great that poor people are often paralysed by huge debt burdens. They may go hungry, or drop their education, rather than do without a television set or a refrigerator. And all that boomerangs back on their ability to raise an adequate income.

Because income does not rise with needs, the poverty gap widens. People feel deprived and cheated, and their frustration makes them bitter and restless. The leadership in the periphery is accused of failing to deliver the goods. This mood is greatly enhanced by the unpalatable experience of corruption. To protect itself against a restive population and maintain order, the elite has to invest more and more of the scarce resources of the country in security and military expenditure at the expense of infrastructure, productive enterprise and welfare. In short, "the West demonstrates day-by-day to the developing countries the standard of consumption available to those in power. It is this which closes the vicious circles of mismanagement. In order to bring the developing countries back to health, not only is it necessary to correct unfair trade structures but also to wean them from their vision of the North's throwaway society."[26]

The collapse of peripheral production

Sadly enough this is not the end of the story. The needs generated in the periphery by the centre economy are all geared to goods which the peripheral population is not able to produce, because the periphery lacks essential factors of production: capital, technology, sophistication, entrepreneurship and organisation.[27] It simply cannot compete. As a result, the little market demand which the periphery can generate is diverted to centre products. Producers of traditional goods and services in the periphery lose their markets and go out of business. Curios, tourism and tropical fruits may be exceptions, but they cannot keep an entire economy on its feet.

The result is that the productive capacity of the periphery deteriorates and the poverty gap grows. A single factory can throw hundreds of traditional artisans out of work. A fleet of drag net trawlers on Lake Malawi can destroy the livelihood of the entire fishing community.[28] The same is true for the international scene. The argument that international competition is healthy because it forces peripheral enterprises to become more efficient does not hold water when there are no survivors among these enterprises.

This observation throws light on the possible effectiveness of Structural Adjustment Programmes:[29] what may work for the semi-periphery under favourable circumstances will not necessarily work for the outer periphery.[30] Just consider the fact that the poorest 42 countries only account for one third of a percent of world exports![31] There is no economic clout anywhere in the peripheral system.

Meanwhile, the market of the centre expands into the periphery. Whatever purchasing power the periphery is able to generate is spent on the goods supplied by the centre. So while the volume of production in the periphery decreases, the volume of production in the centre increases. Granted, the market in the periphery may not be important for the centre because of its insignificant volume. But for the periphery the drain can be quite devastating. Diagram 12-5 shows how the affluence gap grows together with the poverty gap.[32] The concept of "opulence driven poverty traps" begins to make sense.[33]

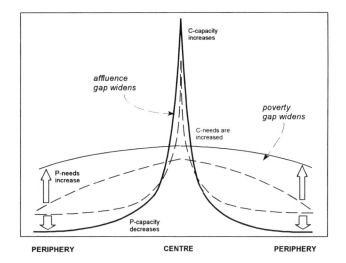

Diagram 12-5
Widening gaps

The scourge of unemployment[34]

According to Gerrard, "the single most important task facing economic theory is to explain the causes of mass unemployment."[35] Is this really so difficult? With the new approach suggested here, I believe we have come close to a basic explanatory model. To satisfy at least some of the needs, whether basic essentials such as food or new needs spilling over from the centre such as radios, the peripheral population has to raise an income. To raise an income, the periphery sells the factors of production it still possesses to the centre: educated brains, raw materials, capital, labour, land. It even sells ecological absorption capacity when it agrees to the establishment of polluting industries or the dumping of toxic waste on its territory.

All this corresponds with what we have described as "suction" in chapter 5. As we have seen, these factors of production are in a "raw" stage and therefore fetch low prices. They include unprocessed materials, undeveloped land, unskilled labour, or skills which lack experience and are not yet integrated into the production process. In comparison, manufactured goods imported from the centre have value added, are often overprocessed[36] and thus tend to be expensive.

Against this background we are able to understand the spectre of unemployment. The argument that the development of labour saving technology and the investment of capital in productive processes create more jobs simply does not hold water. It may be true that product innovation creates more employment opportunities in a sector or a region than process innovation[37] - but at the expense of which other product, sector and region? You may create jobs in factories which build agricultural machinery, but what happens to the rural work force which is displaced by these machines?[38] A burgeoning computer industry may create jobs in the United States, where the hardware and the software are produced, but throw people out of work where the computers are utilised. In other words, the unemployment caused by technological advance is exported. We need to look at the system as a whole!

Let us begin with production. Most discussions on the issue are hamstrung by limited horizons. They take only the nation state or particular sectors of the economy into consideration, not the impact of developments in one section of the total economy on other sections. Most people are also afraid to allow any discomfort to emerge about technological advance. It is the holiest of all cows in the industrial civilisation. But it does not help to put one's head into the sand.

In the centre there is a spiral in which the demand for expertise in a technologically developing economy outstrips the supply and pushes up its price. The result is a high level of remuneration for management, expertise and skills. Skills are again taken over by intellectual capacity. "The revolution that has followed the industrial revolution is not a service revolution but a cerebral one in which value is added not by skilled hands but by skilled minds."[39] A technological elite develops which accumulates both wealth and control, thus economic power. The high cost of this type of labour provokes further developments in labour saving technology, and thus leads to more and more capital-intensive methods of production. In this spiral both capital and labour become progressively more productive, but fewer employees are needed in the process. Those with lesser skills and education increasingly become redundant.[40]

In a situation of rapidly expanding markets, which allows for rapid economic growth, this mechanism is concealed. During the sixties, for instance, there were serious labour shortages in Central Europe because of flourishing internal markets and undersupplied Third World markets. But expanding markets depend on the growth of both needs and purchasing power. As the limits of needs in the centre economy and the limits of purchasing power in the periphery are approached, the market becomes saturated[41] and the economy stagnates. At the same time technological innovation and the concomitant rise in productivity of capital and labour continue to forge ahead. So redundancy can no longer be avoided, even in the centre.[42] Most OECD countries now face serious unemployment problems.[43]

Now let us look at the *periphery*. Because of rising levels of productivity and quality in the centre, caused by relentless technological advance, the peripheral economy cannot compete with the centre. As we have seen, the traditional economy deteriorates, ultimately collapses and sheds its work force in the process. As unemployment increases, there is only one way left for the peripheral population to raise an income: it has to sell its factors of production, in this case its labour, to the centre. The march to the city begins.[44] When a recession hits the centre, however, the rural areas have to function as a safe haven, which reabsorbs migrant workers, without much compensation.[45]

Well, why not? Is a greater and more sophisticated industrial economy not preferable to the small scale, inefficient traditional economy? It could be, if the greater economy was owned by all stakeholders on an equal basis. It could be, if the peripheral population was at least on par with the lower rungs in the centre. The fact is that the industrial economy has no place for the avalanche of work seekers. Whatever their traditional expertise, peripheral workers who drift into the cities are "unskilled" in terms of the industrial mode of production. They cannot compete with the entrenched, highly trained and experienced work force there.[46] If there is a growing supply and a dwindling demand for unskilled labour, the relative price of such labour stagnates or declines. That is how "cheap labour" comes about.

Unskilled labour has no economic clout because it can easily be replaced - if it is utilised at all. Under such circumstances even the power of the trade unions declines.[47] Only unions whose members have become indispensable have power. They push up their demands for higher wages and better working conditions. Labour becomes even more expensive and is, again, partially replaced by capital. So the effect of their action is the emergence of a "worker aristocracy". The result is still higher unemployment.

"Cheap labour", the red rag of the worker movement, is not the worst consequence of this process. More and more people find it difficult to find any job whatsoever, whether permanent or casual, whether earning a living wage or a starvation wage. As Joan Robinson, a prominent economist, has said, "The misery to be exploited by capitalists is nothing in comparison to the misery of not being exploited at all." These people become redundant in the production process, thus as producers, and irrelevant as consumers because they have no income. They eke out a living on the fringes of society in what is euphemistically called the "informal sector". Most of the time the informal sector is nothing but a parasite on the formal sector. What would happen if tourists would not buy African "arts and crafts"! This process is called *marginalisation.*

Marginalisation does not happen all at once. In industrialised countries technology first sucked the work force out of *agriculture* into *mines* and *manufacturing*. The gap in agriculture was filled by technology. Then technology pushed much of the work force into the *services* sector.[48] In the initial stages this may be an enrichment for the society, assuming that the society can afford it. But the new wave of computer technology pushes the work force out of services as well.[49]

Where must they go? They may swell the ranks of the *bureaucracy.* But bureaucracies are not very productive. Bureaucratic procedures seem to become ever more cumbersome and time consuming, and the hidden rationale behind this may be to keep state employees occupied. Studies in the US have shown that bureaucracies can continue to operate even after they have long ceased to serve any purpose at all.

If a society can afford this, why not? It may be one way of spreading purchasing power. But poorer countries cannot afford a burgeoning bureaucracy. Even in rich societies it is a waste of resources and human potential. Much of the wealth of rich countries, shown in their exorbitant GNP per capita figures is in fact wasted on unnecessarily cumbersome and often useless bureaucratic processes.

Another outlet for both surplus capital and labour has been the *armaments* industry.[50] But apart from the problematic nature of this kind of product the arms industry has become highly capital-intensive and presents no respite for the marginalised. Where all these capacities cannot absorb redundant labour, the result is unemployment. In short, technology chases labour from one sector to the other until it has no place to go.[51]

In contrast, producers grow from strength to strength. With technological advance, capital has become more and more mobile and flexible. As we have seen in chapter 5, parts of the production process can be taken out of the stream, transported over long distances and located elsewhere in the world where labour is cheap, labour legislation is lax and regulations against pollution are less severe than in the centre.

This may give the impression that peripheral labour is suddenly able to compete with centre labour. But this is a deceptive impression. While centre workers may lose their jobs, only the cream of the work force in the periphery is utilised, and even that only for particular aspects of the production process. Peripheral workers are also badly exploited.[52] When they begin to make greater demands, the whole operation can be moved to yet cheaper labour pools. As a result, labour standards may decline in the First World without raising standards in the Third World. As Daly sarcastically remarked, "developed country capitalists have generously offered to share the wages of the working classes of their countries with the poor of the world."[53]

Alternatively, the new wave of technology may make it more profitable for firms to move their operations back to the centre - this time without increasing the number of jobs there. This new wave consists of flexible automation which allows producers to change rapidly from one product to another to capture small scale, unstable and quickly changing markets.[54] The tedious assembly work in electronics, so far done by female workers in low labour cost countries, is now being automated. It is reported, for instance, that a Swiss watch maker dropped 220 workers in Singapore, moved back to Switzerland and employed 50 workers instead.[55] Depending on the industry, one robot can replace between one and six workers. This means that 200 000 robots could cause the destruction of up to a million jobs. And even the programming of software for the computers driving the robots is now being automated.[56]

Once we have understood this basic mechanism we understand why marginalisation escalates together with the other two processes, growth of the industrial economy and growth of the population. The more sophisticated and productive the industrial economy becomes, the less use it has for unskilled labour. To train the entire work force is costly. Moreover, new forms of production cannot utilise a great number of workers, even if they were skilled. At the same time peripheral population growth swells the numbers of unskilled entrants to an already saturated labour market. As we have seen, poverty enhances population growth. Large families again cannot afford proper education, training, balanced rations, and other prerequisites of upward mobility. And so the vicious spiral continues.

After having looked at the basic mechanism, we are not surprised to find that unemployment grows in the periphery first, but then also begins to encroach upon the centre.[57] Certainly the old cry for economic growth, so that there might be employment, begins to sound hollow. Technological advance is a marvellous achievement of humankind. It could be instrumental in eradicating physical want from the globe. In fact, the opposite happens. But to recognise this, one has to develop a new analytical paradigm. More particularly, one's perspectives must include the population as a whole and not just an elite, the whole economy and not just a sector, the world as a whole and not just a few industrialised nations.

Underdevelopment and overdevelopment

It has long become clear that the problem is not merely the segmentation of purchasing power within an economy, but the interaction between two or more relatively autonomous economies, whether we think in global, regional, national or sectoral terms. In one section of the economy capacity exceeds need, and thus potential supply exceeds demand. Here the constraint to the economy is, most paradoxically, a *lack of need*. This could be paradise, but the economy is not organised around the aim of bringing about satisfaction. Technological advance, thus productivity, continues unabated. To avoid a recession, the modern economy has only one way out, namely to increase need artificially.

In the other section, need exceeds capacity, thus production, thus income, thus purchasing power. Need cannot translate into demand, demand cannot evoke supply, thus production. Here the constraint is a *lack of capacity*.

The curious thing is that we find a lot of *surplus capacity* in peripheral countries as well. Why should this be the case? The flow from capacity to need is blocked, (a) partly because capacity is paralysed by lack of spare parts, energy, expertise, and organisation, (b) partly because of constraints in the catalysts which enhance the flow, such as communications, transport, credit facilities, and (c) partly because there is no demand. Demand cannot materialise because there is no purchasing power; purchasing power cannot grow, because such growth depends on more earnings being channelled to purchasing power; this cannot happen when capacity is not fully utilised, or when earnings flow elsewhere. Earnings that could build up local purchasing power, can indeed flow elsewhere: to a corrupt and squandering state, to foreign investors, or out of the country through corrupt local elites. Thus, in the periphery, we find *a vicious circle of systemic and chronic recession*.

How can this circle be broken? Economists often argue that the local market is simply too small to generate sufficient demand and suggest that only a policy of production for export can get the economy going. This would bring capital into the country, which would lead to growth, which would again lead to employment, thus to purchasing power. The experience of the "young tigers" in South East Asia seems to bear them out. We shall presently see, however, that this path is not feasible for the bulk of Third World societies. So the needs of masses of peripheral population groups would remain unfulfilled. It would also leave the potential market out of consideration which this need could represent, if only there was the purchasing power to make it effective. Apart from inefficiencies in production and distribution, the real problem is how one can get purchasing power into this potential market and how one can make existing capacity respond to this market.

On the basis of these reflections we are able to define two important concepts precisely:
- **Underdevelopment** is a situation in which production is unable to react positively to rising needs, or drops below its former levels.
- **Overdevelopment** is a situation in which productive capacity outstrips need to such an extent that the latter has to be raised artificially by the application of sophisticated marketing techniques to prevent a recession.

It should be clear from our analysis that underdevelopment and overdevelopment are two sides of the same coin and that, to a large extent, the existence of the former is the result of the existence of the latter. Of course, this is only part of the truth. Our analyses in chapter 5 have shown that there are not only *interaction* factors, but also *inherent* factors within the periphery which cause poverty, and corresponding factors in the centre which cause affluence. But the key to both problems is potency, that is power - or the lack of it.

The way forward

The first problem lies in the segmentation of *purchasing power*. To counteract this tendency, the state (or international agreements) could apply equalisation policies. Progressive taxation can siphon off some of the surplus purchasing power of the centre and use it for welfare projects or public goods and services which benefit the periphery. There could be a "negative tax". This means that, while above a certain income level taxation is progressive, below that level households are given subsidies on a progressive scale.

But handouts are never a good solution. Peripheral groups must be empowered to make greater contributions to the process of production. The task is to spread productive capacity more equally. This problem is much more difficult to solve. The bottom line is a policy aimed at securing equality of opportunity. This includes the promotion of education and training, a greater spread of capital ownership, access to resources, access to credit for small businesses, and employment generation. Internationally, industrialised countries should at least lift trade restrictions against less developed countries.

More important is another consideration. Imbalances in the economy are largely the result of imbalances in factor costs. In the modern economy capital is cheaper than labour as a factor of production. Energy derived from fossil fuels is cheaper than body energy of unskilled labourers and animals. There are many reasons for this phenomenon. Oil producing countries exploit their reserves to gain wealth which leads to a glut on the energy market. In welfare states workers have gained entrenched rights which lead to "market rigidities", such as minimum salaries, unemployment benefits, and pension schemes. Apart from that, machines are perfect slaves: they do not come late, they do not get sick, they do not talk back, they do not go on strike.

Some states grant tax concessions and subsidies to new industries to enable them to become competitive. Research and development aimed at labour saving technology and labour saving investments can be deducted from taxable income. Apartheid South Africa granted substantial subsidies to step up the mechanisation of agriculture with the aim of removing black workers from white farms. Currently dividends from capital investments are tax free to encourage savings and capital formation. Developing countries often make tax concessions to attract international capital. Obviously, the shortfall has to be paid by their own populations.

Without doubt there are good reasons for some of these measures. The point is, however, that it is possible and common for the state to manipulate the factor market. The question is whether this is done on behalf of capital, or on behalf of labour and small business. It is time to realise that, in an era of growing unemployment, the support of labour saving technology has lost its purpose. *The rationale of capital is to help people produce, not to throw them out of the productive process.* At least until the balance is restored, tax exemptions should be granted for labour-intensive methods of production, while the utilisation of labour saving technology, driven by fossil fuels, should be taxed. One can even think of subsidising the employment of targeted groups, for instance, the unemployed youth.

The second problem lies in the overshoot in material need fulfilment. In the centre the market is potentially saturated; the practice of raising wants artificially to get rid of surplus production leads to a wastage of resources, pollution of the environment and a deteriorating quality of life. Ecological deterioration may create new needs, such as the need for clean air, clean water, less noise, which may prompt new waves of production, but this is a vicious spiral, not a solution. The only way out of this dilemma seems to be that, in the centre, capacity should zero in on real need and call it a day. All other energies should be channelled into the enhancement of quality, not only of material output, but of social, cultural and spiritual achievements. That this would imply a reorientation of the entire economic system is obvious. But is there a workable alternative?

As stated above, such a moratorium on surplus production could free up the capital needed for development where it is really needed, that is the enhancement of the quality of life in the periphery and the development of a clean type of technology. Where there is an abyss of unfulfilled material need, capacity certainly has to grow. But it must stay within ecologically sustainable limits, otherwise it does not present a lasting solution. Sustainability is hardly attainable when a rapidly growing population also craves inflated First World standards of living. In the periphery, therefore, population growth must subside and people must aim at a healthy but modest level of consumption. Again fulfilment should be sought in social, cultural and spiritual achievements, rather than material abundance.

Whether the implication would be some sort of "steady-state economy" is a contentious issue.[58] The concept itself smacks of stagnation and may evoke emotions of antagonism in a spiritual climate which believes in unlimited progress. The balanced economy of the future should certainly not be understood as static or frozen, but as a dynamic equilibrium. The moratorium is on material greed, inequity, overexploitation, wastage and pollution, not on the unfolding of human potentials, including science, technology, community, sport, creative arts and spirituality. Life in all its richness will not be fettered but liberated and empowered in such a new dispensation. We shall continue with these deliberations in Part IV.

Revision: What happens to needs and potency in both the centre and the periphery when the two entities interact?

Application: Apart from apartheid, what reasons could have led to the impoverishment of South African black rural areas?

Critique: (a) "That the rich are unwilling to share their illegitimately acquired wealth with the poor, whom they shamelessly exploited, is unforgivable. Why do you cover up their sins with sophisticated economic arguments?" (b) "Fact is that Western initiatives have created more jobs in the poor areas of the world than there have ever been there before. If it was not for the irresponsible growth of the population, the people could be reasonably well-off by now." (c) "If you

> want to enjoy the fruits of the technological age, you have to become part of technological production. Nobody who is willing to work and use modern techniques is forced into marginalisation." Comment.

Section III: The case of the "Asian Tigers"[59]

In recent times many economists have become convinced that the road to development has been shown by the success stories of the "young tigers" in South East Asia, namely Hong Kong, Singapore, Taiwan and South Korea.[60] They also point to the fact that recently such success stories have been replicated in parts of India, Indonesia, Malaysia, China, Brazil and Mexico. The World Bank and the International Monetary Fund have formulated directives for Structural Adjustment Programmes based on these experiences.[61] There are even hopes that this new dynamic may help to solve the economic problems of the West.[62]

All this seems to contradict our argument thus far: the most successful developing economies are those who have chosen not to isolate themselves, but to open themselves up to world markets; who have not protected their fledgling industries, but exposed them to international competition; who have not been hostile to international capital, but welcomed it with open arms; who have concentrated not on satisfying internal markets, but on exports; who have concentrated on growth, not on equality. Together with the collapse of Marxism-Leninism in Eastern Europe, and the dismal failure of socialist policies all over the Third World, these success stories have shown that socialist isolation cannot deliver the goods, while capitalist integration can. Or have they really?

To gain clarity on these issues we need to look carefully at the circumstances which have led to these achievements. In the case of Taiwan, which is particularly revealing, these circumstances include: the early development of agriculture under Japanese colonialism; the fostering of networks of local organisations in rural areas and policies aimed at increasing rural incomes; the development of domestic markets;[63] the immigration of the Chinese elite from the mainland after the revolution; a successful land reform; the strong involvement of the United States during the Cold War; an efficient government relatively free from corruption; strong and pragmatic state intervention; a change from import substitution to export orientated industrialisation at the right point in time; utilisation of cheap labour; penetration of First World markets at a time of global boom; a mentality conducive to economic progress, and so on. One should also not forget that these countries have large populations compared to their land area. So they cannot build on their own resources but must export to survive. If they want to export, they have to compete; they have no choice.[64]

What can we learn from these facts? The first observation is that Taiwan did *not apply classical free enterprise capitalism*. The economy was carefully and pragmatically guided by a strong government. Nor was export-orientated industrialisation the only factor which led to Taiwan's success. Its industrial development followed on a prior spate of agricultural development which placed the bulk of the population on a sound economic basis. A successful programme of land reform was applied, which initiated a tradition of egalitarian policies. Taiwan only switched from import substitution to export when its markets began to be too small for its output. Industrial development was spread across the country so that many people could supplement their agricultural incomes by commuting, rather than migrating to the cities.

So Taiwan's success corroborates rather than questions the contention made in this chapter, namely that the problem of over- and underdevelopment emerges because imbalances get out of hand. Where equalising policies are not applied, for instance in Brazil and recently in China,[65] the problem repeats itself: accumulation of wealth by a few and growing relative poverty of the many.

The second observation is that a *rare combination of historic circumstances and policies* made it possible for Taiwan to grow into a semi-industrial country. Because the combination of such circumstances is fairly unique, it is unlikely that a similar feat can be repeated by all the countries of the Third World.[66] As the recent meltdown of Asian economies showed, globalisation, export-orientation and dependency on Triad markets can also lead an economy into serious trouble.[67]

In the third place, the entire approach builds on the assumption of *cheap labour and environmental laxity*. Both welfare and ecological standards must be kept at low levels if the competitive game is to work. Whether in the industrialised countries or in the young tigers, it is not the capitalist class which bears the cost of low standards, but the lower classes.[68] Once the workers claim higher wages and privileges, once the population asks for environmental safeguards, the show is over. As mentioned above, the advantage of low labour costs is also disappearing more and more because of microelectronics and robotics. Enterprises now need to be near to the markets to be able to react flexibly on rapidly changing consumer tastes.[69] Automation thus again makes it more profitable for entrepreneurs to locate industries in the centre, though centre workers will not necessarily benefit from that.

Once low labour costs cease to be an attraction, the assumption that all countries must have sufficient comparative advantages somewhere up their sleeves is rather theoretical.[70] Unfortunately there are economies which are poor in every conceivable respect and there is no reason why a poor country such as Tanzania must have anything to offer to a rich country such as the United States. In a highly unbalanced situation the principle of comparative advantage does not work according to the text books.[71]

The fourth observation is that the success of the Asian tigers depended entirely on *the willingness and the capacity of the giant markets of the industrial Triad to absorb their industrial products.*[72] Basically, four small countries were clever enough to find nooks and niches in the Triad system where they could comfortably settle in without upsetting dominant interest groups too much. But as the economic power of these young tigers expands, and as others are keen to join the club, this is bound to make an increasing impact on the system:

(a) If all Third World countries wanted to engage in export-led industrialisation, world markets for manufactures would soon become *saturated,* leading to global economic decline.[73] There are just not enough international markets around to absorb the products.[74] And if the roots of growing peripheral economies would indeed penetrate the cracks in the rock of the centre and break it apart, the result would not be balanced global prosperity.

(b) Due to their low labour, welfare and ecological costs, newly industrialising countries are *undercutting* First World prices to capture markets from producers located in "high wage sick countries".[75] This implies a loss of First World income.[76] To avoid becoming uncompetitive, centre producers try to avoid labour costs through increased rationalisation, mechanisation and automation. Many pack up and move to low-cost locations. This constitutes a drain on First World capital. The losers are the workers in the industrial economies. They either forfeit their high incomes and welfare privileges, or lose their jobs. In both cases consumer demand declines, which again leads to recessionary pressures. This again undermines the whole project of exploiting First World markets.

(c) It is also not very likely that centre states would allow all this to happen without resorting to *countervailing measures.*[77] The governments of these countries will lose revenue and get into trouble with their high budget deficits. "In order to balance its external accounts, the United States needs to increase its exports or reduce its imports by about $ 159 billion a year. It needs another $ 50-billion-a-year surplus for ten years to pay interest on its accumulated foreign debts."[78] To reduce imports and increase exports, it has to follow the same recipe as the young tigers - thus smothering all chances for the Third World to get into the international market on a large scale.

It is true that great Third World countries represent vast potential markets of their own which could be developed. There must be considerable slack in their local markets which could be picked up to get the process going. But local production for local needs is not in line with the tiger model. "Would it not make more sense for the Third World countries to transform their own resources into products needed by their own people, rather than export them to the North in exchange for consumer goods for Southern elites?"[79]

The fifth observation is that the young tigers *no longer belong to the periphery;* at best they belong to the semi-periphery. As the economic power of these economies grows, they begin to play the role of centre powers against peripheries elsewhere in the global system.[80] And true to the capitalist mentality, they do so without scruples.[81] Production in the outer periphery is not enhanced when production is enhanced in the semi-periphery.[82] Rather, the entire exercise seems to be another rehearsal of the old drama of growing discrepancies, as described above, not a novel solution to the global problems of economic discrepancies and ecological imbalances.

The counter argument is that the emergence of a powerful economic dynamic in a particular region does not deprive its economic environment but provides powerful stimuli for the development of the entire region. As an economic centre moves to higher levels of value added, it vacates positions lower down which can be occupied by the next layer. Japan, for instance, should not produce rice but computer chips, because there are sufficient other nations in the vicinity which can provide rice. Japan as a whole could become the white collar nation of Asia because there are millions of blue collar workers around. But this argument simply amplifies our observations on centre-periphery mechanisms. Looking at the entire system, the beneficiaries stand out in staggered formations, but who are the victims?

The last observation is that a full industrialisation of the Third World is hardly feasible because of *resource depletion and ecological constraints.* Again, the increased flow from resources to waste can happen either at the expense of weaker and more vulnerable groups - say in Western China and the Philippines - or at the expense of the environment, or both.

At the time of writing the Pacific Rim countries are in severe trouble and the argument has lost much of its clout. The point is that the young tigers are not an example which could or should be followed by the rest of the Third

World.[83] But these remarks are not meant to imply that nothing can be learnt from the young tiger experience. It is indeed possible, for instance, to find more nooks and niches in the global market, to develop new products and services, and to build on the comparative advantage of a particular country. But unless total purchasing power expands and reaches the peripheral masses, every new offer of this kind competes for a place in the sun with established products on a market circumscribed by the wants and the purchasing power of the global elites. And the outer periphery is least able to compete.[84]

Revision: To what extent has the Taiwanese experiment shown a new way to prosperity for the Third World as a whole?

Application: Is there anything your own country can learn from the experiences of Taiwan and similar countries in East Asia? Which of their policies would not be applicable to your country?

Critique: (a) "You can afford to paint the Asian Tigers as a paradise because you do not live under that ruthless capitalist and oppressive regime!" (b) "The Asian Tigers have shown the superiority of capitalist freedom over socialist slavery; those who do not follow their example have only themselves to blame for their misery." How would you react to these statements?

Let us summarise

In this chapter we experimented with a new point of departure in economic theory, namely *need* and *capacity,* rather than supply and demand. In section I we saw that economic activity is a circular flow. Goods and services emanate from productive capacity and satisfy need via production and consumption. Money flows in the other direction. Production and consumption are regulated by the market mechanism which balances out supply and demand. Both of these again depend on money power: supply depends on potential earnings, demand on purchasing power. We have seen that this circular flow is more powerful in the centre than in the periphery.

The basic needs of the centre population are largely satisfied, while the productive capacity of the centre constantly grows. The peripheral population has vast unfulfilled needs, but it lacks the purchasing power to buy greater quantities of centre products. To sell their output, centre producers have to *raise the need levels of the centre population* through advertising and aggressive marketing. This again has an effect on the need structure in the periphery because people begin to desire the commodities produced by the centre.

In section II we explored the consequences of these developments for the peripheral economy. When the meagre income of the periphery is spent on centre products, *peripheral producers lose their markets.* Both the increase of needs and the decrease of productive potential increase the poverty gap in the periphery. This mechanism again induces the periphery to sell its raw factors of production to the centre to obtain an income. But technological advance makes the centre less and less dependent on the factors of production offered by the periphery. One of the more disconcerting results of these developments is the growth of unemployment, particularly in the periphery, but recently also in the centre. On the strength of this analysis we defined *underdevelopment* as the inability to respond to an increase in needs with an increase in production, and *overdevelopment* as the necessity to raise needs artificially to dispose of surplus production.

In section III we considered the counter-argument which builds on the success stories of the "young tigers" in South East Asia. Looking at the *case of Taiwan,* we saw that part of its success was its commitment to greater equality within the economy; that the historical circumstances which made its success possible, were unique and can hardly be repeated; that the bulk of the Third World could not follow their recipe without destroying the centre markets on which it depends; that those who rise to centre status also act as centres in relation to the rest of the periphery, and that the whole exercise is based on the assumption of cheap labour and ecological laxity. Far from dismantling our argument, therefore, the case of Taiwan only reinforces the demand for greater balance in the world economy. The following chapter will increase its urgency even more.

Notes

[1] For a discussion of the change from need to wants or utility see Simons 1996:25ff.
[2] "Economics is the study of how scarce resources are allocated among competing ends." Rufin & Gregory 1983:24f. Cf Samuelson & Nordhaus 1989:25.
[3] Of course, economists do not say that they are irrelevant: "Scarcity is the consequence of the mismatch between (our unlimited) wants and the ability of the economy to meet these wants." (Rufin & Gregory 1983:57). It is also assumed that wants are implied in the demand function, while capacity is implied in the supply function. The point to be made, however, is that if these two functions are not analysed further into their constituents (as we do here), the foundational importance of both needs and capacity, and thus the critical role of differences in economic power, are lost in the argument.
[4] Samuelson & Nordhaus 1989, the "classical" textbook in economics, does not have the term "need" in its index. It is used on p 26 in inverted commas and mentioned in the same breath with "wants". Rufin & Gregory say that "the law of demand shows that the

everyday concept of *need* is not a very useful concept in economics." The reason is that it implies an "absolute necessity for something" (1983:58). Well, certainly there are absolute necessities in life, are there not? But the definition of the concept is also unnecessarily restrictive. The fact that I could use a piece of wire to tie the exhaust pipe to the chassis of my car does not imply that there is no need for the correct type of nut and bolt. Instead, economists use the concept "wants" which Rufin & Gregory define as "the goods and services that consumers would claim if they were given away free." They also say that "Economics is based upon the principle of *unlimited wants.*" (1983:57). The fact that "collectively, we all want more than the economy can provide" is simply taken for granted and no distinction is made between, say, the need of a homeless person for fire wood on a winter night, and the want of a playboy for a Porsche. One cannot help to remark that, in a textbook which is otherwise exemplary in offering precise definitions, this is shoddy reasoning at best and unconscious deception at worst.

[5] See Nürnberger 1988:86-94 for some detail. Cf. chapter 3 section III of this book.
[6] Rufin & Gregory 1983:36ff; Samuelson & Nordhaus 1989:55ff.
[7] See the diagram in Samuelson & Nordhaus 1989:42.
[8] Junne in Caporaso 1987:120.
[9] "Many of these (developing countries) are mere consumers of the technological revolution, inasmuch as they mainly purchase military hardware and consumer goods for their small middle classes. Most are simply bypassed by the process of technological change ... although they also suffer the consequences of the techno-economic restructuring of the world system through the relative downgrading of their competitive capacities." Practically the whole of Sub-Saharan Africa and many Latin American countries are in this position (Castells & Tyson in Purcell 1989:22).
[10] "It will become increasingly difficult for countries that fail to reach this level to obtain the hard currency necessary to import the capital goods and know-how critical to their development process." The large majority do not have this potential, so they "seem destined to lag increasingly behind the OECD countries and their new economic partners, the NICs." (Castells & Tyson in Purcell 1989:21).
[11] This is not based on economic necessity; it is due to the Western capitalist culture. In the highly successful Japanese economy, for instance, one finds greater community awareness: life time employment, a wage system based on seniority, enterprise unions, and so on. "It is primarily a community of people, rather than a piece of property belonging to the shareholders" (*United Nations Development Programme* 1993:37f).
[12] See Meeks 1989:173ff on consumerism, going back to Veblen.
[13] "Business interests dictate that they supply the maximum amount of goods and services for the longest possible time and the highest possible profit." Goods and services must be promoted and consumption encouraged. "On promotion depends production; on production depends employment, and on employment depends the capacity to consume. Thus the economic system is a closed circle." Goldring et al 1993:4. I would rather say that it is a spiral, not a circle.
[14] Galbraith spoke of producer sovereignty, rather than consumer sovereignty, which is an overstatement of a valid contention (opposed by Scott Maynes 1976:263ff.)
[15] It is not a law of nature that this should happen. Post-war Germany operated according to a social contract by which gains made from rises in productivity were used to build up mass purchasing power, to which production could again respond with higher levels of production. The result was the German economic miracle (Fröbel et al 1981:11).
[16] For detail see Galbraith 1958; Tawney 1948; Hirschmann 1977.
[17] Packard 1985.
[18] In 1993 the budget deficit in Germany was 3.3% of GDP, 3.4% in the US, 5.8% in France, 7.7% in the UK and 9.6% in Italy (Froyen 1996:447). "During the 1980s the United States spent much more than it produced through a series of mounting budget deficits, constraining future economic growth and the choices of future generations in order to maintain a high level of present consumption" (Fukuyama 1992:123).
[19] Attitudes towards luxuries have changed considerably during the history of economic thought. On the one hand luxuries led to specialisation, the development of arts and

crafts, the opening of markets, which ultimately benefited society as a whole. Since Adam Smith economists usually thought they represented a diversion of capital and labour from social necessities and that they impeded progress. Keynes (in his *General theory of employment, interest and money* 1935-6) argued that luxuries were desirable if they produced jobs which would otherwise not be available. Another argument was that capitalism might be outrunning its investment possibilities; thus any investment possibility is good. (*Encyclopaedia Britannica* 1986 vol 17, pp 966f.) Looking at the facts we now have to return to the classical position.

[20] In economics this law is usually applied to single consumer goods only, but it is certainly also valid for the total volume of possible consumption.

[21] *United Nations Development Programme* 1993:7.

[22] Heilbroner 1985:56.

[23] Here are a few possibilities: In chapter 5 we have mentioned the rising costs of imported machinery, or "capital goods". Another cause of inflation in peripheral countries is *overspending by the state*. Because there are so many unfulfilled needs in a peripheral country and people expect more from the state than its financial resources allow, Third World governments generally spend more than the country can afford. Again demand is higher than supply. Paradoxically governments of rich countries, such as the United States, also have vast budget deficits, but this is not our topic at present. Third World regimes also want to remain popular by creating jobs for their supporters in unproductive bureaucracies. Governments have the possibility to make up the shortfall between their incomes and their expenditures by *printing more money*. But if that happens the same value is just represented by more pieces of paper and each of the papers has less value. If the inflation rate is 20%, a ten Rand note is still called a ten Rand note a year later, but it is worth only 8 Rand, and 6.40 Rand the year after. In some Latin American and Near Eastern countries this practice, amongst other causes, has often led to galloping inflation.

[24] There must be "reasonable distribution of productive assets (particularly land) so that people do not come to the market with totally unequal buying or selling power"; credit facilities for the poor; markets which are open to all; sufficient physical infrastructure; information; a liberal trade regime; "a legal system that encourages open and transparent transactions..." and so on (United Nations Development Programme 1993:4).

[25] Ed Yardeni believes that there is nothing wrong with the free enterprise system as such, only with the corruption of those who become rich and do not plough back their gains into worker incomes to build up purchasing power (quoted by Mike Bygrave in *Mail & Guardian,* July 31:21).

[26] Von Weizsäcker 1994:101.

[27] "Developing countries are shut off from technology that would allow them to produce even the most basic manufactured goods. There is nothing to sell except what they have in abundance: raw materials and primary commodities" (Purcell 1989:2f).

[28] Cf Ferguson A E, Derman B, Mkandawire R M 1993: The new development rhetoric and Lake Malawi. *Africa* 63/1993 1-18.

[29] "Due to prevailing uncertainty and disorganization, ambitious restructuring programmes are thwarted from the start by the actual inability of their target beneficiaries to clearly appropriate the strategies" (Scholtes 1993:49). For an overview of the Structureal Adjustment Programmes see Mosley in Fontaine 1992:27ff.

[30] "The liberalizing brand of structural adjustment fashionable in the 1980s appears to have more relevance to richer than to poorer LDCs and least of all to those that are in absolute decline" (Mosley in Fontaine 1993:43f).

[31] Tussie & Glover 1993:240. "The magic of the market-place does not work for those who cannot enter it, namely the rural poor in developing countries" (S Nanjundan 1993: The re-emphasis on small enterprises: a review article. *Industry and Development* 33/1993:112f).

[32] With the global division of labour fully installed, "some will specialise in profits, the others in losses" (Eduardo Galeano, quoted by von Weizsäcker 1990:117).

[33] Van Marrewijk & Verbeek 1993.

[34] The following section is adapted from my pamphlet with this name, Nürnberger

1990:14-17.
[35] Gerrard 1989:10. Classical theory explains unemployment in terms of the theory of market equilibrium either as voluntary or frictional unemployment. Keynes explained it as deficiency in aggregate demand (Gerrard 132ff). Our analysis is akin to this approach but goes much further.
[36] Just take any modern gadget and subtract all non-essential features and you will get considerable savings.
[37] Cf Katsoulacos 1986, Goddard and Thwaites 1986:107ff.
[38] Subsidised mechanisation of white agriculture in apartheid South Africa has displaced over a million black people before 1980. 90% of farm workers were engaged in physical work, only 2,2% in mechanical or managerial work (Nürnberger 1988:123).
[39] Liston & Reeves 1988:265 quoting from the Economist, Aug 22, 1987.
[40] For the inverse correlation between labour productivity and employment cf National Productivity Institute 1988: *Productivity Focus 1988 edition*. Pretoria: NPI, p 14.
[41] It is curious how economists can argue that there can be no market saturation because human appetites are unlimited. This argument is wrong on two counts: human needs and wants in toto are not unlimited, and purchasing power is not unlimited.
[42] Caporasi 1987:199ff.
[43] In four years, 1990-1993, the number of unemployed in OECD countries rose from 26 to 36 million, 19 of which in the European Union (Van Bergen 1993:97).
[44] For rural-urban migration and unemployment cf Thirlwall 1994:102.
[45] For an analysis of the South African migrant labour system, which utilises rural labour but leaves the social cost to the rural areas concerned, see Nürnberger 1988:24ff.
[46] "The basic proposition ... is that capital-using technological progress, which increases the marginal productivity of capital more than that of labor, increases the demand for capital relative to labor and so tends to raise the growth rate of the capital-labor ratio." Abramovitz 1993:221f.
[47] Cf Junne in Caporaso 1987:75.
[48] According to Bade employment in the West Germany in 1982 was: 64% in services, 30% in production, 6% in agriculture. In 1939, in contrast, agriculture accounted for roughly 28% and services for about 34% (1986 199f). According to Liston & Reeves (1988:11f.) agriculture, forestry and fishing in the United Kingdom accounted for only 1.8% of GDP in 1986. Cf Abramovitz 1993:234.
[49] "Present trends suggest that knowledge inputs may be displacing capital, land, and labor as the primary defining feature of the production process in advanced industrial capitalism" (Mytelka in Caporaso 1987:43).
[50] For detail see Tinbergen 1987.
[51] Jones 1982 1ff.
[52] Fröbel et al 1981.
[53] Herman Daly in *Population and Environment* 15/1993:66f.
[54] Junne in Caporaso 1987:71ff.
[55] Junne ibid:81.
[56] Junne ibid:82f.
[57] Unemployment in the European member countries of the OECD increased threefold, from 3% in the mid-1970s to about 10% in 1992. In the developing countries it is much worse. According to the United Nations there are currently 700 million unemployed in the Third World and the figure is expected to rise by another 750 million within 20 years (Van Bergen opus cit 1993:68). "In Sub-Saharan Africa, not a single country had single-digit unemployment figures throughout this period." Moreover, "employment has consistently lagged behind economic growth ... in both industrial and developing countries." While the economies of France, Germany and the United Kingdom grew impressively between 1960 and 1987, employment actually declined. Taking 1960 figures as base (= 100), the Gross Domestic Product in Germany grew to 173 in 1973 and 222 in 1987; employment declined first to 96 then to 85 for the same years (United Nations Development Programme 1993:35). In Japan and the United States employment figures rose, but not nearly as high as Gross Domestic Product. With the exception of East Asia, the work force is also projected to outgrow employment all

Chapter 12 - A more fundamental approach | 331

over the world up to the year 2000 (United Nations Development Programme 1993:35-37). Finally, a telling example from South Africa: the labour index in the coal mines rose from 100 in 1950 to 165 in 1980, while the capital index rose to 436 during the same period (Nürnberger 1990:30).

[58] For the concept see Daly 1991.
[59] This section is adapted from Nürnberger 1998, Chapter 6. Cf Korten 1990:74ff.
[60] An example for this stance: "We are witnessing the triumph of economic integration, of the open model, of development based on Northern technologies and Southern labor ... By implication we are also witnessing the failure of autarkic and import-substituting alternatives ... The process is not taking place in, or significantly benefiting the poorest countries in the developing world; but it is taking place everywhere else. And it potentially can happen anywhere ... Each country must specialize in line with its abilities, opportunities, and preferences. Each must seek to serve a wider market than its own." (J K Galbraith in Purcell 1989:227-9).
[61] Mosley in Fontaine 1992:27ff.
[62] "The hope here is that ... the success of the emerging NICs, as measured by their ability to buy and sell abroad, will open a huge pool of demand, large enough both to balance the U.S. trade deficit and to spur industrial country growth rates ... It is far from certain, however, that this is what will take place." (Purcell 1989:2).
[63] Korten 1990:54.
[64] Reynolds 1988:6.
[65] In China rapid economic development led to severe discrepancies between the favoured South-Eastern coastal regions and the North Western interior.
[66] The young tigers, and their immediate emulators, may also remain the "successful exceptions" (Keesing 1988). "The developing countries' share of global value added edged upward only 3 percent between 1965 and 1980, but ten NICs (Newly Industrialising countries) recorded almost 75 percent of the gain." Seven NICs accounted for 60% of the South's manufactured exports (Junne in Caporaso 1987:91). On the "tigers" as exceptions see also Castells & Tyson in Purcell 1989:21f.
[67] Haber 1989.
[68] For the costs of the economic miracle in South Korea to the workers see Ogle 1990.
[69] Purcell 1989:4.
[70] For a description and critique of the theory of comparative advantage see Scholtes 1993:43-51.
[71] "However valid comparative advantage may be as a logical exercise, it is irrelevant in a world dominated by international mobility of capital in pursuit of absolute advantage" (Herman Daly in *Population and Environment* 15/1993:66f). Refer also to Caporaso 1987:1ff; Daly & Cobb 1989:209ff; Thirlwall 1994:366f; Anzuck 1982.
[72] "Triad Power (that is the US, Japan, Europe) constitutes the fundamental core of the world economy. Any additions (e.g. the NICs) can exist and prosper only on the basis of their access to the markets, technologies, capital, and capital goods generated in the core" (Castell & Tyson in Purcell 1989:29).
[73] "Economists have long warned that the widespread adoption of export-oriented policies by developing countries, especially in manufactures, could saturate world markets, or at least the tolerance of protectionist lobbies" (Gover & Tussie 1993:240). On the negative impact of the NICs on the industrial nations see Lovett 1987:xvi ff.
[74] Korten 1990:56.
[75] Lovett 1987:xvii.
[76] The share of the United States in world production dropped from 25,7% in 1967 to 20,8% in 1989, that of Europe from 25,9% to 22,2%, that of Japan rose from 5,6% to 7,8%. The share of Asia without Japan increased from 11% to 19,8%, and that of South America from 7,1% to - 7,5%. Van Bergen, A et al: De vierdaagse werkweek. *Elsevier* 49/1993:68.
[77] Purcell 1989:6ff. Reynolds 1988:8.
[78] Purcell 1989:9.
[79] Daly in *Population and Environment* 15/1993:68.
[80] Purcell 1989:4.

[81] "Many countries that are industrializing and/or striking good bilateral deals with countries in the North see little common interest with the smallest and least developed." Tussie and Glover 1993:240.

[82] On a recent trip to Tanzania I witnessed an example of this: after trade liberalisation imposed on Tanzania by the Structural Adjustment Programme, the streets of Dar es Salam were awash with shacks selling cheap industrial commodities. Apart from a few oranges and coconuts, apart also from a few curios sold to First World tourists, there were hardly any indigenous products to be seen.

[83] Cf W Cline: Can the East Asian model of development be generalized? *World Development* 10, Febr 1982.

[84] According to Gerd Junne the new wave of industrialisation, that is automation, will hit the "four tigers" least, because they have the means to beat the West with its own means, but the "second layer" which "tried to follow the example of the 'four tigers' ... may be hit very hard." They may have to fall back on their traditional role as raw material producers. "The next wave of technological development, however, might undermine this postion as well. New production processes will need less raw materials and fuel," because of recycling, the development of local energy resources etc. (in Caporaso 1987:72).

13 A wider frame of reference

Entropy and evolution

What is the task of this chapter?

In Part III we are dealing with a new approach to economics. We have taken two steps so far. In chapter 11 we suggested that a paradigm shift is overdue in economics as a science. In chapter 12 we demonstrated that the explanatory power of economics changes dramatically if one shifts the foundations of economic theory from supply and demand to the deeper lying levels of need and capacity. In this chapter we take another, even more drastic step, namely to place economic realities into the framework of insights gained by the *natural sciences*.

Why do we do that? In the first place because multi-dimensional problems need interdisciplinary approaches. The natural, social and human sciences define their problems, develop their paradigms, conduct their research and come to their conclusions in relative isolation from each other. The exponential growth of knowledge makes ever greater specialisation inescapable. But specialisation creates blind spots and blinkered perceptions which may not only harm the discipline concerned but also become dangerous for humankind at large. The application of the tools developed in one science in the field of another can produce startling results. In fact, economics has often claimed that its basic insights are applicable to virtually all other spheres of life. One can explain party politics, patterns of courtship, or church attendance in terms of supply and demand on the market place.

But economic reality can, in turn, be explained by the natural sciences. The theories of entropy, evolution and acceleration, developed in the natural sciences, are believed by some theorists to be so fundamental that they are capable of providing an interpretative framework for reality as a whole, including its economic and ecological dimensions.[1] In fact, we shall see that the application of these theories to the economic and ecological dimensions of reality leads to insights that challenge economics to rethink its most cherished assumptions. In this chapter we shall give a brief description of the three theories and reflect on their economic and ecological implications.

Section I: The law of entropy

Physics has come up with two "laws of thermodynamics" which describe the behaviour of energy, the basic substance of reality. The *first* of these says that the total amount of energy in a closed universe remains constant. Energy can be transformed from one form into the other, say from motion into heat, from heat into light, from light into electricity, from electricity into motion and so on, but it can never be lost in the universe as a whole. So all processes in the universe as a whole constitute a zero sum game. This does not mean that it cannot be imported into, and exported from, a subsystem of the universe. For instance, the Earth imports vast amounts of energy in the form of sunlight and exports vast amounts of energy in the form of heat radiation into outer space. we shall soon come back to that.

The *second* law of thermodynamics says that processes in the universe have a tendency to move from a state of high density energy to a state of low density energy, from high concentrations of energy to low concentrations, or from imbalance to balance in the distribution of energy. If one compresses gas in a container and connects it to a container in which there is a vacuum, the gas will tend to spread equally throughout the space of both containers.

High density energy is energy caught up and organised in constructs such as tree trunks or crude oil; low density energy is energy which is released into space without form or organisation. When a piece of wood burns up, for instance, the transformation of energy stored in the wood supplies us with light and heat. But when the burning process has taken its course, most of the energy stored in the wood has escaped into the atmosphere in the form of heat from where it is radiated back into outer space.

Because anything that we perceive as reality in the universe is the result of energy concentrations in various forms of organisation, we can also say that the universe as a whole has a tendency to proceed *from order to dis-*

order. In the very long term it is literally breaking up, falling apart. The word entropy is used in physics to indicate the degree of disorder caused when energy is released and dissipated.[2]

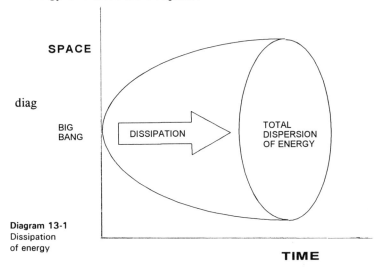

Diagram 13-1
Dissipation
of energy

Physics offers a plausible explanation for this tendency. Beginning with the "big bang", reality proceeds from an infinite concentration of energy to an infinite *dissipation of energy* (diagram 13-1).[3] How the original concentration came into being, that is, what happened before the big bang, has not been established by science. What will happen after complete dissipation is also not clear. Some scientists believe, for instance, that the force of gravity may pull reality back into a giant black hole, that is, an infinite concentration of energy, to complete a cycle. But all that is not relevant to our deliberations because by then humanity, nature, our planet, even our milky way will all have ceased to exist.

What is clear, however, is that as long as there are imbalances in the distribution of energy in the universe, the general tendency towards greater balance will continue to lead to processes in space and time. When the bottle breaks, the water flows out and disperses until it has reached a consistent level within the confines of its new environment. As we all know, water rushing from a higher to a lower level can be very powerful. It can wash houses away during a flood, or drive hydro-electric power stations. But once it has found its equilibrium in the lake below, it has no such power.

In the same way, completely dissipated energy (or maximum-entropy energy) is powerless energy. Effective power is only generated by a flow of energy from higher concentrations to lower concentrations. As energy is distributed more evenly in the universe it loses more and more of its dynamic. When all energy has dissipated it will have reached a state of balance and thus a state of complete powerlessness. Dispersed energy cannot hold anything together; it cannot give form to anything; it cannot move anything. At the end of the process, therefore, there will be no structure, no process, no contrasts of any sort. Reality, as we know it, will have ceased to exist. The very concepts of space, time and power make sense only between the big bang and the point of total dispersion.

Because anything that has a shape, order or direction constitutes a concentration of energy, the dispersion of energy is tantamount to the destruction of this shape, order or direction. This means that whatever is built up also breaks down again. When any process is in operation within reality a less structured situation is always the more probable outcome of such a process than a more structured situation. The sun burns up, the continent erodes, the body decomposes. Even social structures, cultures and convictions have a tendency to disintegrate. "All processes, even the process of intellectual discovery, operate in accordance with the principles of thermodynamics".[4]

Construction implies deconstruction

The process moving from total concentration to total dissipation is not regular but irregular. It resembles the waves caused by a stone thrown into a pond, which continue until the process fizzles out. Obviously the irregularity in the universe is infinitely more complex than the waves in the pond. But none of these structures and processes appear at random. The direction of the movement of energy from lower to higher concentrations in some places and from higher to lower concentrations in others follows physical and mathematical laws. One of these laws is the law of gravity, which says that when concentrations of energy come closer to each other they attract each other and conglomerate. Such clusters of energy appear in endless varieties of sizes, shapes, speeds and relationships. These irregularities of the entropic process form the structures and processes of reality as we know it.

Thus, within the general tendency of dissipation of energy (destruction) we have specific processes of concentration of energy (construction). Let us call all organised concentrations of energy, whether in the form of structures or in the form of processes, *constructs*. These constructs are composed of concentrated or "low entropy" energy, while dissipated energy is called "high entropy" energy.

Wherever such a cluster of energy emerges and evolves it attracts low entropy energy from the environment and feeds high entropy energy back into the environment. So a higher level of concentration in the construct implies, at the same time, a lower level of concentration (that is a greater degree of dissipation) in its environment. In a refrigerator, for instance, water can cool off and turn into ice. But this can only happen because the atmosphere outside the refrigerator is warmed up. And the process of cooling and condensing also consumes concentrated energy. The result is that, on the whole, more energy is dissipated than concentrated. The upshot of these observations is that every construction exacts a price in the form of deconstruction and this price is paid by the environment.

Moreover, the overall cost (deconstruction) is always higher than the overall benefit (construction). Why that? If it is true that the cosmic process moves from total concentration to total dissipation of energy, it is only logical to assume that every concentration has to be compensated for by a greater degree of dissipation somewhere in its environment. This means that where constructs emerge, evolve and are maintained, a greater degree of deconstruction has to take place. Expressed in different terms, a greater degree of order somewhere in the system implies a greater degree of disorder somewhere else in the system. Stronger constructive forces "eat away" at weaker ones. But as the weaker ones disappear, the stronger ones also become victims of the process. Ultimately all concentrations "dissolve" or "wear away". In in the end there will be only "chaos" or "nothingness" left.

Therefore, entropy is generated in two ways: (a) any construction implies deconstruction in the environment of the construct and (b) the new construct is also bound to dissolve in time. The upshot of these observations is that, in terms of the existence of constructs, processes in the universe as a whole constitute a negative-sum game. Relevant questions are (a) whether efficiency gains can reduce the generation of entropy and (b) whether the unavoidable entropy can be channelled into directions which do not harm human interests.

This is where the economic relevance of the law of entropy comes in. Economists love to say that "there are no free lunches!" Somebody or something somewhere in some way always has to pay the price. This raises important questions:

(a) Who has to pay, other humans or the rest of nature?
(b) If humans, do those who pay the price also reap the benefit?
(c) If nature, can it absorb the impact without degenerating or collapsing?

The energy balance on the planet

All this sounds depressing. But it is not as bad as it seems. We have said that the Earth is not a closed system. Large quantities of energy are constantly imported from the sun into the biosphere. The biosphere is the complex system of organic life located in the Earth's crust, the ocean and the atmosphere. Biological processes such as photosynthesis catch this energy, store it, and use it up. When it is used up, energy is released into the atmosphere and radiated back into outer space.

For all practical intents and purposes the energy pool of the sun is inexhaustible, because it is so vast that it will be sufficient for millions of years to come. The "rubbish bin" of outer space, where the entropy of the Earth is dumped, is even greater. For life on earth as a whole, therefore, the entropic process should pose no immediate danger. The energy derived from the sun can be used to maintain and increase the process of construction on earth for a very long time and the dissipated (high entropy) energy can constantly be radiated back into outer space.

But time imposes a limit to this process. The problem is that while there may be no limits in volume, there are indeed limits in flow.[5] A tap can flow endlessly if the supply of water flowing into the tank equals or exceeds the water flowing out. The processes through which the energy derived from the sun is transformed into constructs by the biosphere are relatively slow. In contrast, the process of deconstruction can be very fast. A large tree, which took thirty years to grow to its present size, can be burnt to ashes within a few hours. If the process in which constructs are deconstructed outstrips the process of reconstruction, the entire system begins to deteriorate. This happens, for instance, when forests are burnt down.

At the same time, the process in which dissipated energy is radiated into outer space can be impeded, for instance by clouds. We know that at night the atmosphere cools down less when it is overcast than when it is clear. Radiation into outer space can also be impeded by a shield of gases in the atmosphere. Then the speed at which high entropy energy is created may become greater than the speed at which the Earth can get rid of it. The production of carbon dioxide, for instance, is believed to be the cause of "global warming" or the "greenhouse effect".

So while reflecting on entropy we have come across the two basic ecological problems of today: *over-exploitation* of natural resources and *pollution*. If exploitation and pollution are light, nature can deal with them. The process in which constructs emerge and flourish at the expense of their environments takes place in concentric circles. Photosynthesis caters for a very large pool of primitive life on which successive layers of more sophisticated forms of life can flourish. Plants live on humus produced by bacteria, antelopes live on plants, lions live on antelopes. Human beings live on both plants and animals.

In nature this system regulates itself. The environmental substratum of any construct must be large enough to absorb the impact of the construct. Therefore the growth of the pool of lower forms of life, on which higher forms of life subsist, must be fast enough to compensate for the destruction imposed on it by respective higher forms of life. Conversely, the number of constructs on any level and the demands each makes on its environment must be small enough not to destroy the viability of the lower level of constructs on which they all depend. When there are more antelopes than grass to feed on, they begin to die until the balance is restored.

The upshot of these considerations is that the numbers of higher level creatures and their average consumption must stay within the limits of (a) the power of lower level creatures, upon which they feed, to regenerate themselves, and (b) the capacity of the "sinks" to absorb the pollution. Predatory organisms cannot multiply beyond the limits imposed by the volume of their nutritional base. Once any species moves beyond this threshold it is in trouble. Ecologists call this "overshoot".[6]

Humanity as a destructive species

It is this dynamic balance which is being upset by modern civilisation. By virtue of their superior intelligence, human beings are able to go beyond scavenging in their environment as hunters and gatherers to survive. They are able to design and use tools to break down raw materials and put some of the parts together again to make usable artifacts. The rest is discarded as waste. The artifacts also end up in waste. That is why the impact of humans on nature has always been higher than that of comparable animals such as monkeys. Apart from nature, entropy can also be offloaded on other human beings. Through raiding, conquest, colonialism, imperialism, exploitation and discrimination, the more powerful engage in their construction at the expense of the less powerful. We can observe the following phenomena:

(a) By dismantling natural checks on their proliferation, without introducing cultural checks, humans have opened the sluice gates to rampant *population growth*. In terms of the biological world as a whole, this is a "cancerous" development in the literal sense of the word.

(b) Humans continuously enhance their powers to extract and process resources by means of *technology*. They use these rapidly accumulating powers not simply to secure the long term flow of the prerequisites of human life, but to exterminate fellow creatures and exploit the support base on which they thrive. Although the aim of technology is construction, the dismantling of higher forms of being into their constituents is evolution in reverse, thus destruction.

(c) By overpopulation and the use of technology humans *spread their entropic waste* throughout all layers of the biosphere.

(d) Humanity has not learnt to differentiate between various levels of reality, but *subjects all of them to the same processes* of competition, domination, exploitation and pollution. Not only fossil fuels, minerals, plants and animals, but also other humans are utilised for their own ends.

The human race has impacted its social and natural environment ever since human life began. But the process has gradually accelerated. In recent times it has begun to gallop. To get an impression of the growth of the total impact, one would have to multiply the relative advance in technology between two points in time by the growth of the population during the same period. Indications are (a) that we are busy *depleting* natural resources which cannot be renewed, notably crude oil; (b) that we *overexploit* natural resources which could regenerate themselves if given a chance, such as fish stocks and forests, and (c) that we *overburden* the capacity of the environment to absorb the entropy or waste generated, for instance when we emit toxic gases from smoke stacks. And this we do faster and faster.

If one visualised nature as an organism, or as a community intent on defending its own life and prosperity, one would have to concede that the human species has become the worst vermin our planet has ever been subjected to. Many people are concerned about the survival of dolphins, whales, elephants and black rhinos. But what about all the other creatures? It is estimated that about 50 000 species will become extinct annually by the year 2000.[7] Modern humans cynically assume that they are entitled to subject and utilise whatever they encounter for their own purposes. Much worse than large swarms of locusts or smallpox epidemics, which only have limited targets, the aggressive self-interest of humanity resembles the lava of a volcano which kills, destroys and buries whatever gets in its way.

As we have argued again and again, population growth and economic growth cannot continue for ever on a limited planet. Sooner or later the entire process will have to spiral down again until it fits into the limits set by nature, limits which are also getting more narrow as the process continues. The result will be incredible suffering for millions of people - apart from other creatures on earth. Is all this inevitable?

Section II: The theory of evolution

While entropy says that there is a general tendency towards destruction and decay, evolution says that life on this planet has begun in the form of simple organisms and has been unfolding ever since into more complex forms of life. Early developments of life have been slow, but evolution has been enhanced by sexual differentiation. Hereditary factors (or genes) come in pairs of which one is usually dominant, the other recessive. Because of the multitude of different genes and the seemingly endless ways in which they can combine, the variety of particular species and specimens has become incredibly great.

Organisms which possess combinations of hereditary factors which enable them to survive better in particular environments out-compete others which are less resilient; hence some multiply, some die out. This pattern is called the *survival of the fittest*. It does not preclude the possibility that organisms can cooperate and survive and prosper together. We call that symbiosis. There are also certain "jumps" in the evolutionary process, called mutations. If a changed factor leads to greater capacity to survive within a particular environment, the new variety again has an advantage over its competitors.

Evolution theory has been the particular domain of the biological sciences. They have investigated the underlying mechanisms in great detail and used them, for instance, to generate new breeds in agriculture. But just as the law of entropy is not confined to physics, the theory of evolution is not confined to biology. The human sciences have applied it (sometimes with problematic intentions) to the development of social, political and economic systems, to cultures and religions, to science and technology. The physical sciences in turn have come up with theories concerning the evolution of matter. They have shown that the elements which make up physical reality have all been put together over long periods of time from elementary particles. So today the view that reality evolves in time forms the basic framework for understanding reality as a whole.

The relation between entropy and evolution

We have seen that there is a tendency in reality towards dissolution and chaos (entropy) and there is a tendency for reality to develop into ever more complicated forms (evolution). How can this apparent contradiction between the theories of entropy and evolution be resolved? As mentioned above, the physical sciences have found that the entropic process is characterised by wave-like irregularities caused by the force of gravity, which lead to temporary buildups of energy, which again force the entropic process into roundabouts. It is these localised concentrations of energy in constructs which form the substance of the theory of material, biological and cultural evolution. So the two theories do not contradict but complement each other.[8]

In spite of their growing complexity, however, construction cannot eliminate the general trend towards destruction. The spectacular turns of a river meandering through a hilly landscape represent the shortest possible distance to the sea. Obstacles make the river's path longer, but eventually they too will be eroded by the water. So the generation of entropy is the overriding principle; evolution must be seen within its context. Chaos theory may throw light on the relation of growth in complexity to entropic dissolution, but we cannot pursue this further here.

Because evolution takes place within the context of the entropic process, growth in complexity creates concentric discrepancies, reaching from the most advanced to the most deteriorated. It is no accident that there are vast stretches of empty space between galaxies, solar systems and planets. But the phenomenon can also be observed in the social, economic and political spheres of reality. Wherever there is construction, there is deconstruction. "Rising stars" in the entertainment or sport industries cannot help but cause the relative decline of the status of others in society. The rise of an empire leads to the dissolution of previously sovereign states. The growth of competitiveness of one economic entity leads to the deterioration of the competitiveness of others.

We have seen, however, that the biosphere is a self-regulatory system which, at least for time spans relevant to human reflection, is able to make up for the loss. There is more plankton in the sea than necessary for fish to survive. More seed is produced by plants than necessary to reproduce their species; so birds can feed on them. It is only human civilisation which upsets this dynamic equilibrium. However, to understand the full impact of modern developments on nature we need to look at another facet of the entropic-evolutionary process, namely acceleration.

The acceleration of historical processes

The growth in complexity, due to the evolutionary process, is subject to acceleration, that is, it moves faster and faster. The higher developed an aspect of reality has become, the faster its evolution - and its demise. The social sciences express this insight with the theory of exponential growth.[9] But the phenomenon characterises the entropic-evolutionary process of reality as a whole (diagram 13-2). The evolution of *physical matter* has been extremely slow. The big bang is believed to have happened some 15 billion years ago.

Since then all the elements, galaxies, suns, planets, continents and rock formations have evolved. *Organic* evolution was faster. *Cultural* evolution was still faster. The latest and fastest evolution is that of *science and technology*. Even this development began slowly and picked up speed. Today a computer is obsolete within a year or two. In his book, *Future Shock,* Alwin Toffler has provided us with a powerful and popular depiction of this acceleration.[10]

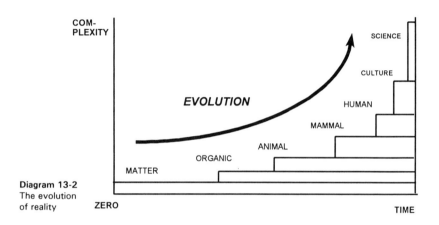

Diagram 13-2
The evolution of reality

The nature of economic evolution

Economic development has generally been perceived in terms of evolution.[11] In fact, the very concept of "development" suggests a process of unfolding. What has not always been recognised is that *evolution takes place in the context of the entropic process.*[12] We have seen in our earlier analyses that economic development leads to staggered discrepancies between economic centres and economic peripheries, even though these centres shift and become more differentiated over time and even though the economy as a whole may grow. In the inner core of the centres the economy grows exponentially while population numbers remain constant or decline. Semiperipheral countries, such as the Asian tigers, may have higher growth rates, but from a lower base. In the outer peripheries the economy tends to stagnate while the population tends to grow exponentially - at least for the time being. In the centres we find growing complexity and integration; in the peripheries we find growing dissolution.

The entropy caused by the rapid evolution of the industrial economy is deposited in its social and natural environment. As early as in chapter 2 of this book we have seen that the flip side of the "creation of material wealth" in economic centres is either the impoverishment of economic peripheries, or the depletion of natural resources, or both. Because evolution always takes place in the context of the entropic process, there simply is no third alternative - except where growth means growth in quality rather than quantity.

An example of the entropic consequences of economic growth in the centre economy is the growth of *unemployment* in peripheral contexts. We have discussed this phenomenon fully in chapter 12. Unless the entropic process is balanced out through deliberate countervailing measures, economic growth leads to increasing marginalisation. But the generation of entropy in the *natural realm* is, in the long term, much more serious. Pesticides, antibiotics, fertilisers, fire arms, in fact all the manifold tools with which humankind controls nature, are weapons against nature.

One result of the growth of the competitiveness of humans in relation to other creatures is the exponential growth of the human population. Plants and animals are simply eradicated in the process. Seen in this light, economic development is tantamount to an acceleration of the *entropic deconstruction of the eco-system.* This happens for the sake of temporary (and in many cases problematic) life enhancing powers which benefit humanity at the expense of its non-human competitors in the eco-system. It also benefits the present generation of humans at the expense of future human claimants to the resources of the eco-system.

As mentioned above, the ecological impact is roughly commensurate with the *level of development multiplied by population numbers.* The growth of depletion and pollution per person among the poor may seem to be minimal in relation to that of the affluent, but their greater numbers, their concentration in informal settlements and overpopulated rural areas, and their inability to clean up their environments can make the overall impact even greater.

Entropy generated in the economic realm impacts the ecological realm and this again impacts the future economy. *Fossil energy* consumption is a case in point. We have seen that toxic emissions created by the utilisation of fossil fuels begin to reach dangerous levels. But that is not the worst problem. Technology can devise means to control pollution as long as the energy to do so is available. The greater problem is that these deposits are irreplaceable and that their quantity is strictly limited. They represent low entropy energy accumulated over long periods of time through the process of photosynthesis in which solar radiation is gathered up and transformed into carbon.

The astounding productivity levels currently reached in agriculture and manufacturing, which are leading to population numbers and standards of living unparalleled in human history, are totally dependent on this diminishing resource. Affordable and ecologically safe alternatives, which can reach the magnitude required for the sustained economic growth of a growing population, are nowhere in sight. It is often not realised that alternatives, such as solar energy or tidal energy, presuppose a high level of sophistication. And their manufacture also demands energy.

Growth in production thus does not represent growth in the wealth of humankind, as generally assumed, but the acceleration of its depletion. The throughput from resource base to waste is accelerating. Schumacher has taught us two decades ago that we are using up our "natural capital".[13] There is no way out of this problem. If we match increases in production with increasing efforts to combat pollution, we do relieve the sinks, but cleaning-up operations and recycling efforts also add to the energy bill.

Whether the increasing scarcity of fossil fuels will price them out of the market within half a century or within three centuries is not really the point. The point is that, seen in the context of the vast sweeps of history, an industrial civilisation based on fossil fuels can at best constitute an episode. Our descendants will have to make do without them. And the crunch will be felt first by those who possess the least economic clout to compete in a tightening market situation.

In the long run, however, even the affluent will not escape the consequences of oil wells running dry. Sooner or later humanity as a whole will have to adjust their lifestyles and their numbers to match the pace at which the biosphere can transform solar radiation into low entropy energy.[14] By that time the capacity of nature to do just that will have been substantially reduced. Humankind may well have to face an accelerating downward spiral.

These adjustments can hardly happen without sacrifices and hardships. Whether economic growth in the centre leads to the collapse of the peripheral economy or not depends on the capacity of the periphery to outpace the impact by its own economic growth, thus by increasing its own entropic impact on nature. Whether both growth processes lead to the collapse of the ecosystem or not depends on the relation between the rate of this economic growth and the rate at which the respective natural environments can absorb the impact and regenerate themselves.

This fact places the issues of social justice and intergenerational justice onto the agenda with increasing urgency.[15] As we have seen in chapter 11, economics, any science for that matter, cannot be value-free. Every science is built on assumptions and assumptions are co-determined by interests. The failure to account for one's choices is rooted in a conscious or subconscious attempt to abdicate from responsibility, if not to legitimate the ruthless pursuit of private or collective interests.

The role of technology

Social, economic and political developments are empowered by scientific-technological evolution. Technology has evolved from the simple tools of the stone age to the marvels of modern technology. Technology is nothing but interference in the evolutionary and entropic processes:

(a) On the one hand it introduces *catalysts* to enhance the speed of dissolution to gain more energy for construction. For instance, crude oil is burnt up to generate electricity. In more general terms, technology dismantles higher forms of being to gain access to their components as building blocks for new constructs. This is evolution in reverse, unless cars and sky-scrapers are seen as products of the evolutionary process.

(b) On the other hand it introduces carefully worked out *obstacles* which steer the entropic process in particular directions to achieve certain constructs which seem to be useful or desirable. One can, for instance divert a river for irrigation purposes.

(c) One can also *brake* the process of disintegration, for instance, by putting cement between bricks.

Seen from the vantage point of short term interests, technology has led and is still leading to vast improvements in the human condition. But seen from the vantage point of long term survival and prosperity, technology is currently leading to devastations of staggering proportions. Nature has been destroyed in the wake of any civilisation, the more so the more advanced it has become. Examples are the once forested mountains of Greece which were denuded by Phoenician fleet builders and the collapse of the Mayan civilisation in Central America through overpopulation, depletion and warfare.[16] But what we are doing today defies all comparisons.

Some observers have come to the conclusion that the technological enterprise is to blame for the human predicament and should be abandoned. But this is romanticism. The fact of the matter is that we can no longer do without technology. Indeed we need it even for finding ways out of the current impasse. Technology is a tool, not a demon. As stated above, technology can also be used to brake and channel the process. *So far,* increasing control over nature has induced human beings to multiply and prosper at the expense of plants and animals. *So far,* the majority of human beings continue to struggle, while a minority reap huge benefits. But there is no rule of nature which says that this must continue. While it is not possible to avoid entropy, it is possible to reduce and rechannel it. But this would presuppose a new mindset. The following rules seem to be decisive and, in the long run, inescapable:

(a) The entropy generated by the economy *must be channelled out* of the human realm into the non-human realm, which forms the natural biological support base for humanity, that is, plants and animals.

(b) The rate of exploitation of the biological resource base must be reduced to the level at which it *can regenerate itself.* This means that we must slow down our economy to the speed at which solar energy can be converted into biological life.

(c) Similarly, the generation of pollution and waste must be reduced to the level at which natural "sinks" *can absorb the impact.* For instance, we may not emit more carbon dioxide into the atmosphere than can be absorbed by the biosphere.

(d) The rate of depletion of resources which cannot be renewed must be reduced to the rate at which *alternative resources* can be developed. The depletion of fossil fuels, for instance, should not outpace the development of other energy sources, such as solar energy.

(e) This implies that humankind must reduce both its *numbers* and its levels of *consumption.*

(f) Because the burden of this demand cannot possibly be laid upon the poorest, but must be shouldered by those who overconsume, greater *equity* is not only an economic but also an ecological necessity.

(g) We must reach a more *balanced pattern of need satisfaction.* There are vast areas of human need which are severely undernourished in affluent societies. The task is to shift satisfaction to areas which do not impoverish, deplete, destroy and pollute.

(h) The constructive process as a whole must be channelled into *life-enhancing,* rather than life-inhibiting and life-destructing directions. The survival of rain forests is more important than the repayment of international debts. The wellbeing of fellow creatures is more important than the ambitions of warlords.

(i) Technology must be used to reduce the costs of construction and the speed of destruction by utilising resources *more efficiently* - but without stepping up the volume consumed - and by enhancing the quality of satisfaction achieved by smaller inputs. Technology must help us to change *from quantity to quality* in all spheres of human activity, but particularly regarding material consumption.

(j) All this presupposes a fundamental *change in perception and attitude* among masses of ordinary people, beginning with the academic, commercial and political leadership. This also means that the ethical dimension must be placed on the economic and political agendas more forcefully than ever.

Revision: Summarise the theories of entropy and evolution and apply them to the economic process.

Application: Do you think that the order and prosperity created by modern economic achievements in your country implies the simultaneous creation of disorder and poverty among less fortunate population groups?

Critique: (a) "Your assessment of human nature is out of touch with reality. Everybody is guided first and foremost by self-interest and the economy can only function on this basis. Whether this is due to evolution or original sin, as the bible maintains, makes no difference to the fact." (b) "This chapter is a good example of what can happen when you confuse different sciences. The science of physics deals with a dimension of reality completely different from that of biology, biology from that of economics, economics from that of ethics, ethics from that of religious studies, what ever that may be." (c) "You have disregarded the law of gravity which balances out the law of entropy in that energy always tends to reconglomerate." How would you respond to these statements?

Section III: Objections

Issues of justice could be scorned or neglected by economics only as long as its basic assumptions could be taken for granted. This is no longer the case. Let us look at some of them.

1. Economics assumes that the pie as a whole *can grow indefinitely*. If everybody's share grows, thus the argument, the issue of distribution is irrelevant. But is it really? If a pie doubled from 100 to 200 units, a constant share of 10% would double to 20 units and a constant share of 90% would double to 180 units. Thus the advantage of the rich over the poor would rise from 80 to 160 - with multiplier and all. Higher standards of living among the rich again pull up the need levels of the poor and the latter are worse off than before. Rising discrepancies in income also have serious social and political consequences.

The argument also flies in the face of the principle of equality of opportunity which is part and parcel of the assumptions - and the efficiency - of a free enterprise system. It is not true that a more equitable distribution stifles the rate at which growth can be achieved for the benefit of all, quite apart from the sustainability of such growth. Equality of opportunity opens up access to material and human resources, thus bringing more people into the range where potential entrepreneurial initiative and technological competence can translate into effective initiative and competence. "The notion of a trade-off between growth and equity, which helped to entrench anti-growth policies in socialist economies and anti-equity policies in conservative ones, has been further discredited by the many economies that consistently outperform the rest on both counts: Costa Rica, Indonesia, Japan, Korea, Malaysia ... and the Scandinavian economies".[17]

2. Another argument is that *technology* and the *market mechanism* will see things right: as a resource is being depleted, the market will signal scarcity by higher prices, technology will be activated to overcome the problem by designing means to exploit alternatives, and so on in eternity. No - not in eternity! If I have five bank accounts and switch from one to the other as they are being depleted, I will finally reach the last cent in the last bank account.[18]

3. Another argument says that *efficiency gains* due to technological development lead to a reduction in the use of natural resources. Clem Sunter, a member of the South African business elite, maintains that the industrial nations are "busy designing them out of the system."[19] Presumably the cars of the future will be composed of some sort of ethereal substance which you cannot even see or touch any more?[20] The idea behind this contention is that resources can be substituted by capital. But capital is either a factor of production (capital invested in machines, warehouses, and so on), or it is a symbolic representation of rights (financial capital). An increase in chain saws leads to an increase in timber cut down, but it does not lead to an increase in trees.[21] And financial capital is a bubble which can burst at any time.

It is true that materials can be recycled, that toxic effluents can be reduced, and that existing pollution can be cleaned up to some extent. But it is also true that this again devours capital, energy and materials in increasing quantities, rather than adding to the stock.[22] While recycling forestalls the premature abandonment of a resource, thus reducing the rate of waste, it does not add to the total volume of a resource, nor does it subtract from total entropy. In fact, recycling deepens the levels of exploitation and increases total entropy.

While the fact that technology has indeed increased efficiency and reduced the material content of gadgets and energy consumption per unit considerably over the last decades must be applauded, the total material volume of such gadgets and the total volume of energy utilised still increases exponentially as ever greater proportions of the world population come to produce and consume them. Even in highly industrialised countries, where technology has increasingly pushed people out of the production process into the service and communications sectors,[23] the material sectors, agriculture and industry, are still growing in absolute terms.[24] In fact, the service sector could not survive without the primary and the secondary sectors.[25] People in industrialised countries use up to ten times more metals on average than people in non-industrialised countries, and "the average person in an industrial market economy uses more than 80 times as much energy as someone in sub-Saharan Africa."[26] At high levels of affluence, throughput of energy and material may level off, but as long as the rest of humankind is surging ahead to reach First World standards, the need for materials and energy will snowball.[27]

4. Another assumption is that this kind of progress is *universally accessible*. Even the poorest nations could, if they so wished, reach the levels of production and consumption currently prevalent in industrialised nations. But this is clearly not the case.

> Consider the following simplified calculation: the per capita fuel consumption in South Asia was 0.23 tonnes per year in 1988, while in North America it was 8.05. That is 35 times as much![28] South Asia consumed 243 million tonnes (3% of the total), North America 2192 million tons (27.2% of the total). If only India, with a population roughly three times as large as that of the US and Canada combined, industrialised up to North American levels, it would consume three times the amount of fuel that North America does. This would push up the total from 8.058 billion tonnes to 14.6 billion tonnes, that is up 82%.
> Add to this the composite growth of India's population and the composite growth of the US economy, say 2% each, then fuel consumption of these two regions alone would reach about 18 billion tonnes per year 20 years later. At that rate the two regions would deplete the currently known reserves of 124 billion tons in less than 7 years. The emissions of toxic gases would rise by 740% over these 20 years.
> Now add China with 1.2 billion people, Brazil, Indonesia, Africa and the rest of the Third World - each with their population and economic growth rates - plus the continuing growth of other industrial nations.
> It should be clear that this simply cannot happen, however optimistic one would like to be. The argument makes the economically powerful feel good about what they are doing; it is ideological legitimation plain and simple.

5. The assumption of infinite growth which is accessible to all people is built on two other assumptions, namely that *the Earth is an open system* where resources are abundant and mutually substitutable, and that an egalitarian system exists in the sense that there are *no intrinsic limits to human initiative* in a free economy. But what would happen if the Earth in fact

turned out to be a closed system in which resources were strictly limited and not indefinitely substitutable? What would happen if increasing competition for decreasing resources led to the survival of the fittest and the elimination of the weakest rather than to the infinite growth of the pie?

We mentioned above that the Earth is, in fact, not a closed system. However, the acceleration of the process imposes time constraints which can reach thresholds of virtual closure. To the extent that the velocity of the throughput of energy in the economy is greater than the velocity of (a) natural construction through photosynthesis and (b) natural disposal through decomposition and radiation, the system closes up. So the classical arguments are no longer valid and no magic wand will pull future generations out of their predicament.[29]

6. The most potent argument has always been that human nature is *inherently selfish* and cannot be changed. This untested assumption poses in "scientific" garb when it is argued that evolution itself is to blame. To survive in the early stages of its evolution, it is maintained, humanity had to give absolute priority to self-interest. To gain greater power over the environment the ego was extended to include the family and the tribe.[30] Everything else was treated as a potentially dangerous and potentially useful environment. If dangerous it had to be kept at bay or eradicated; if useful it could be subdued and utilised. So the human being is conditioned by the evolution of the species to be self-interested. If this were true, the appeal to responsibility for society, for humankind as a whole, even for future generations, would remain a futile undertaking. Ecological concerns could only grow as far as immediate self-interests were involved. This argument is both fatalistic and fallacious:

(a) We should not *confuse culture with nature*. All creatures are subject to the survival imperative, yet only humans destroy their habitat.[31]

(b) *Higher levels of evolution supersede lower levels.* Evolution is an ongoing process that does not stop somewhere on the way. Technology, for instance, makes up for physical deficiencies. Systems of meaning, values, norms, institutions and procedures make up for character deficiencies. The appeals of ecologists to conscience, reason and political will are neither utopian nor idealistic but pragmatic and realistic.[32] It is the dream that growth can go on forever which is utopian and irresponsible. If people do not change by their own free choice, a changed situation will simply impose itself on humanity.

(c) Evolution proceeds much *faster at higher levels* than at lower levels. While the genetic pool of humankind may be relatively fixed in the short term, assumptions, values and norms adapt very rapidly to changing situations. Cultural change is vastly accelerated by widening horizons due to the information explosion and growing interlinkages. Even if one conceded that the human race was conditioned to be inherently selfish, history teaches that the "extended self" grew from clan to tribe, to nation, to humanity.[33] We are now on the threshold of the potential inclusion of future generations and the natural substratum of human life. The only question is whether we want to embrace and support this growing awareness or thwart it for the sake of short term profit and shallow enjoyment.

The problem is, therefore, not the enslavement of the human race in its evolutionary fetters but an example of cultural failure. Traits which enhance our powers over reality have advanced very rapidly; traits which would have enabled us to control our appetites have lagged behind. But if this were only a matter of cultural lag, it could be made up sooner or later. In fact, it is a matter of *cultural degeneration*.[34] While science and technology have been part of the accelerating evolutionary process, communal cohesion and responsibility have been abandoned to an unmitigated entropic process.

This failure is partly due to a set of flawed philosophical assumptions which have granted legitimacy to base desires at the expense of higher values. But philosophical reflections are potent only because they are used by more powerful forces in society for their own purposes. Rampant selfishness is not a characteristic of human culture as such; it is the specific problem of the liberal capitalist version of the technological civilisation. Here the culture of moderation and responsibility is consciously broken down by sophisticated and ruthless appeals to the human lust for control, excitement and enjoyment with the sole goal of higher profit.[35]

A corollary is the lack of courage and foresight, on the side of politicians to whom the health of the society is entrusted, to recognise and counteract this rampant abuse. Not only do they share the ravaging culture, but they also have to pander to it because, in a democratic society, power depends on the votes of the average consumer. We have built sophisticated weapon systems to conquer the lands of others and keep aggressors in check, we have developed complex legal systems to contain discontent and maintain public order, but we have not developed means to counteract cultural entropy and enhance cultural evolution in the sphere of human accountability.

7. Occasionally economists become irritated enough by ecologists to go on the offensive. If humans are just a part of nature, it is argued, "why should insects, bacteria, intestinal parasites, and HIV viruses not have rights equal to those of human beings?"[36] Those environmentalists who preserve nature for the enjoyment of humans are simply deemed to be *hypocritical.* An extreme fringe may be consistent in its defence of the rights of nature, but then "the consequence of this belief is an indifference to mass starvation in

countries like Ethiopia, since this is simply an example of nature paying man back for overreaching ..."[37]

This argument obviously cannot hold water. In Ethiopia and similar countries nature suffers at least as much as humans, and humans suffer because nature is being destroyed. There is no nature there which has the power of "paying back"; there is only a common process of dying at the hands of humans. Ecological responsibility does not deny that there is a hierarchy in nature and that humans are situated at the peak of this hierarchy. It just realises that the respective lower levels of the hierarchy provide the foundations for the respective higher ones. If the foundations are destroyed the whole edifice comes tumbling down - and neither nature nor humanity gains.

The most critical task of economics today, therefore, is to demythologise its own assumptions. Prominent economists have tried to do just that for quite some time, though without making a substantial impact on big business, the state, the production of young economists and the massive research projects undertaken to the benefit of these giant stakeholders.[38]

How to deal with entropy

The identification of development with evolution as such is not wrong. But evolution has to be seen *in the context of the entropic process*. There are two ways in which entropy can overwhelm evolution, depending on whether entropy is channelled out of the human realm altogether, or shifted to other sections of the human population. In the latter case order (progress and prosperity) can indeed be achieved in one part of the system but only at the expense of greater disorder (deterioration and misery) in another part of the system. Entropy deposited within the realm of the human species particularly affects populations which have been subject to imperial domination, which are unable to compete because they have been left behind in the acceleration of science and technology, or which cannot defend themselves because they have not yet been born. This throws new light on the centre-periphery mechanisms analysed in Parts I and II.

So to reach equity entropy must be channelled out of the human realm. Unfortunately, at least for humans, not all entropy can be deposited in the rest of nature. This raises ethical concerns. Every gain implies sacrifices, and those who have to make the sacrifices are not necessarily those who reap the benefits. This may be in order as long as there is a *balanced give-and-take:* the farmer sweats to supply food to the industrial worker; the industrial worker sweats to supply agricultural machinery to the farmer. But alas, we are not in a paradise of social equity!

But we have to go beyond moral precepts, strong as they may be in their own right. Purely economic considerations show that the most cost-efficient society is one in which *both the benefits and the sacrifices are distributed equitably*. Why should this be the case? The law of diminishing marginal utility says that higher units of consumption lead to declining additions in satisfaction; the law of diminishing marginal productivity says that higher inputs lead to diminishing additional output. This means that the more unequal a society is, the greater the sacrifices of the less privileged and the smaller the benefits to the more privileged. Thus, as the level of equity declines, the cost-efficiency of the social system as a whole deteriorates. That is, the degree of entropy increases, and the evolutionary gain decreases.

The other possibility is that entropy is channelled out of the human subsystem. In this case humans reap the benefits and nature bears the cost. This cannot be avoided if human beings are to survive. However, if the entropy, which is channelled out of the human realm into the realm of its natural life support system, is *greater than the capacity* of this system to absorb the impact and regenerate itself, the human species endangers itself. In other words, there is a definite threshold up to which the quantitative volume of the material culture of humanity can grow. Beyond this threshold the process is reversed and entropic destruction takes over.

Let us summarise

In this chapter we attempted to place our deliberations on the global economy into the context of the behaviour of the natural world, as described by the natural sciences. The application of the tools of the natural sciences, such as the theories of entropy, evolution and acceleration, to the economic realm can yield dramatic insights. The *law of entropy* says that, because energy is involved in a long term process of dispersion, there is a universal tendency from order to disorder. Any construction implies a greater degree of destruction elsewhere in the system. Life on earth is made possible by the fact that plants process the energy imported from the sun into low entropy energy, which is transformed into high entropy energy by other living creatures and radiated back into outer space.

The *theory of evolution* says that there is a tendency in the universe to form more complex forms. This process led to the emergence of the great variety of phenomena found in the natural world. Emergence and evolution proceed in stages: the physical world, the plant and animal kingdoms, humanity and the rich variety of cultural achievements, culminating in science and technology. Emergence means that a higher form of organisation ensues which is more than its constituent parts, yet it cannot exist without its constituent parts. Each stage of the process takes a shorter time to mature and the entire process is subject to *acceleration*.

The evolutionary process towards greater complexity must be seen within the context of the entropic process towards dissolution. Because any construction implies a greater amount of deconstruction somewhere in a closed system, every benefit implies a cost. The entropic cost of economic progress can be channelled either to other members of the human family, or to the natural support base. In the former case we have to achieve *parity* in the distribution of sacrifices and benefits, otherwise the economy becomes inefficient - quite apart from the ethical demand for equity. In the latter case the exploitation of nature has to remain *below the threshold* where the natural environment can no longer absorb the impact and regenerate itself through importing energy from the sun.

We maintained that the problem did not lie in the development of technology as such, but in the kinds of *human motivation* which utilise these powers in beneficial or detrimental directions. We strongly denied that human selfishness and human narrow-mindedness were implanted in our psyche by the evolutionary process and had to be accepted as unalterable fate. Culture can always supersede and correct nature. But in our civilisation, the dominant ideology of liberal capitalism has led to severe retrogression in this crucial dimension of culture. If we are serious, we have to rediscover the fundamental importance of convictions, values and norms for a responsible economy. Humanity simply cannot afford selfish and destructive patterns of behaviour any more.

In practical terms humanity must learn
(a) to strive for more equitable participation in the economic process,
(b) to reduce its numbers,
(c) to be more modest in its expectations,
(d) to live below the threshold of nature's ability to absorb its impact and regenerate its substance,
(e) to make production more cost-efficient,
(f) to change from quantity to quality,
(g) to avoid harm to life where possible.

Notes

[1] This should not be a strange idea to economists. For a long time main line economics has believed in the "alleged scientific supremacy of physics over all other scientific disciplines" and desired "so to construct the field as to emulate the scientific status of physics." The problem is, however, that "many economic theorists appear to be unaware that the scientific philosophy of modern physics has changed radically ..." (Drakopoulos 1994:333). The natural sciences themselves have discovered that the am-

bition of an exact and valueless science is problematic because all human insight is governed by assumptions and subject to distortions.

[2] The concept "entropy" was first used in physics by Rudolf Clausius. Its meaning is ambiguous. According to one explanation it is a combination of "energeia" (the Greek word for activity, power, efficaciousness) with "tropos" (the Greek word for direction, tendency, character, condition) or with "trope" (the Greek word for change of direction, thus transformation); cf Capra 1982:60. It is usually used to indicate the result of a process going in a particular direction, in this case disorder or chaos. But it is also often used to indicate the tendency or movement towards that condition of disorder. In this essay we used the former, more common meaning. The second law of thermodynamics says that "the entropy of the universe tends to a maximum." (Brennan 1990:200). This definition reflects the fact that Ludwig Boltzmann gave statistical expression to the law of entropy by adding the concept of probability (Capra 1982:61). This means that the tendency towards chaos can be reversed, but the likelihood of such a reversal is smaller than the normal process.

[3] See Hawking 1988 for a full discussion.

[4] Townsend 1992:98.

[5] See Daly 1991:91ff.

[6] Meadows et al 1992:1ff.

[7] Von Weizsäcker 1994:105.

[8] Sachsse 1984.

[9] It has been developed for the social sciences, amongst others, by Meadows et al (1972), Mesarovic and Pestel (1975) and Meadows et al (1992).

[10] Toffler 1983.

[11] An influential example is Rostow 1990.

[12] The plain fact is that classical economics overlooked a fundamental characteristic of earthly reality. And the reason is not hard to find: " ... perhaps the answer has something to do with the negative implications of the entropy law for any ideology based on continuous growth" (Daly 1986:321), thus for an ideology legitimating collective self-interest. See further Seifert et al 1992; Daly & Cobb 1989:194ff; Rifkin 1990; Sachsse 1984 (who still shares the optimistic liberal point of view in economics). The relevance of the phenomenon of entropy for economics was highlighted by Georgesku-Roegen 1971 (see also his contribution in Daly 1973:37ff); it was further developed by Daly 1973. Its relevance has been questioned by Young 1991. Daly (1992) and Townsend (1992) have convincingly dismantled the arguments. Young has also explained why economists have not taken up the clue. One reason is the idea that it is covered by the law of diminishing returns. This is, as Young concedes, wrong because the phenomenon of lower marginal returns is due to rising deficiencies in other factors of production, which can be augmented, while entropy signifies a loss of total stock in the process of production. The idea that prices on a perfect market discount entropy is also wrong, even if all costs are internalised, because entropy is not a function of the relative scarcity of single resources, but the fate of all resources, and the market is not sensitive to total stocks available.

[13] Schumacher 1973.

[14] "Barring some major breakthrough in fusion or fission energy, we will in the future have to live largely on the current flow of solar energy rather than on the accumulated sunshine of paleolithic summers" (Daly & Cobb 1989:204).

[15] For the concept of intergenerational justice see Brown-Weiss 1989.

[16] *Newsweek,* June 12, 1995, p 15.

[17] World Bank 1992:137; cf. Fields 1980:122f.

[18] J L Simon, an economist particularly committed to the classical tradition in his trade is indignant over the prophets of resource scarcity: "It is natural for economists to recognise that forces arise to increase supply in consequence of diminutions in supply and increases in prices" (in *Journal of Population Economics,* vol 6/1993,146f). The author has hardly recognised the irony hidden in his words: as forests are chopped down, logging is accelerated to keep profits rising; when the forests are gone, the profit motive turns to the next exploitable resource! How can one seriously condone and legitimate

this kind of ravaging with elegant economic theory?
[19] Sunter 1987:37.
[20] Sunter's formulation may have been a slip of the pen, but there are supply-side economists who seriously think along these lines. George Gilder (1981): "The United States must overcome the materialistic fallacy: the illusion that resources and capital are essentially things, which can run out, rather than products of the human will and imagination which in freedom are inexhaustible." Or Julian Simon (1982): "You see, in the end copper and oil come out of our minds. That's really where they are." (Both quoted in Daly & Cobb 1989:109).
[21] This is conceded by Young, who otherwise attacks the contention that entropy leads to scarcity: "It clearly will not be possible to reduce R [resources] to an infinitesimally small value simply by allowing K [capital] to increase rapidly enough." (1991:171). In fact, improvements in efficiency "alter the rate of entropic change of the system without increasing the availability of resources" (Townsend 1992:98).
[22] "The economic limit stems from the physical fact that enormous amounts of energy, as well as other materials, are required to recycle highly dispersed matter. Even in the case of energy ... it always costs more energy to carry out the recycle than the amount recycled. So recycling energy is always uneconomic (regardless of the price of energy!)." (Daly 1991:93).
[23] Jones 1982.
[24] Meadows et al 1992:36.
[25] Meadows et al 1992:78.
[26] *World Commission on Environment and Development* (Brundtland) 1990:14.
[27] Meadows et al 1992:80.
[28] Simpson 1990:95.
[29] Kenneth Boulding pointed out the difference between a cowboy ethic, in which lone individuals can roam and plunder the infinite prairie, and a spaceship ethic, where a closely interdependent group of people has to make do with what is available on board (Korten 1990:33ff).
[30] This can be seen not only in ancient Hebrew and contemporary African traditions, but also in old Germanic traditions. See Russell 1994:120.
[31] "Excessive aggression, competition, and destructive behavior are predominant only in the human species and have to be dealt with in terms of cultural values rather than being 'explained' pseudo-scientifically as inherently natural phenomena" (Capra 1983:302).
[32] For such appeals see Schumacher 1973; Meadows et al 1972; Daly 1973; Mesarovic & Pestel 1975; Laszlo et al 1977; Barney 1980; Daly & Cobb 1989; von Weizsäcker 1992, 1994, and many others.
[33] Cf Chapter 7, Section III.
[34] Cf Chapter 11, Section II.
[35] Cf Marshall Berman: "Their secret - a secret they have managed to keep even from themselves - is that, behind their facades, they are the most violently destructive ruling class in history ..." (Quoted by Heilbroner 1985:136).
[36] Fukuyama 1992:298. The author argues that, if one denies that humans are humans because they have moral choice and are not simply the unwilling product of natural processes, one has to extend equal rights to all creatures. I agree with the emphasis on moral choice, but we must take decisions for, not against nature.
[37] Fukuyama 1992:298.
[38] Examples are Schumpeter, Keynes, Galbraith (1958); Boulding (1970); Schumacher (1974); Heilbroner (1985, 1988), Holland (1987), Davidson & Davidson (1988); Lester Thurow (1983); Lutz & Lux (1979); Karl Polanyi; Henderson (1978); Naqvi (1993); Gerrard (1989), Etzioni (1988), geographer David Smith (1979, 1994), Krugman (1993), and many others.

PART IV

The transformation of social structures

14 Agents of change

What is the task of this chapter?

In Part I we have sketched the social-structural problems of humankind at the turn of the millennium. In Part II we have explored the assumptions, values and norms underlying these social structures and processes. In Part III we have called for a new economic paradigm. Where do we go from here?

Sensitive people cannot help but be afflicted by the immense human suffering caused by poverty, opulence, conflict and the destruction of the biosphere. They cannot subdue their anxiety concerning the future. They cannot help but be appalled by the extent of ignorance, error, narrow horizons and detrimental motivations found in the traditional and the modern worlds. Sensitive people will want to do something about these scourges of humankind and do something practical. What should be done? How can it be done? What can be done by whom? These are the questions which part IV seeks to address.

The present chapter will reflect on possible *agents of change* and their range of competence. The next chapter will draw up the *agenda*. The last three chapters will deal with *public policy*. Because our approach strives to be multi-dimensional and comprehensive, the original manuscript included a chapter on creative change in personal life styles on various income levels. This material has, in the mean time, been published as a separate booklet and we do not need to repeat it here.[1]

We begin with levels of influence and competence from international organisations down to individuals. Special emphasis will fall on the power of primary groups. Then we reflect on the relation between convictions and institutions, ending with the role of the church and the nation state.

Levels of power and competence

In the next chapters we shall design a policy model which is meant to establish the perimeters of the task confronting us in very general terms. To be practically useful it must be broken down into successive layers of sub-tasks assigned to various levels of competence.[2] Ultimately a new economy consists of an infinite number of single but closely interrelated acts, arrangements and processes. To clarify this issue the reader might want to draw up a grid which correlates the levels of responsibility mentioned in this chapter with tasks indicated in the policy model and fill in examples as we go.

We have seen that power is concentrated in some parts of society more than in others. Not to overwhelmed by a task deemed greater than our capacity, we must become aware of the widely differentiated system of power and responsibility which makes up the structures of society. We can safely assume that every citizen has some influence somewhere in the system. We all have our particular potentials, and we all have our particular constraints. Even the most powerful leaders of the most powerful nations have their frustrating limitations.[3] Even the most inspiring innovators cannot move a passive and fatalistic populace.

So there is no justification for anyone to be a spectator. Nobody should continue life as usual and wait for the "authorities" to act. Leaders are leaders only in as far as they are able to articulate, consolidate and channel the potentials and powers of their respective constituencies. Ultimately it is the body politic which has to move, otherwise there is no movement. And the body politic consists of highly diverse sets of statuses and roles in society, all of which are, at their distinctive levels, of equal importance.

So let us gain at least a superficial impression of the full range of levels of competence in modern society.[4] At the highest level there are *international organisations* such as the United Nations, the International Court of Justice, the Commonwealth, the European Community, the North Atlantic Pact Organisation, the Organisation of African Unity, the International Labour Office, the World Bank, the International Monetary Fund, and many others.

Their spheres of competence include issues such as the resolution of international conflicts; the regulation of international trade; conventions on basic values and their implementation, such as human rights and environmental protection. There is no reason, for instance, why GATT agreements cannot include binding codes of conduct or processes which compensate for international economic imbalances.

The constraints of these bodies are mainly related to the collective self-interests of their members and the sovereignty of the nation state. Opposition to an important step forward, such as limitations on whaling, cannot be outvoted, and there is no clout to enforce the agreements once they have been arrived at. These constraints do not need to be permanent. In fact, international cooperation has already been institutionalised to an amazing degree if one compares the present situation with that of a century ago. The goal is not an almighty superstate but an evolving negotiated social contract based on human rights, equity and sustainability.

The next level is the *nation state*. It operates through its legal systems, its bureaucratic organs and the policies of the current administration. Its sphere of competence includes the achievement of democratic procedures, the rule of law, containment of corruption, the enforcement of justice, the resolution of internal conflict, the educational system, research, the moulding of the economy by means of monetary policy and taxation, and the construction of the necessary economic, social and human infrastructure. Its constraints are numerous: lack of financial resources, the power play of interest groups, administrative incompetence, corruption, the dead weight of popular lethargy, external enemies, international crime syndicates, and so on. None of these constraints is an inescapable fate.

At another level we find the *private sector of the economy*. Here we can distinguish various spheres of influence ranging from the chief executives of great multi-national corporations[5] to small undertakings, workshops and professions; the formal and informal sectors; various social partners, such as employers' organisations, trade unions, consumer organisations, and the state. The spheres of competence found in the private sector are extremely varied. They include research and experimentation with alternative products and methods of production; ethical marketing practices; internal democratisation; commitment to codes of conduct; setting professional standards; developing corporate values, and exposing business practice to public scrutiny.

Constraints are mainly related to competition, whether inter-factor (between labour and capital), intra-sector (between different firms producing similar products), or international. The weakest "partners" in the free enterprise system are the marginalised, the unemployed, the unorganised, the economically dependent. There is no reason why competition cannot be mellowed by cooperation and augmented progressively by a common commitment to save and restore the resource base, attain a higher quality of life for the society as a whole, and increase the level of participation of all able bodied citizens in the productive process, rather than pursuing individual profit at whatever cost.[6]

Then there are *the mass media.* Their sphere of competence includes information, analysis, exposure of public evils, political advocacy, and the maintenance of the collective value system. Their constraints can include control by an authoritarian state; prescriptive owners; their dependence on short-sighted and devious advertisers; the desire of the masses for sensational news; the degree of literacy; different cultures of authority (elders, experts or popular idols); different cultures of information and communication (reading, watching, or gossiping); the purchasing power of the population, and the availability of transistor radios and television sets. Again none of these constraints need to be permanent.

Then there is the great variety of *voluntary groups and non-governmental organisations,* such as activists, pressure groups, and religious communities.[7] Their spheres of competence include experimenting with new life styles; pioneering new approaches; filling in welfare gaps left open by the state; channelling the motivations of committed individuals into practical actions, and providing acceptance, belonging and encouragement within a community of like-minded people.[8] The importance of group loyalty can be observed both in international agencies[9] and local groups. Latin American base communities, for instance, have become vehicles for communal self-expression, self-enhancement and advocacy. Even hard-nosed economists begin to discover the desirability of a sense of community in industrial and commercial undertakings.[10] One of the tasks of the state is to create the environment in which non-governmental organisations can flourish. The constraints of voluntary organisations are mainly linked to the fact that, in modern societies, they are dissipated and pluralistic. Networking, dialogue on fundamentals and cooperation may progressively overcome some of these constraints.

The particular sphere of competence of *religious communities* is to cater for the transcendent dimensions of life: a system of meaning, assurance of one's right of existence, and authority. They offer a vision of what ought to be; they critique and confirm values, norms and goals; they offer acceptance and belonging; they build up self-confidence. Their main constraint is the privatisation of religion in a pluralistic and secularised society. A religion which does not go public loses its contact with reality and its impact on society.

Finally we come to the level of individuals. The character of the enterprising individual must be transformed from the death-defying patriot of yesterday and the greedy hoarder of today into the responsible citizen of tomorrow. The place where individual and collective awareness and responsibility can grow is *the primary group*. Primary groups are families, colleagues, sports teams, friendship groups, boards, councils, or cabinets. The constraints of primary groups seem to lie in their minuteness and vulnerability under modern conditions. But if one takes into consideration that they operate at all levels of influence and competence and that they are able to develop linkages, or to coagulate into greater networks, it becomes clear that they can have a considerable social impact, apart from their indispensable role of grounding the individual. Let us pursue this further.

The primary group as the pivotal power unit in society

We have said that social change comes about when a critical mass within the population wants to bring it about. To build up such a critical mass on the motivational level is the work of countless individuals who generate, dissipate and live out new insights in the context of their primary groups. Many institutions can provide a platform for this task: research institutions; the media; primary, secondary and tertiary education; religious communities; business organisations and unions; political parties and non-governmental organisations; the courts and the police force, and so forth. But in the end it is people who have to be motivated to take up the initiative and become active.

When speaking of "the people" as initiators and sustainers of transformative processes we should avoid two pitfalls, the individualistic and the populist view. As mentioned above, single individuals are powerless against the structures of society, even those individuals who are placed in positions of authority. It is commonplace that systems reward and promote their most loyal supporters. One gains power by getting to the top of a social structure or to the front of a social process. Those who obstruct the system are neutralised; those who do not cooperate are marginalised.

On the other hand it is not true to say that it is the "masses" that bring about change. An amorphous mass of individuals would resemble a heap of sand which is unable to move, or a stampeding herd of antelopes which has no sense of direction. It is the primary group which is the locus of initiative and commitment. A society derives its coherence from the operation of primary group loyalty on an infinite number of levels. Groups do not only operate on "grass roots" level, for instance, in the form of Latin American base communities, but wherever people relate with each other.

The most basic primary group is the family. But there are many others: peer groups at school; platoons of soldiers; colleagues in university departments; golf clubs; neighbourhoods; boards of executives; work teams on the shop floor; party caucuses; cabinets, etc. Because traditional structures, such as the extended patriarchal family, the tribe, even the work team,[11] are breaking down, primary groups are increasingly dependent on the voluntary cooperation of committed individuals.

Social research has revealed that primary group loyalty is the most powerful motivating and restraining force in human society. A soldier does not die for his country, or his sovereign, or an ideology, but for his regiment, or his immediate comrades. Factory workers are not easily lured by incentives which undermine peer solidarity.[12] The reason for this phenomenon is the fact that humans cannot live without a sense of acceptance and belonging, thus without a sense of their right of existence. Belonging is granted most effectively by primary groups, but it is granted by such groups subject to their own criteria and conditions. That is the root of their power over the individual. The values such groups embody and the type of relationships they engender is absolutely crucial for the maintenance or the change of cultures, whether one thinks of the extended family in traditional societies, boards of directors of multinational corporations, or street gangs. It is here where a culture is primarily based. And it is from here that change can be generated.[13]

To protect, enlarge, re-institutionalise and re-orientate *the family,* in particular, is one of the greatest challenges of the future. There is no iron law which says that the present nuclear family cannot be extended again to a three or four generation family. But such extended families will have to grant both freedom and belonging to the individual. Unfortunately the family often has had a restrictive, rather than an empowering effect, especially in traditional societies. Yet this no law of nature. While in Africa ancestor veneration seems to prevent innovation, in Taiwan the deceased are given sacrifices to empower modern entrepreneurs. Female and adolescent emancipation does not have to imply the breakup of the family. Rather, the internalised structure of the family must be transformed in such a way that it becomes liberating and empowering for all its members.[14]

Female emancipation and participation are particularly important. Apart from the demands of equity, apart also from the economic imperative of liberating the vast hidden potentials of a discriminated and oppressed group, women tend to have internalised the typical family values of cooperation and mutual concern, as opposed to the individualistic masculine values of competition and domination.[15] With women in leadership positions these values can be fed more readily into the system at all levels of competence.

Our argument is that the primary group has considerably more clout than a lone individual. Obviously primary groups are susceptible to the convictions and commitments of their members. It is true that the kind of influence an individual exerts in a primary group is subject to the ongoing power play within the group. This again is determined by the distribution of the gifts of discernment, argument and leadership within the group, and the ongoing responses of the group to its changing contexts. In a primary group there is a real chance that those who expose facts and offer valid interpretations can shift the power within the group in their direction. A committed individual can make a greater impact in a primary group and through a primary group in society than anywhere else in the social system.

Primary groups in turn have considerably more potential to develop a sense of direction than masses. Depending on its position within the society, a primary group has more chances of making an impact on the society as a whole than any other social formation. In the end it is councils, boards, cabinets, faculties, and the like, who call the shots. Those interested in social change, therefore, must seek to convince individuals, inspire them to stand for their convictions within their primary groups, and motivate these groups to take up the issues concerned in their spheres of competence and influence.

A further dimension of social cohesion is that primary groups are interlinked with each other in a variety of ways. Some of these linkages are purely functional. Employer representatives, for instance, formally negotiate with unions about wage increases. Other linkages come about through the fact that every individual belongs to a variety of primary groups which partially overlap. A cabinet minister may make friends with a union leader or a general at the local golf club. An academic can encounter a lorry driver in a house church. The potential significance of these linkages cannot be overestimated.[16] Constraints imposed by one group can be transcended in another. Challenges experienced in one group can be passed on in another. Such linkages can also be consciously developed. Primary groups which have been won over to a course of action can establish *networks* between themselves, and such networks can ultimately develop a dynamic of their own which can have a considerable impact.

Again two pitfalls must be avoided. Some say that only the *top structures* of society can make a difference, because they command all the power. From this view follows that ordinary people have to sit and wait for the "big guys on top" to act. The result is that their considerable potential is not utilised. Others maintain that the privileged will not want to change because it is against their interests; it is *those who suffer* who have to take the initiative, organise themselves, overthrow the system and liquidate the elites. From this follows that the considerable potential of the elites cannot be utilised.

Both positions are ideological. They are also dehumanising. They also do not take social reality into account. It is not true that elites cannot come to their senses, nor is it true that grass roots communities cannot exert power. Our analysis of social reality shows that all people at all possible levels and in all possible dimensions of society can be, and should be, involved. Nobody should be excused from accountability, least of all the powerful. Appropriate structures and checks and balances must be set in place to guarantee that.

But the motivations of all these parties must be changed. The poor would not be poor, if they had the will and the power to change the system. In as much as they acquire power, they are in the same boat as the elite. To assume that the poor and powerless are not guided by self-interest, especially when they cease to be poor and powerless, is both naive and self-contradictory. If their consciousness has not been changed from self-interest to communal responsibility the system will not change for the better.[17] Motivations again depend on convictions. This brings us to the next point.

The importance of convictions

Change happens in any case. The question is only whether humans are masters or victims of change. To be in charge of their destinies, humans have to get their act together. To get their act together, they need visions, convictions and motivations. Humans have reached an evolutionary stage where they can no longer rely on their instincts. Their superior powers can be used for short-sighted and selfish gain, but they can also be used to attain and maintain healthy balances in society and nature. To go beyond their private lives and their immediate life worlds and envisage the comprehensive wellbeing of the whole of reality demands a high vantage point.

Humans must be able to fathom the discrepancy between *what is* and *what ought* to be. What ought to be is expressed in terms of values and norms. These again depend on a system of meaning. A system of meaning presupposes some sort of perception of the ultimate source and criterion of reality. Empirically such an entity cannot be shown to exist, but, as we have argued in chapter 7, human beings need to transcend experienced reality towards its ultimate foundations. "If there was no God, we would have to invent one."

The human capacity to transcend experienced reality is manifest in the realm of religion in all its forms. In spite of the backwardness, waywardness and harmfulness of many religious and ideological convictions, the realm of convictions as such is critical for human life. It empowers humans to disentangle themselves from shallow expediency. It liberates them to reach out beyond the limits of their immediate experience towards the vast horizons of time and space. Convictions create an awareness of the whole. This in turn enables them to see, interpret and organise reality in its wider contexts.

Awareness of the discrepancy between what is and what ought to be causes restlessness and dissatisfaction with present situations and trends. It is not only rooted in unfulfilled personal needs and desires, but in a sense of the inadequacy of the whole of experienced reality. Awareness of what ought to be prompts humans to develop a vision and to deal critically and constructively with the world in which they live. Convictions are thus able to liberate humans both *from* reality and *for* reality.

Moreover, convictions posit an authority to which humans are accountable, rather than elevating humanity to absolute status. Human interests are not unimportant, but they must be shown to be legitimate. Religion typically transcends all limits. It forms the basis for universal, rather than petty responsibility. It prevents people from indulging in the mentality ascribed to Louis XIV of France: "After me the deluge!"

Faith can motivate humans to explore possibilities which might be hidden in dimensions of reality beyond their reach, notably in the future. It provides them with meaning in the face of meaninglessness, assurance in the face of failure, and the authority to take their lives into their own hands where the powers of fate seem to be overwhelming. The visions of dedicated individuals have led to such great movements and institutions as the Geneva Convention of 1864 and the Red Cross/Crescent, the League of Nations and the United Nations, the Declaration of Human Rights, UNESCO, UNICEF, FAO, WHO, also to resistance against dictatorial regimes and the formulation of alternatives, the development of a socially conscious free enterprise system, and to countless NGOs and their inestimable contribution to the alleviation of suffering and a more fulfilled life.[18]

In short, convictions are quite indispensable. The only question is what type of convictions they are and whether they are part of the problem or part of the solution. Not all convictions are leading in the right direction. Human convictions are subject to human limitations and human depravity, even those deemed to be based on revelation. Committed believers, including myself, should acknowledge historical relativity and human fallibility in the realm of truth, stop being apologetic about their religious commitments, and subject them to critique and reconstruction.

The "redemption" of motivations

Convictions ought to generate motivations, but that does not happen automatically. A great number of people have no socially effective motivation at all. There is the typical *cynic* who only finds fault and questions the human potential to achieve anything good at all. There is the typical *victim* who incessantly complains about the injustices of others. There is the typical *bystander* who does not trust in his/her ability to make a difference. All these attitudes are not very helpful.[19] The first problem is, therefore, how to galvanise people into action; to make them enthusiastic; to impart on them a sense of self-confidence, calling and authority.

The second problem is the direction of the motivation. As mentioned above, change is inevitable in the long run; the question is what kind of change it will be. Leaders and primary groups must be committed to the vision of comprehensive wellbeing. They must also be ahead of their times. They must channel change in the right direction. They must also ease in this change to avoid abrupt and thus disruptive change. The idea of total revolution - raising everything to the ground and building afresh on the ruins - is not only unrealistic, but also self-destructive.

Neither leaders nor primary groups automatically move in the right direction. Clan loyalty may be an obstacle to progress. Gangsters and mafias are also built on group loyalty. "The activities of peer groups range all the way from a sort of guerrilla warfare against management to enthusiastic cooperation with it".[20] Apart from ignorance, outright folly and sinister motivations, groups are held together by common interests. They are not immune against collective selfishness and its ideological legitimations. Therefore motivations must be "redeemed" before they can become positive and effective.

Pure altruism does not seem to be the right approach for a society-wide impact, not even from a Christian point of view.[21] A more potent concept is enlightened self-interest: it is in my own interest to work for the good of the social and natural environment of which I am a part and on which my own life and prosperity depend. So enlightened self-interest boils down to a widening of horizons. That ingroups isolate themselves from outgroups is a well-known phenomenon. To overcome this tendency, the walls which separate ingroups from outgroups must become lucid and permeable. Or rather, the boundaries of ingroups must become progressively more inclusive. Ultimately people must discover that the welfare of every individual and group depends on the welfare of the entire social and natural context in which it is located. The most comprehensive horizon is represented by what we called the "vision of God".[22]

Groups also have a tendency to restrict themselves to narrowly defined objectives. When you play football, you do not talk business. When you are in church, you do not enter into politics. These restrictions must be pried open to allow for more holistic and natural relations within the group.

Opinion leaders can hook fundamental concerns to instances where public awareness flares up, such as the Chernobyl disaster, or the dying of forests in Europe. They can also create such an awareness where alarming facts are readily available, for instance, unemployment statistics. One can also use existing lobbies, take up their legitimate concerns and challenge them to suggest solutions to the deeper lying problems. The concern of "pro-life" groups for the right of a fetus to develop into an adult, for instance, is quite legitimate. But the problems which have led to the demand for the right to abortion cannot simply be dismissed. If you do not want abortion you must make contraception acceptable, safe and available to all. Pro-life groups should also widen their social concern to include the provision of basic needs such as safe streets at night to prevent rape. And what about applying the pro life principle to delegitimate violence and war!

The importance of institutions and positions

What we have said about convictions and motivations does not imply that we opt for *idealism*. It is simply not true that, once the beliefs of people change, social structures will follow suit. Institutions tend to maintain their own stability and develop their own dynamics. Therefore social institutions must be tackled simultaneously, and with the same vigour, as convictions and motivations. Of course, the reverse is also true: a change of social structures does not automatically lead to a change of attitudes, as the *materialists* believe. So how are they related?

It is inappropriate to separate convictions and social structures into neat compartments, to assign convictions to some sort of private spirituality without institutional embodiment, while considering institutions to be nothing but secular and practical arrangements. Neither exists without the other. Collective patterns of behaviour and social institutions are both based on beliefs, assumptions, values, norms, goals and procedures. Mental structures are internalised by those who have been socialised into the system. What is perceived to be normal becomes normative, and what is perceived to be normative guides collective behaviour. The validity of these mental structures is anchored in the collective subconscious and socially embodied in mandated statuses and roles. As such these structures develop a stability and a dynamic of their own which is largely impervious to new insights.

Any institution therefore embodies a collective spirit. No institution could operate without some sort of minimal consensus which includes beliefs, assumptions, values, norms, goals and procedures. If institutions are to change, collective consciousness has to change. Private opinions remain private opinions until they are collectively internalised as valid and become operational in the normal procedures of the institution. It is for this reason that a "change of heart" does not automatically lead to a change of attitudes; a change of attitudes does not automatically lead to change of actions; a change of actions does not automatically lead to change of socially accepted patterns of behaviour, and the latter do not automatically lead to new institutional structures and processes.

These facts have frustrated many prophets and reformers. Those without recognised statuses and roles cannot reach the levers of power, while those who are indeed in control have been rewarded with their positions and privileges for being the system's most loyal supporters and servants. Incumbents are the most unlikely to rock the boat. This insight has convinced early Marxists that revolution and the liquidation of elites was the only way to achieve radical change.[23] But it would be wrong to conclude that assumptions, values and norms are powerless and insignificant. Once a "critical mass" of people has gathered around a cause, social change can be surprisingly rapid and smooth. It is the goal of social action to reach, and go beyond, this threshold. In such a situation it is of critical importance that the incumbents of positions of power are ahead of their times and do not cling to the past.

The critical mass is constituted not by numbers but by accumulated social power. Obviously a ruling elite has more power to make changes than people on grass roots level. But they again have to convince the people that the changes are in their own interest, or at least unavoidable, so that they become legitimate in the perceptions of the population. Conversely, a popular movement has to mobilise considerable numbers of people to constitute social power and convince the leaders that the changes are in their own interest, or unavoidable. But these two extremes are only opposite poles of a whole system of linkages. To make an impact, therefore, we have to work in all directions, from the top down, from the bottom up, and criss-cross over the entire system.

The strengths and weaknesses of the Church

Against the background of these reflections we can assess the potentials of the church in social change.[24] The church has important *assets* not easily attained by secular organisations in peripheral areas, such as:
(a) spiritual and moral foundations which can be mobilised to generate vision, motivation and responsibility,
(b) access to the most deprived grass roots communities,
(c) a traditional focus on the family as the most basic unit of society,
(d) members in all kinds of secular professions and primary groups with all kinds of spheres of influence scattered throughout the fabric of society,
(e) a potential network of cross-cultural relationships which can be activated relatively easily,
(f) an international network of communications.

On the other hand the average church has typical *weaknesses* which need to be overcome, such as:
(a) a spiritualised concept of salvation which neglects social concerns,
(b) a traditionalist orientation which looks backward into the past rather than forward into the future,
(c) an inflexible orthodoxy which spiritualises human needs and offers stereotyped spiritual recipes,
(d) a hierarchical, often authoritarian leadership structure,
(e) a lack of social-analytical skills,
(f) a tendency to withdraw into cozy and homogeneous ingroups which shun challenges and conflicts,
(g) an atmosphere which does not attract the youth, the men in their prime, or leading intellectuals.

What is needed, above all, to make the church relevant in economic and ecological terms is the awareness that sustainable development is derived from its peculiar *vision* and therefore part of its *mission*. Other prerequisites such as openness, social concern, service, an empowering leadership and the acquisition of skills follow from that.

The crucial role of the nation state

The importance of the primary group at the motivational level is matched by the importance of the nation state at the institutional level. In spite of the growing internationalisation of economic processes, the nation state is still the most powerful agent of change, whether to harm or to enhance the economy. This is why the discussion on the choice between a free enterprise system and a command system has lost none of its relevance.[25] State imposed systems have floundered in Eastern Europe and Africa. But in some of the newly industrialising states of South East Asia the state has played a critical and economically beneficial role. All seems to depend on what the state actually does. The following approaches have proved to be counterproductive in the past and need to be abandoned:
(a) The model of *communal sharing*. We have inherited this model from preindustrial societies where it served its purpose well. It offered social cohesion, cooperation and security. Many peripheral areas are still in this stage of development and cannot afford to do without traditional institutions. The model also seems to express the core value of any acceptable social ethic. But in its traditional form it also undermines individual responsibility and paralyses economic initiative. Under modern conditions, injudicious sharing can lead to humiliation, dependency and parasitism.[26] Moreover, because of its inflexible and hierarchical status and role struc-

ture it cannot be retrieved for a society which has undergone the process of emancipation. Communities of the future have to build on equal dignity, freedom, responsibility and sharing in the productive process.

(b) Seemingly on the opposite extreme is what could be called the *elbow model*. Individuals, communities and societies use their respective muscles to squeeze themselves into beneficial positions. Inevitably they do so at the expense of weaker ones. They can do so only as long as the stronger ones choose to tolerate their behaviour. Capitalism as a whole is characterised by this approach and the success of the "young tigers" in South East Asia, so often taken as a blue print to be followed by other developing nations, is a good example. But it offers no humanity-wide solution.[27] Moreover, it is precisely this approach which has led us into the present predicament.

(c) The *central planning* model (or the command economy), practised by Marxist-Leninist regimes, was a response to the ravages of the capitalist approach. Its dramatic failure in recent times has exposed its basic flaws. The most important of these is the assumption that freedom, responsibility and prosperity can be generated by means of authoritarian and totalitarian impositions. Development can only emerge and prosper where free individuals, operating in the context of voluntary grass roots communities, can develop their gifts and their initiative for mutual benefit.

(d) Again this does not mean that the blundering along of the *free market* model is the answer. It should be clear by now that the free market is a fiction in most concrete cases. Where it does operate, it leads to rapidly increasing imbalances, thus negating its own principle of freedom.[28] While it is true that the state should not run the economy, it should make "free enterprise" and balanced need satisfaction possible by levelling the playing fields. "To the extent that competition is self-eliminating we must constantly reestablish it by trustbusting."[29]

When dealing with the nation state, therefore, we need to make a fundamental assumption: it is *the people* who are responsible for their collective wellbeing; the state is nothing but their instrument. This is important both for the political and the economic realm. Authority is vested primarily in the people, not in the state. So the initiatives should emerge from the people. The state should create the necessary space for these initiatives to flourish, provide the necessary infrastructure, and supply the necessary empowerment to make them effectual.

The concept "people" again should not denote "the masses", that is, a collective of isolated individuals who can be manipulated by left wing or right wing elites, but the integrated network of communities. Moreover, the people are more than a bunch of competing interest groups. Fundamental is the awareness that the wellbeing of the whole is the precondition for the wellbeing of the parts. An explicit or implicit social contract should emerge, to which the legal system should give expression. It is with these assumptions in mind that we speak about the tasks of the nation state.

> *Revision:* Spell out the role of *primary groups* in social change on various levels of influence and competence.
> *Application:* Assuming that you have been motivated by this book, how would you try to make an impact?
> *Critique:* (a) "Social processes are driven by collective self-interest and public power, not by personal anxiety, misguided morality, or hysterical activism. Calm down and let society sort itself out!" (b) "If we were all filled with the love of Christ the world would be changed overnight." (c) "Only the state has enough power to change the economy; only the state can act in the interests of the people as a whole." How would you respond?

Let us summarise

After having analysed social structures and patterns of collective consciousness in Parts I - III, Part IV makes concrete suggestions on how the entire syndrome of economic and ecological imbalances could be overcome. The present chapter dealt with possible agents of change, while the following chapters will reflect on policy directions.

We began with various *levels of competence,* from the international community down to the family. Neither the "big guys on top" nor the "grass roots communities" should be absolutised. We ascribed pivotal importance to the network of *primary group loyalties* throughout the system as agents of change. We saw that multiple relations exist between primary groups operating at many levels of society and catering for many dimensions of life. Their collective potential to make an impact is vast, provided their motivations can be transformed.

We continued, therefore, with a reflection on the relation between *convictions, motivations* and *institutions*. Healthy motivations are indispensable. But the power of institutions should not be ignored. In institutional terms it is the *nation state* which has the greatest power to redirect social processes. However, the state should be seen as the agent of the "people", whose task is to liberate and empower individuals and communities and to level the playing fields. It should not constitute an authority structure in its own right which tries to run the economy for the people.

Notes

[1] Nürnberger 1995.
[2] Cf Duchrow 1987:141ff who identifies (a) political institutions, (b) international economic insitutions such as GATT, IMF and World Bank, (c) national and transnational firms and banks, (d) trade unions, (e) science and technology, (f) the media and private individuals as consumers, savers and borrowers.
[3] "When the people refuse to acquiesce, the power of the tyrant evaporates" (Korten 1990:27).
[4] See the table in Duchrow (1994:300-301). Korten offers a useful distinction between the "prince" (the state), the "merchant" (private enterprise), and the "citizen" (civil society, including NGOs, such as people's organisations and voluntary organisations (1990:95ff). The interplay between these agents overcomes the old capitalism-socialism divide: both state and business have their role to play, but the emphasis is now on grass roots community initiative and empowerment against big government and big business.
[5] For a reformation of the "spirituality" of the corporation see Stackhouse 1987:130ff.
[6] Stackhouse believes that the corporation has a dynamic and a character of its own, which needs to be reformed apart from the attitudes of those that serve it (1987:132ff). But that can only be done by primary groups.
[7] For a more detailed treatment of voluntary organisations see Korten 1990:91ff; for the change of their economic paradigms see pages 114ff, for particular targets of such groups see pages 147 and 167. Drucker argues that the nanny-state fails, while the future belongs to "citizenship through the social sector" (1993:152ff).
[8] For their different roles see Korten 1990:185ff.
[9] Korten 1990:197ff.
[10] Drucker 1993:152ff, 156ff.
[11] Drucker 1993:157.
[12] "New norms cannot be effective unless they are also accepted by the group" (Horton and Hunt 1976:160).
[13] "Molecular revolutions do occur; that is, revolutions started by social actors who, like molecules, are organized in groups, in a community, in laboratories of thought and action ... Without the courage to take the first few steps, they would be unable to move ahead, and the possibility of a great change would never open up" (Boff 1995:27).
[14] Cf Korten 1990:169ff.
[15] Korten 1990:169
[16] "Informal cliques that cut across formal work assignments may lead to cooperation and smoother functioning of the the organization" (Horton & Hunt 1976:160).
[17] "Consequently, raising the consciousness of power holders of the nature and consequences of power relationships to impress upon them their stewardship responsibility is as important as carrying out consciousness raising exercises among the powerless that help them discover the sources of their own inherent strength" (Korten 1990:168).
[18] Küng 1997:220; for the last mentioned item, Küng 1997:196ff.
[19] Block 1993:221ff.
[20] Caplow 1971:449.
[21] There is a very definite dialectic between self-assertion and self-denial in the biblical faith which conventional theology has often overlooked. 2 Cor 8:9 indicates the relation: the rich and powerful moves down to enrich and empower the poor and powerless (self-denial). The poor and powerless move up (self-assertion) to fill their rightful place - which is at the side of those who are able to move down to enrich and empower the poor and powerless!
[22] Chapter 7, Section II.

[23] Marx and Engels conceded that a liberal democracy could bring about bloodless change (Nürnberger 1998:54).
[24] Cf liberation theology; recent Catholic social teaching; Protestant social-ethical awareness enhanced by the ecumenical movement, as well as socially conscious Evangelicalism, e.g. Phiri et al (eds) 1996; Samuel & Sugden (eds) 1987; Sider 1993; Sine (ed) 1983. Yamamori et al (eds) 1996.
[25] Cf Nürnberger 1998.
[26] See chapters 8 and 9 for detail.
[27] See chapter 12, Section III, of this book, or Nürnberger 1998, chapter 6.
[28] See Nürnberger 1998, chapter 7.
[29] Daly & Cobb 1989:49.

Drawing up the agenda

What is the task of this chapter?

Parts II and III dealt with the transformation of collective consciousness. Part IV deals with the transformation of social structures. In the previous chapter we dealt with possible *agents of change*. In the present chapter we set up the *parameters* for more detailed policies suggested in subsequent chapters. Section I summarises the kind of *mindset* which we need to bring about fundamental change as it has crystallised out during our discussions so far. Section II sketches five *priorities* on the economic agenda. This is followed by a model which depicts the *major tasks* to be tackled in centres, peripheries and in their field of interaction. To prompt a sense of realism we add a model of *obstacles* in the way of these goals.

Section I: A new mindset

A social system should be changed if it is inefficient or counterproductive in terms of the needs, values and goals of a society. As we have seen in chapter 6, the two dimensions of a social system, social structures and collective consciousness, interact in a complex system of feedback loops. Wherever human agency is required, a system of meaning and a source of motivation are indispensable. In this section we summarise the kind of mindset we need for the reconstructive process, as it crystallised out in Parts II and III.

In the 20th century the debate has centred on two questions, (a) whether the economic system leads to *economic growth*, and (b) whether it brings about economic *equity*. The first is the concern of liberalism, the second the concern of socialism.[1] *Sustainability* is a third concern which is slowly gaining ground. Liberalism is currently in the dominant position. Probably the most important advantage of this system is that it evokes, liberates and empowers "free enterprise" in all its forms and opens the way for the optimal development and utilisation of all the potentials an individual, a group, or a society can muster.

While free initiative is the prerequisite of optimal development in all its forms, the goals targeted by free initiative can be advantageous or counter-productive. Theoretically, initiative is a morally neutral, goal-directed psychological dynamic; its goals can cover the entire spectrum from self-serving ruthlessness to self-effacing service. But human beings are not automatically altruistic and the ruthless pursuit of self-interest can be socially harmful. In fact, selfishness has plagued the human race in all phases of its history. That is why cultures have devised psychological, social and legal means to keep its harmful effects under control.[2] Especially when the survival of a society was in the balance, which has been the rule rather than the exception in early cultures, private motivations could not be allowed to run wild.

For this reason free initiative has been curtailed deliberately and systematically in most *traditional* societies. "In every pre-capitalist society we find acquisitive activity disliked or despised."[3] African traditional cultures are a case in point. They have developed a system of meaning in which equilibrium within the community - stretching backwards to the deceased and forward to the not-yet-born - is absolutely paramount. The acquisition and accumulation of resources, as well as the development and deployment of particular gifts for private purposes, are not only viewed with suspicion, but considered to be positively dangerous for the viability and cohesion of the community. Private gain is taboo in all its forms. What I achieve is never my own; economic potential is a collective, not a private asset. At a time when communal survival was of the essence, profound wisdom lay in these assumptions.

In *modern* society all this has changed. With its growing mastery over nature, increasing food security, rising prosperity and the development of democratic institutions, emerging capitalist societies have progressively relaxed constraints - not only on initiative, but also on the pursuit of self-interest.[4] Liberalism locates self-interest in a particular individual, nationalism in a particular collective.[5] Both perceive cut-throat competition not only as acceptable behaviour, but as a positive value, even at the risk of open conflict. State authority is supposed to prevent the contenders from eliminating each other physically, but not economically.[6] From infancy, Westerners are conditioned to compete, to excel, to outperform others - and to consider what they have acquired as their own by right.[7] So is communal and ecological responsibility simply outdated?

We have seen in Part I that, while centre economies have soared, marginalisation has increased. As a result, there are huge pockets where communal survival continues to be the paramount consideration. Moreover, accelerating development may lead to the decreasing availability of natural resources. If that is true, the struggle for survival may engulf ever greater sections of humanity. In such situations individual selfishness is fatal.

Does this imply that humanity should revert to the communal fixation of roles which inhibit personal initiatives as found in traditional societies? Can one argue that, because in traditional societies self-serving attitudes are discouraged and controlled, they were ahead of modernity in terms of cultural evolution? This would be a wrong conclusion. When power-seeking, status-seeking, avarice or plain initiative are socially and psychologically repressed, they are not necessarily overcome. As any profound encounter with such communities reveals, selfishness reappears in all kinds of covert and destructive guises. An imposed traditional social structure would also lead to intense frustration, listlessness and lethargy among the emancipated. Freedom is indispensable for economic progress.

Moreover, it is only in an environment of freedom that genuine responsibility can grow. In our endeavour to overcome modernist individualism and selfishness our task is not to go back to the fetters of traditionalism but forward to a liberated communal and ecological responsibility. Neither the mindset of traditionalism nor the mindset of modernity is able to support the combination of economic development, social equity and ecological sustainability which constitute the imperatives of the emerging global situation. We need to move beyond both alternatives.

Let us summarise some of the main characteristics of the type of mindset which would be needed.[8]

(a) The most fundamental prerequisite of an appropriate mindset is to gain *a comprehensive picture of reality* in terms of space, time and power. The average consciousness even of the educated elite, has lagged behind contemporary scientific insight. The following dimensions can be distinguished:
- Reality must be seen in terms of *physical* space and structure. Our planet is a tiny part of a giant universe. This universe is constructed according to

principles which we cannot ignore without causing irreparable harm. In chapter 13 we have highlighted the theories of entropy and evolution which seem to be particularly relevant in this regard.
- Reality must be seen in terms of *social* space and structure. Humankind is in the process of forming a single social network. As we have seen in chapter 12, what happens in one section of global society has repercussions for all others. Many analyses still suffer from the assumption that national economies are isolated entities with equal potentials to develop and prosper. The situation is even worse when it comes to economic policies and business behaviour.[9]
- Reality must be seen in terms of *time, process and acceleration*. Short term decisions and arrangements must be seen in the context of their long term cumulative effects. To ignore the phenomenon of acceleration is dangerous for the future.
- Reality must be seen in terms of *power discrepancies*, both between different sections of society on the one hand, and between humankind and nature on the other. The not yet born are the weakest partners in the constellation. For an economy to be balanced and sustainable, power concentrations and concomitant marginalisations must constantly be redressed through countervailing strategies.
- The prospects of *organisational breakdowns* of highly sophisticated organisational systems at local, regional and global levels must be faced realistically.[10] There is wisdom in creating local and regional systems with reasonable degrees of viability and self-sufficiency. If power is balanced, these systems can interact freely without creating dependencies.

(b) We have to perceive human reality in its *concentric contexts:*
- Material needs have to be seen in the context of *total human needs*. Unavoidable sacrifices in the material sphere of life can be compensated for by satisfaction in other spheres.
- The individual has to be seen in the context of *the community*, the community in the context of the *society*, the society in the context of *humankind*, contemporary humankind in the context of the *sequence of generations*.
- Humans again have to be seen in the context of *nature* of which they form an inextricable part and upon which their continued existence depends.[11]

(c) Humankind has to develop a *vision of comprehensive wellbeing* for the entire system of concentric contexts. Inevitable trade-offs have to be built into the vision. Comprehensive wellbeing, to be comprehensive, must be optimal, or balanced wellbeing, rather than maximised wellbeing of one

part at the expense of other parts. When cancer cells thrive, they destroy the body on which they feed.
- Comprehensive wellbeing as ultimate goal implies the targeting of *specific* deficiencies in wellbeing, occurring in any dimension of reality, until thresholds of optimal satisfaction are reached.

(d) Humankind has to overcome *fatalism:*
- *Metaphysical* fatalism of traditionalism, rooted in the concept of dynamistic power, has to be overcome by confrontations with scientific insight and a dynamic understanding of history. Part of this task is to dismantle the dependency syndrome. This does not mean that positive elements of the traditionalist world-view cannot be retrieved, especially where modernity has its own blind spots.
- *Religious* fatalism, which diverts human motivations and expectations from current needs and predicaments towards an imagined and reified otherworldly realm, must be replaced with genuine transcendence which puts pressure on human, social and natural reality to move towards what it ought to be.
- *Moral* fatalism which says that human beings are irredeemably selfish, individualistic and predatory must be replaced with the determination to expose and overcome asocial motivations and behaviour.
- *Economic* fatalism, which accept the inevitability of misery, and posits economic growth as an indispensable and inescapable necessity, must be replaced with a realisation how recent this preoccupation is, how short-lived it will be in the context of the large sweeps of history and how important human solidarity is in global economic terms and in the long run.
- *Ecological* fatalism, emanating from extrapolations of current trends into catastrophic scenarios, must be replaced with the realisation that the future is open and the flow of history is unpredictable. Far from providing grounds for complacency, this openness provides a source of hope and inspiration to avert looming dangers.

(e) The collective experience of humankind seems to suggest that it is important to *unleash and empower the initiative* of the individual, the group and the community, while state tutelage has proved to be counterproductive. The role of the state is to make free enterprise equally accessible to all citizens by preventing the powerful from marginalising the powerless (capitalism) and refraining from running the economy on behalf of the population (socialism).

(f) Humankind must learn to *combine freedom with responsibility*. In traditionalism responsibility is taken to the extreme, but there is little freedom. In modernity freedom is taken to the extreme, but there is little responsi-

bility. Both these versions are contradictions in terms. There is no true responsibility where there is no freedom, as the operation of both traditional societies and state planned economies demonstrate. And there is no true freedom without responsibility as the enslaving character of callousness in Western societies shows.[12]
- Responsibility entails a definite set of *secondary values,* such as concern, care, reliability, punctuality, efficiency, precision, non-corruptibility, customer service (or citizen service in the case of the bureaucracy), etc.

(g) Humankind must develop a new set of *criteria* according to which the right of existence is bestowed and authority is allocated. Both the criteria of progeny in traditionalism and the achievement of wealth and power in modernity must be recognised for what they are: evolutionary outdated and counterproductive cultural relics pertaining to past sets of circumstances. If humankind is to have a future, collective selfishness and its ideological justification must be superseded by concern for universal wellbeing.

(h) Human *motivations* must be changed:
- from concern for the *survival* and prosperity of one's own group to concern for *justice* for all groups;
- from concern for justice to *concern* for the weak and vulnerable;
- from concern for humans only to concern for the endangered *natural world* of which humanity forms an inextricable part.
- The historically latest stage, ecological responsibility, has become so overwhelmingly important in the recent past, that it has to be accorded first priority. It also commands logical priority because all other economic goals become unattainable and therefore irrelevant without the maintenance of natural resources and the living biosphere in which humankind is embedded.

Elsewhere I have argued that this progression is part of the evolution of biblical ethics.[13] It can also be argued that it is part of the evolution of human culture in general, and that the current abandonment of our cultural heritage in favour of producer megalomania and consumer avarice is a retrogressive phenomenon. It can also be argued that humankind has to go through the above four stages again and again and that - contrary to the assumptions concerning "economic man" - there is nothing extraordinary or supernatural about it. It is just the magnitude of the task which has increased through the globalisation of economic processes and the rapid growth of technological power.

> *Revision: What are the reasons for drawing up a list of characteristics of a new collective consciousness? Refer to various facets in the analysis of mindsets offered in this chapter.*
> *Application: Imagine the situation in a board room where the directors of a corporation discuss possible options in the rationalisation of their operations, including mechanisation, retrenchment of workers, exploration of new markets, the disposal of toxic waste, and so on. How could the proposed mindset begin to take shape in such a situation?*
> *Critique: (a) "What is the use of such idealistic demands? They only demonstrate, once again, the fact that moralists are out of touch with economic reality. People just do not function in this fashion. If you think, on the one hand, of the billions of poor and uneducated people around the world, who are primarily concerned about their next meal, and if you think of the entrenched power of self-interested elites on the other, the practicality of the new mindset you propose is pure fantasy." (b) "Why make things so complicated? If only we were all motivated by the love of Christ, the world would be a better place for us all." Try to respond to these two objections.*

Section II: A new economic agenda

Five priorities on the agenda

We now move from assumptions and motivations to practical politics. On the basis of our analyses so far we can draw up a new economic agenda (diagram 15-1).[14] It includes five points in the following order of priorities:

(a) We have to find ways of protecting and enhancing the *natural habitat* on which all life on earth depends. The concern for ecological viability has to be upgraded from a hardly recognised sideline to the most fundamental problem in economics. The key issues to be tackled here are the impact of industrial growth and population growth on the environment. The target is a reduction of the volume of economic throughput from resource base to waste. This volume must remain below the threshold where nature can no longer absorb the impact and regenerate itself. This does not imply bringing the economic process to a halt, but only heeding the speed limits.[15] It also does not imply a lowering of standards but a change from quantity to quality. Because consumption levels of the poor cannot be reduced much further without serious harm to their health and happiness, it is the more affluent who have to bear the burden of frugality. But the poor have to sacrifice inflated expectations.

(b) We have to find ways of securing the material prerequisites for the *life and health* of all human beings. The deceptive expectation that development will create superabundance must be replaced by a recognition of the fact that the extent of misery is growing and will continue to grow exponentially in the future. The key issues here are a lack of efficiency and output in poor societies and the wasteful utilisation of existing means of production by the elites.

(c) We have to find ways of making sure that both the material wealth generated by human creativity, and the sacrifices necessary for its generation, are *distributed equitably*. This implies a change from moralist complaint, capitalist apologetics and socialist flaws to the recognition of the fact that equity is an economic and ecological imperative. Economically, vast discrepancies in income and wealth are "irrational" because resources are offloaded where marginal utility is lowest and withheld where it could be highest. Ecologically, discrepancies lead to the dumping of entropy within the human realm rather than in the natural support base.[16] To achieve this goal one must move from a concern for equalising consumption to a concern for equal participation in production.

(d) We have to find ways of caring for those who are *unable to make a contribution,* even under conditions of equal opportunity. This implies a change from private charity and socialist welfare programmes to a comprehensive social contract which institutionalises concern for the life chances and the empowerment of the weak. Such a change presupposes a collective will based on the realisation that all humans are dependent and vulnerable and need each other.

(e) We have to find ways of *balancing out need satisfaction* in all dimensions of life. Motivations must change from the emphases on profit and pleasure to comprehensive quality of life. To achieve this end, misleading and compulsive marketing techniques must be controlled or neutralised. Room must be made for neglected sources of satisfaction without falling into the trap of cultural impositions. Surely we do not want a totalitarian state which tells its citizens what to believe, strive for and consume.

If the goal is to serve humanity, this should be the broad agenda of economic policies, business practice, economics as an academic discipline, and personal economic responsibility. Whether one looks at the financial reports of enterprises, economic research and training, public economic policy, or popular economic behaviour, this does not appear to be the current agenda of economics and certainly not in this order of priorities. To push some of these concerns into the existing framework will hardly do. We have to go to the roots and think from scratch.

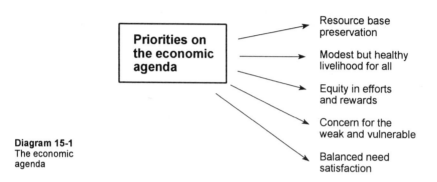

Diagram 15-1
The economic agenda

A basic policy model

In chapter 5 we developed a model of causation which distinguished between factors located in the centre, in the periphery and in the interaction between the two respectively. It also distinguished between structural mechanisms and volitional factors rooted in collective consciousness. To address the syndrome as a whole we have to overcome each of these causes.

Two insights must serve as points of departure. The first is that *the industrial economy in the centre is not sustainable* in the long term. Therefore it also cannot provide the model for peripheral development. The second is that the centre economy is shrinking relative to the total population, while *the marginalised section of the population is growing.* So the goal of integrating the entire peripheral population into an expanding centre economy is unattainable. We should admit this fact and abandon current economic self-deception.

This again implies that the peripheral population must develop an alternative economy. It should be geared to the production of goods and services which satisfy local needs by local means. While peripheral countries can exploit whatever comparative advantage they may possess to fill niches on world markets, their participation in international trade should be highly selective. Free trade as a means of forcing one's way into the Triad - the model of the "Asian tigers" - is not a viable option for the poorest of the poor. On the contrary, they may get seriously hurt in the process.

This means that we have to think in terms of two separate paths to sustainable development, one in the centre and one in the periphery. Ultimately these two streams have to merge into a common, balanced and sustainable economy. But until they merge, the relationship between them must be carefully monitored and active steps must be taken to level the playing fields. This also applies to centres and peripheries within rich and poor countries respectively. The principle can best be demonstrated in the case of semi-peripheral countries. Here the central sector should be integrated into the world economy as far as it can muster comparative advantage in the global market, but no further. If it cannot survive competition it should not produce for export but address the needs of the local population. The argument that its products could be imported more cheaply from world markets does not hold water if the country concerned does not have sufficient export potential to finance such imports.

If a country cannot produce certain levels of sophistication and quality in commodities and services, it simply has to do without them, except perhaps for a few strictly defined necessities. Necessities do not include the luxury wants of the elites. Neither import substitution nor export-led industrialisation should be subsidised, because this can only be done at the expense of the peripheral sector. In fact, it is the central sector which should yield the tax revenue with which the viability of the peripheral sector is to be enhanced. If it cannot survive without state subsidies, it is not viable and should not be propped up artificially. The centre economy in such countries should also not be financed by taking up foreign loans, except in special cases and under highly controlled circumstances. Not only the governments concerned but also the lending agencies bear responsibility in this regard.

The social structural dimension

On the basis of these considerations we can distinguish the following tasks (diagram 15-2):

(a) **In the periphery** the main problems are population growth, ecological deterioration and a level of production which cannot cope with the needs of a growing population. To overcome these problems the peripheral population needs empowerment:

(i) The quantitative *growth of the population* must be arrested to make the growth of the productive capacity and the quality of life of the existing population possible.

(ii) The quality of its productive capacity must be enhanced through holistic *socio-economic development* which grants participation in the productive process to all able-bodied citizens and retains the capacity of the natural environment. The goal should be, as Korten vividly explains, the transformation of the caterpillar into a butterfly, not the growth of a greater caterpillar.[17] Transformation must include beliefs, values, community organisation, economic structures, political structures, technology and international relations.

(b) **In the centre** the problem is the growth of productive capacity beyond reasonable needs, the resultant artificially induced growth of demand, and the impact of growing throughput on the resource base and the natural environment. To overcome these problems the economic dynamic of the centre must be redirected, not arrested. Production must change from quantitative output to the enhancement of the *quality of life*.[18] This implies a change of motivation from profit and pleasure maximisation to communal responsibility and a commitment to "generational equity":

(i) Concerning consumption, centre demand must reflect *balanced need satisfaction*. The population should regain its courage to control, through its democratic institutions, the commercial abuse of human weaknesses to create pseudo-needs. This can be done, for instance, by the progressive taxation of advertising and its substitution with computerised information on available products. Supply will then respond to reconstructed demand.

(ii) Concerning production, the society should cease to grant legitimacy to minority profit maximisation through technological advance and capital accumulation. Priority should be given to the creation of meaningful *employment* for the majority and to the health of the *natural environment*. The current distortion of the factor market has to be overcome by progressive taxation of energy, capital-intensive methods of production and the generation of waste and pollution.[19] If a subsidy or tax relief is granted at all - which is not a good thing in principle - it should be granted to labour-intensive and resource-saving production rather than to capital, energy and technological sophistication.

(c) The **interaction between centre and periphery.** Here the problem lies in the fact that the greater power of the centre leads to relative or absolute advantages for the centre and relative or absolute disadvantages for the periphery due to structural mechanisms and the self-interested use of power. To address the problem economic potency must be balanced out:

(i) *Demand and supply should be relocated* from the centre, where the market is essentially saturated, to the periphery where there are abysses of unfulfilled needs. Purchasing power must be built up in the periphery, not through handouts, but through favourable terms of trade and the growth of the productive capacity of the peripheral population. The centre dynamic should be partially harnessed and redirected to help bring about the prerequisites of this development in the periphery.[20] The debt burden of countries whose economies are paralysed by debt servicing must be written off, and new loans should be applied for and granted with utmost circumspection. Care should be taken that it is the periphery's self-generated, self-sustained and ecologically sustainable development that is built up, not an extension of centre offshoots which impose development patterns of the centre.

(ii) The *abuse of power must be monitored and curtailed* through national and international legal instruments. To render the abuse of power socially unacceptable, basic human values must be retrieved and reinforced in the population.[21] Structural mechanisms which benefit the centre should be neutralised with countervailing processes. Centre populations should refuse to accept wealth due to the flow of financial and material resources from the periphery to the centre, and reverse this flow by compensatory payments based on institutionalised agreements. It should be obvious that all this means that the network of multinational corporations, which has begun to control the global economy, has to be subjected to international controls. If they are to remain, or become, acceptable members of the world community, they have to transform their own motivations, assumptions and goals.[22]

Two further dimensions of critical importance cut across the centre-periphery model. The first is **armed conflict.** Greater determination must be invested in conflict resolution. The current pattern that great nations intervene only when their energy supplies are in danger, reveals an irresponsible and unacceptable attitude.

(a) Most conflicts are caused by discrepancies in wealth and power. There is the urge to dominate and the resentment of being dominated - both in economic and political terms. Competition for power and resources again leads to the abuse of ethnic sentiments to legitimate and mobilise collective interests. *Redressing economic imbalances* and demythologising ethnicity cut out major sources of conflict. The resolution of conflict again set resources free for survival and development.

(b) A determined international commitment must be generated to bring about the *democratisation* of Third World countries.[23] Conflicts should be resolved through courts of law rather than through violent means. Conflict resolution through negotiation is better than fighting it out. But negotiation is less satisfactory than due processes of law because negotiation allows partners to exert pressure according to their different levels of power and get away with socially unacceptable behaviour.

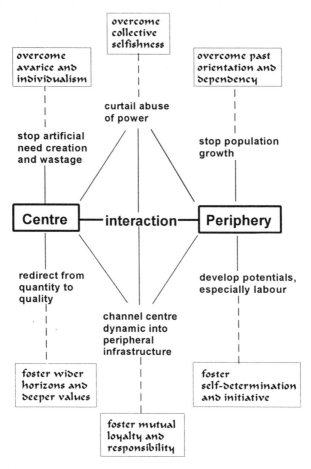

Diagram 15-2
The basic
policy model

(c) If we are serious with our commitment to peace, democratisation cannot be restricted to the nation state. Democrats who insist on the autonomy and sovereignty of the nation state are inconsistent in their philosophy. Nor can democracy be confined to the political sphere and leave *economic* power differentials out of account. Ultimately there is no substitute for a gradual but determined move towards the horizontalisation and democratisation of national and international economic relationships.[24]

(d) The *arms trade* should be subjected to tight international controls. Balanced arms reduction should be negotiated between partner states in conflict zones under international supervision. The arms trade should be tightly controlled by an international legal framework. Humankind simply cannot afford wars any more. Efforts must be made to disarm the population in conflict-ridden societies, especially militias and criminals. Obviously this presupposes legitimate and responsible governments.

The last and most fundamental aspect is the **preservation of nature.** Here the problems are depletion of non-renewable resources, over-exploitation of renewable resources, and pollution of the bio-sphere, all of which are caused by industrial growth, population growth and the growth of destructive conflict. The goal is to maintain and enhance the wellbeing of the natural world, which is our only resource-base. Three aspects need attention:

(a) We have to extract and utilise *non-renewable* resources, such as fossil fuels and scarce metals, with utmost restraint, cut out wastage as far as possible and look for alternatives. Exploitation levels should remain below the rate at which alternatives are being found and developed at any given point in time.

(b) We have to reduce the exploitation of *renewable* natural resources, such as fish and forests, to below the limits of the capacity of nature to regenerate itself.

(c) We have to cut down on *pollution and waste* to well below the level of the capacity of natural sinks to absorb the impact.

The dimension of collective consciousness

At the level of collective consciousness the task in the centre is to overcome the destructive attitudes of *individualism and materialism*. Ideologies which legitimate the blind pursuit of self-interest and instil a desire for material overconsumption, must make way for wide horizons, long term planning, self-control, freedom and responsibility. This cannot simply be left to preachers of morality. Unredeemed selfishness must be contained by institutions which make accountability obligatory in all walks of life and on all levels. It is not only the task of religious communities but also of parliaments, political parties, the legal profession, the media and the educational system.

In the periphery we need a kind of *spiritual empowerment* which liberates the population from traditionalist fetters and the dependency syndrome, on the one hand, and from the lure of Western consumption patterns on the other. Individual and communal initiatives must be enhanced without leading into the trap of individualist selfishness.

The problems caused by traumatisation through *cultural clash* and imperial domination must be tackled. As we have seen, cultural self-assertion may lead to the defence of traditionalist traits which present obstacles to progress, such as the authority of the ancestors, the subordination of women, or the right of diviners to determine agricultural practices. The arrogance of the technological civilisation and the denigration of other cultures are out of place. But if one wants to enjoy the fruits of progress, one has to link up with the set of assumptions, values and norms which makes the generation of progress possible.

In the interaction area we have to overcome collective selfishness and foster *mutual responsibility*. The awareness must spread that the "space ship Earth" is becoming too small and too vulnerable to allow destructive competition and violent conflict. Humankind must begin to appreciate the value of mutual assistance and cooperation in a "world growing together".

Obstacles to change

Formidable obstacles have to be overcome if these goals are to be achieved. Some of the more important are summarised in diagram 15-3. The vertical axis in this model represents obstacles located in the sphere of *convictions,* assumptions, values, norms and goals. There is an intricate interaction between accepted systems of meaning and their respective normative systems on the one hand and the vital interests of the main actors in society and their ideological legitimations on the other. We have seen that "popular wisdom" is a powerful but problematic mediator.

The horizontal axis of the model represents constraints which are not based on mindsets but on *structural and technical factors.* It includes ignorance, wrong information and narrow horizons on the one hand, and the rigidity of existing structures and technical constraints on the other.

(a) *Ignorance.* Available information is plentiful, but it is only accessible to a tiny elite of academics and experts. Neither the masses nor the great decision makers seem to possess the theoretical knowledge and practical experience necessary for the transformation of society. If these issues are not taken up by mass media, educational institutions and popular movements we are wasting our time.

(b) *Narrow horizons*. Social research has revealed that even major decision makers in the world today have extremely narrow horizons. Ethnocentric motivations and short term interests prevail.[25] The masses at grass roots level, who struggle to make ends meet, place ecological considerations and social accountability on the back burner.

(c) *Structural rigidities*. Social structures are always unwieldy and difficult to change. People have internalised certain beliefs, obligations, collective expectations and a myriad of procedures and conventions embedded in vast networks of institutionalised relationships. It is difficult enough to change the course of a large business undertaking, let alone that of a national economy. Countries which had made up the former Soviet Union, for example, found out with dismay that one cannot simply switch from socialism to capitalism and expect it to work. Nor could the United States easily adopt socialism, even if it wanted to.

(d) *Technical problems*. Social change and economic development are, to a large extent, beset by technical problems. How do you harness the energy of the sun in a simple sun cooker? How do you boost crop yields, yet retain the fertility of the soil? How do you make labour-intensive production competitive? How do you reallocate land in a feudal situation such as North-Western Brazil? How do you overcome the consequences of worsening periodic droughts in Ethiopia? How do you deliver food aid to refugees in war torn regions such as Somalia or the Southern Sudan?

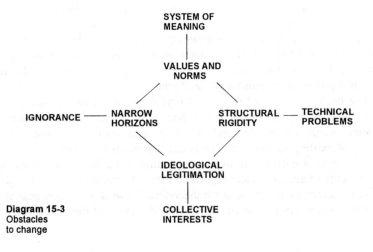

Diagram 15-3
Obstacles
to change

Chapter 15 - Drawing up the agenda | 393

There are close linkages and feedback loops between all these problems. A technical problem is never only a technical problem because the prevalent system of meaning defines what can be perceived as "nothing but a technical problem" in the first place. The kind of solution which would be considered to be appropriate also cannot be taken for granted. The system of meaning again is largely distorted by collective interests. Conversely, structural changes transform the system of meaning as well as collective interests. To some extent relevant information can overcome inappropriate assumptions, but information is again interpreted on the basis of assumptions. Assumptions can also enhance or inhibit the free flow of information. In short, we have to look at the whole network of feedback loops if we do not want to be trapped into blinkered perceptions and one-sided policies.

Revision: Summarise, with the help of diagrams, the main tasks of social change as well as the obstacles confronting such efforts.

Application: Imagine that you were called to be a government consultant on upgrading the periphery in a semi-industrialised country such as Brazil or South Africa. What would be your priorities?

Critique: (a) "These sweeping generalisations are of no practical use at all. It is the detail which matters." (b) "If you think that an academic, sitting in his little office, can find a quick-fix for a whole society you are deluding yourself and others." (c) "Your economic agenda goes far beyond anything that the real economy could ever achieve; what we need is a set of realistic guidelines. Respond to these statements.

Let us summarise

Part IV deals with social change. The previous chapter indicates potential agencies of change; the present chapter sets up general parameters, while the following chapters deal with public policy. In Section I we summarised the kind of *collective mindset* which has to underpin any initiative towards structural change. Our view of reality must become as inclusive as possible; place material needs into their concentric contexts (total needs, community, society, humankind, nature); develop a vision of comprehensive wellbeing; target specific deficiencies occurring in any dimension of reality; overcome metaphysical, ideological, scientific and moral fatalism; acquire a value system which accords the right of existence on the basis of human dignity, not performance, and combine freedom with responsibility. The value system which motivates thought and action should progress from survival to justice, from justice to concern for the weak and vulnerable, and from concern for humans to concern for an endangered natural world.

Section II offered a *new economic agenda*. Its five priorities are sustainability, adequacy for all, equity, concern for the weak, and balance in the satisfaction of needs. We then constructed a *model of targets*, which is based on the model of causes developed in chapter 5. In the periphery the productive capacity must be enhanced and population growth must be arrested. In the centre consumption must be reduced to authentic needs and production must be channelled from quantity to quality. In the realm of interaction productive capacity must be relocated from surplus areas to deficiency areas, the abuse of power must be curtailed and the consequences of asymmetrical interaction must be counterbalanced.

The problems of conflict resolution and ecological sustainability cut across the centre-periphery model. Economic power concentrations must be overcome because they lead to entropy being offloaded on weaker sections of humanity rather than on the natural support base. And the impact of humanity on nature must be reduced to below its capacity to absorb waste and regenerate itself.

At the level of *collective consciousness* individualistic materialism in the centre, traditionalist fetters in the periphery, and collective selfishness in the interaction between the two must be overcome. We added a model depicting different kinds of *obstacles,* including convictions, interests, ignorance, structural inertia and technical problems. With this broad overview in place we can now launch into more detailed policy considerations.

Notes

[1] For the following, cf Nürnberger 1986, especially chapter 7.
[2] Daly & Cobb 1989:89ff.
[3] Heilbroner 1985:109.
[4] For a short summary of the value system of "cowboy economics" see Korten 1990:43ff.
[5] For an interpretation of nationalism as an expression of the collective desire for greatness, rather than economic gain, see Fukuyama 1992:266ff.
[6] For more detail see Hirschman 1977.
[7] Note that I do not distinguish here between desire for material possessions and what Fukuyama calls *thymos,* that is the longing for recognition or prestige, because, beyond basic needs, material acquisitiveness precisely functions as means to gain recognition and prestige in the capitalist society.
[8] Cf the useful ideas of Capra 1992:81ff.
[9] "The Earth is one but the world is not. We all depend on one biosphere for sustaining

our lives. Yet each community, each country, strives for survival and prosperity with little regard for its impact on others." World Commission on Environment and Development 1990:27.

[10] The repercussions of a rapid spread of gloom on global stock markets, as experienced in 1998 with the sudden deterioration of the Pacific Rim, and the impact of the turn of the millenium on global computer networks are relatively mild forms of such breakdowns. The chain reactions leading to two world wars in the 20th Century and the catastrophe which could have ensued if a nuclear attack had been triggered during the Cold War are more formidable. But these examples could all be dwarfed by global organisational breakdowns in the future if the present trend towards highly sophisticated integration and interdependence continues.

[11] See Cobb 1992:87ff for anti-anthropocentrism.

[12] For detail see Nürnberger 1998, chapter 7.

[13] Cf Nürnberger 1988:301ff.

[14] Wogaman 1986:58ff sets the following priorities: (a) adequate production, (b) equity and security, (c) employment and educational opportunity, (d) conservation, (e) a new world order which overcomes the gap between rich and poor. Note that I begin with the ecological question.

[15] Meadows et al 1992:99.

[16] Cf Sider (1977:52).*

[17] Korten 1990:133.

[18] See Power 1988.

[19] For "green taxes" see Turner, Pearce and Bateman 1994:166ff.

[20] For international aid and its problematic features see Korten 1990:136ff.

[21] Cf Küng 1998:234ff.

[22] According to the economist H Hesse, ordinary people have an interest in being protected from exploitation and domination. Economic changes open up new avenues for the demons of avarice and ambition and moral norms must be adjusted accordingly. Where asocial behaviour surfaces in a free society, it must be blocked; if the group pressure of collective morality is not strong enough to do this, state laws must be imposed (Küng 1997:238f).

[23] Korten 1990:171ff.

[24] Korten 1990:173.

[25] Laszlo 1977.

16 Policies for peripheries

What is the task of this chapter?

In chapter 14 we drew up an agenda for action. In the following three chapters we shall deal with institutions and policies aimed at redirecting social processes. The field to be covered is immense. To keep our discussion within manageable proportions we simply offer a few pertinent examples of what can be done, first concerning developments in the *periphery* (chapter 16), then in the *centre* (chapter 17), then concerning the *interaction* between the two (chapter 18). These suggestions try to give expression to the overall vision and to prompt creative imagination, exploratory thought and experimental action. The detail has to be worked out by experts and practitioners in the field through interdisciplinary consultation and cooperation.[1]

Some of the examples offered are already being implemented; some are widely discussed; some seem to be revolutionary, some even utopian. But looking back from where we should be in half a century's time, even the wildest dreams may appear to have been inadequate. Is it realistic to expect such revolutions to happen? The recent history of societal transitions in Eastern Europe and South Africa has taught us that what seems to be unthinkable today may be self-evident tomorrow. Moreover, the radicality of the vision of what ought to be should not be watered down by feasibility considerations, because it establishes the direction in which we need to move. Visionary thought is necessary to pave the way for, and give direction to, incremental action in real-life situations.

Overview

It should be obvious by now that the contents of the following three chapters are intimately connected. A brief restatement of the problem and the suggested solution may be useful to guide the reader through the detail. Economic imbalances are the essence of the problem. In the periphery they include a reduction of deaths without a reduction of births, and consumption of modern consumer goods and services without the means of producing such goods and services. In the centre they include a productive capacity which outstrips genuine need and leads to artificial need creation, and the resultant exploitation and pollution of natural resources without their regeneration. The interaction between the two is marred by asymmetrical power relations. The suggested solutions can be summarised as follows:
(a) In the *periphery* population growth must be arrested and the economy must change from stagnation to balanced and sustainable development.
(b) In the *centre* the economic dynamic must be redirected from growth in quantity to growth in quality; from ecological destruction to environmental protection; from marginalisation of the periphery to its empowerment.
(c) *Between centre and periphery* the abuse of power must be curtailed and countervailing processes must be institutionalised to neutralise structural mechanisms which benefit the centre at the expense of the periphery.

The respective *collective mindsets* have to change accordingly: in the centre, individualism and avarice must make way for comprehensive horizons and collective responsibility; in the periphery, past-orientation and dependency must make way for self-determination and initiative; in the relation between the two, collective selfishness must make way for mutual loyalty and responsibility for future generations.

As a result of these measures, the wasteful depletion of non-renewable resources, the overexploitation of renewable resources, the destruction of the environment through pollution and the buildup of conflict potential should make way for a sustainable and peaceful future for all.

Realistic goals

Before we begin with a discussion of policies concerning the periphery, we have to debunk unrealistic expectations. The goal of the periphery should not be to aspire to the levels of production and consumption found in the economic centres because this is neither feasible nor desirable.

It is not *feasible* because the environment cannot yield the resources and absorb the impact. Long before the entire Third World could reach First World standards, fossil fuels will largely have been depleted. As they become scarce, their prices are bound to rise to levels which make their use for large scale energy generation uneconomical. So far there is no alternative in sight. Even if virtually unlimited new deposits were discovered, which is not likely, the ecological impact of pollution would become prohibitive. No amount of indignation, frustration or rationalisation can make these realities go away. The sooner humankind accepts the inevitable, the sooner it will be willing to contemplate alternatives.

Industrial overdevelopment, as found in the centre, is also not *desirable* because it does not cater for a balanced satisfaction of needs. In many respects the quality of life is lower in modern cities than in traditional societies. By abandoning First World standards, Third World societies could demonstrate to the rest of humankind that it is quite possible to lead healthy and happy lives without the economic rat race, environmental degradation and spiritual superficiality of the industrial civilisation.

If humankind became modest and restrained, the following *proximate goals* should be attainable within a few decades without destroying the biosphere: employment for every able-bodied person; a site and service scheme which allows each family to build a modest home and upgrade it in time; a reasonably balanced ration, clean water supplies and microbe-based toilets for everybody; a solar cooker, perhaps even a solar hot water system, for every family; sufficient protection against the cold; primary health care, simple home cure education and a system of regional clinics; primary education for every child and access to higher education for the gifted; a transistor radio for every family and a television set for every community, administered by local schools, to cater for communication, cultural upliftment and participation in regional and national affairs; access to adequate sports facilities for everybody; a bicycle or an animal drawn cart per family; public transport for going to school and to work where necessary.

If we could achieve these goals, material living conditions would be adequate and the great majority of humans could attain rich and fulfilled lives. In fact, the quality of life of the peripheral population would have improved dramatically. It would also have surpassed the standards of living humankind had to contend with throughout its long history before the 20th Century.[2]

The first goal of peripheral rehabilitation should be to decrease the vulnerability of those most exposed to recurring hazards. There are people who have built their huts on river banks which are prone to be swept away by floods. There are people crammed in slums which can turn into infernos when strong winds sweep fires through them. There are people who do not have access to clean water and become vulnerable to epidemics. There are people living in fragile environments which are prone to drought, erosion, deforestation and desertification, and who become vulnerable to malnutrition and famine. This is where peripheral policies should begin.

The first aspect: population control

We shall now mention a number of factors which need to be taken into account. The sequence is not meant to indicate any order of priorities. Let us begin with population growth. This is a highly emotional issue and the first requirement is that we remain as factual as possible. The indisputable fact that centre populations consume too much, for instance, in no way diminishes the other indisputable fact that peripheral populations grow too rapidly. The sooner we face up to *both* these facts the better for the poor and for the future of humankind as a whole.[3]

Secondly, we must distinguish between private morality and public necessity. Personal or communal issues, such as family planning, female emancipation and human rights are important, but we are here concerned with another order of magnitude. Overpopulation is a matter of the survival and prosperity of a whole country, ultimately of the human race as a whole.[4]

Thirdly, where rapid population growth occurs it is the result of a developmental imbalance: due to modern medicine, more hygienic conditions and more balanced diets the death rate has fallen dramatically while the birth rate has not been reduced accordingly. The problem has been created by the introduction of Western manipulative science into one side of the equation and not into the other. Those who reject the scientific-technological control of the birth rate - whether on account of traditional cultures, or religious convictions, or human rights considerations - should be consistent and also reject the scientific-technological reduction of the death rate.

As a fundamental policy guideline we demand, therefore, that strictly equal amounts be allocated in the budgets of states and relevant non-governmental organisations to the improvement of health care (lowering the mortality rate) on the one hand, and family planning measures (lowering the fertility rate) on the other. In the long run, overpopulation is more destructive of human life and wellbeing than any epidemic.

Contraception

How can fertility be controlled? For a culture sensitive to human rights and democratic freedoms it is self-evident that services should not be imposed but offered by the state and freely accepted or rejected by the citizens. But this presupposes that the population is capable of taking informed decisions and have the means to implement them.

The first factor is located at the level of collective consciousness. There are considerable cultural and emotional obstacles to contraception in Third World countries, often fuelled by religious sentiments and traditions. Freedom of choice implies the reconstruction of religious convictions and cultural assumptions. Here churches and religious communities have to play their indispensable role - apart from educational institutions, political parties, the mass media and voluntary organisations.

It is a sad fact that some religions, including some churches, are part of the problem in this regard, rather than part of the solution. Some argue that contraception deprives human beings of their chance to live. Thus contraception is taken to be akin to abortion. Obviously this argument is fallacious. Human beings which have never been conceived do not exist and cannot be deprived of their lives.

Some traditionalists resist female emancipation and contraception on the grounds that they are Western impositions, that male authority is part of their cultural identity and that their children are evidence of their virility. But culture should serve human communities, not enslave and harm them. It is self-interest which is legitimated by such arguments, rather than a precious cultural heritage. Fathers (and mothers) should base their self-esteem on the quality of the life of their children, rather than their numbers.

The position of the Vatican is that contraception is unnatural, thus sinful, because it obstructs the purpose of the sexual act, which is defined as procreation. This argument is also untenable. Apart from the fact that procreation is not the sole purpose of human sexuality, human nature is such that we have to transcend nature. Primitive stone tools and modern inventions, such as chemical fertilisers, the use of tractors, global food trade, and modern medicine, are all as unnatural as contraception. Scientific-technological interventions into nature always have detrimental side-effects and these need to be controlled. But the argument that contraception is unnatural is irrelevant because nothing we do as humans is natural.

Assuming that the internal obstructions to free and responsible decision making have been removed, family planning should be voluntary. The Chinese have managed to arrest population growth by draconian means. Because this nation comprises a quarter of humankind, their success is of great significance for humankind as a whole. Yet the inhuman means to achieve this goal are problematic to say the least. As people concerned about human rights we must insist on freedom of choice.[5] But families must also be enabled to implement their decisions. There are at least four preconditions which must be fulfilled before freedom of choice can become effective:

(a) *Information*, including sex education, creation of an awareness of the global problem of overpopulation, and conscientisation concerning human rights and responsibilities.
(b) The *availability* of harmless and affordable contraceptives.
(c) *Old age security:* parents should not generate a large offspring for fear of being abandoned in their helpless years.
(d) *Female emancipation.* Women must be entitled to refuse pregnancies without fear of victimisation. A modern state has the duty to create the institutional framework for female emancipation and gender equality in all spheres of life, including procreational behaviour.[6]

These requirements tend to be fulfilled as the prosperity of a population grows. Because we have to break out of the vicious circle between population growth and misery, however, we cannot wait for this to happen. Just as food supplements or medicines are introduced where the body cannot fend for itself, the state has to take measures where the society cannot cope on its own.

Abortion

In some societies the legality, and availability, of abortion has become a substantial factor in the reduction of the growth rate of the population. Here individual morality is made subservient to public necessity. In what follows I will not address the issue of individual morality, therefore, except for saying that where abortion is the lesser evil, it must be legal and possible to choose the lesser evil, otherwise the greater evil will occur - such as back street abortion, infanticide and suicide. However, our present argument is located at the level of public necessity, rather than individual morality.[7] It says that the appropriate means of reducing the birth rate is contraception, not abortion.

I found it helpful to compare abortion with warfare. According to current social norms, a war can be waged in self-defence against the violence of an aggressor. Soldiers are expected to die for the wellbeing of the community, though they may be personally innocent. Enemy soldiers can be killed, though they are equally innocent. Neither the one nor the other is considered to be murder.

Abortion, just like warfare, is killing human beings, though in the case of abortion it is killing humans-in-becoming.[8] The problem is not, therefore, whether abortion is killing (potential) human beings, but whether killing human beings is justified and under which circumstances. If the state is entitled to lead the cream of its youth into death on the battle field, when it considers this to be in the interest of the society, there is no reason why a state should not allow mass abortion, if an avalanche of births threatens the survival and prosperity of the nation.[9] If moralists insist on "just war" criteria, they should draw up similar criteria for legalised abortion.[10]

However, what if the greater evil can be avoided without resorting to the lesser evil? In my opinion neither war nor abortion should ever be justified, because both are avoidable. The necessity of resorting to military might is due to the failure of conflict resolution through peaceful means: democratic decision-making, institutional equity, legal procedure, negotiation and diplomacy. Under modern liberal presuppositions war should be outlawed as an option.[11]

Similarly abortion is the result of violence or neglect. Healthy marriages, sexual discipline, responsible family planning and effective crime prevention should obviate the problem. It is only when the fabric of a society has deteriorated beyond certain thresholds that the question whether abortion is the lesser or the greater evil arises.

The problem is largely caused by the carelessness with which our modern culture treats socially harmful developments. War and abortion are both condoned violence. They represent the tendency of our culture to choose the easy way out, to take shortcuts, to try and get away with it, whether others are hurt or not.

Just as warfare can be avoided through equity and diplomacy, population growth can be curtailed through sexual discipline and contraception. Those who are *against* abortion must be *for* sexual discipline and contraception, just as those who are *against* war must be *for* equity and democracy, because that is where the solution is to be found.

The second aspect: self-reliant and sustainable development

Underdevelopment is due to a lack of balance between production and consumption. That is the root of the problem. Third World populations have been lured into the propensity of consuming the type, the quality and the quantity of goods and services made possible by a modern industrial economy without having the capacity, the sophistication, the organisation and the mentality to produce such goods and services themselves. This includes medical services which reduce the death rate.

If peripheral populations had ample financial resources they could simply purchase all these goodies. This is the situation in some oil-rich countries, but even here it is a relatively short lived, thus deceptive situation. To live on borrowed money, true to the commercial slogan "enjoy now, pay later", is no solution because what will come "later" is poverty, not abundance. The other way out, to beg for handouts, is both dehumanising and unsustainable. So there is really no choice: in the long term a society cannot consume more than the equivalent of what it is able to produce. The simple lesson is that no society can continue to live beyond its means for long. To break even it has to reduce its consumption and raise its production. There is no other way.

If all this is true, there is no escape from the fact that peripheral countries and regions should base their economic policies on local self-sufficiency and sustainability.[12] However, to use a suggestive picture again, this means the transformation of the caterpillar into a butterfly, not the growth of a greater caterpillar. The policy directive called for is to *develop local skills* and exploit *local resources* to fulfil *local needs*. Tanzania's Ujamaa experiment did not falter because Nyerere's principle of self-reliance was wrong, but for other reasons.[13]

To the extent that the basic needs and the basic means of their populations are largely located in rural areas, such countries should build up their agricultural potential before even thinking of urban-industrial development. Under no circumstances should the only potentially viable sector in the country, that is agriculture, be milked by the tax man to subsidise an industrial production which has not proved its viability.

There should also be no enhancement of production for export at the expense of local need satisfaction. Apart from international debt servicing, to which we shall come in chapter 18, the hard currency ostensibly needed by the country is invariably needed by the rich and powerful, not by the peasants. If the rural majority cannot enjoy modern luxuries, why should they pay for the consumption of such luxuries by urban elites![14]

Agricultural production should be boosted not by introducing imported fertilisers, machinery and fuels, but by allowing realistic food prices to generate the incentive to utilise existing agricultural potentials and develop the productivity of local methods and tools. State intervention is necessary to enhance this objective, for instance in the areas of research, soil conservation, afforestation, transport, and so on.

To achieve agricultural self-sufficiency, local food prices may not be lowered artificially for the benefit of urbanites and at the expense of peasants. This was a common mistake made by Third World governments, especially in Africa.[15] The result was rapid urbanisation and the collapse of rural production. Urban elites have to adjust their patterns of consumption away from centre standards towards the possibilities the country can offer.

To escape dependence on South Africa, for instance, Zimbabwe raised food prices and achieved a dramatic rise in food production, became self-sufficient and had enough for export. In 1994 Zimbabwe was the biggest supplier of maize for the UN World Food Programme. It is highly significant that not only commercial farms but also tribal subsistence agriculture responded positively to prices.[16]

The example of Taiwan is revealing in this regard. Here a flourishing agriculture soon reached its natural limits. But the developmental dynamic generated on the agricultural base spilled over into secondary occupations such as spinning, weaving, sewing, carpentry, forgery, leather works, trading, road building, transport and so on, all servicing local needs. Workshops could then develop into small industrial enterprises which produce value added, but still for local consumption and based on local skills and resources. In this way the country-side developed evenly and the great migration to the cities was avoided.[17] Finally, Taiwanese producers launched out into niches in the global markets.

The idea of import substitution, which has led many developing countries into stagnation, was not wrong in principle. But one should not try to produce Western type manufactures locally with imported capital, raw materials, expertise and machinery. This only leads to dependency, because the local population cannot generate these means of production. It also leads to inefficiencies of scale, because the local market is, in most cases, too limited to reach sufficient volumes for making such a production competitive.

The "Asian tigers" switched to aggressive export-led industrialisation and Triad market penetration, when import substitution reached its limits. If this road is not open to all Third World countries, as we have argued in chapter 12, the lesson to be learnt is that import substitution does not go far enough. Instead of translocating Western means of production to Third World countries to produce Western manufactures, these Western products themselves should be substituted by products which can be produced locally without importing machinery, expertise and raw materials.

This does not imply that poor economies should be isolated from the world. Even the poorest country needs foreign exchange for essential goods and services which it cannot produce itself. It means, rather, that they should regain sovereignty and dignity in their international relationships. They should develop their local capacities, build on their strengths and develop the pride of doing without things they cannot afford. It is a process of inner and outer emancipation. Expenditure on imports should not go beyond the genuine earnings of exports. Nor should imports be allowed to undermine local producers.

We also do not suggest that it is wrong for a poor economy to try and find niches of genuine comparative advantage and exploit them fully - with value added as far as possible. Apart from foreign exchange earnings, the exploitation of niche markets both at national and international levels can lead to economies of scale. But comparative advantages in niche markets are constantly changing. Therefore one needs alertness, motivation, information and innovation. Governments can assist upcoming entrepreneurs to detect and utilise potential comparative advantages by providing training, market research, cheap and accessible channels of communication, credit facilities and backup insurance in case of product failure.

If neither isolationist policies nor free trade is the answer, we have to reappraise the merits of a two-tier approach. There was a time when the phenomenon of a "dual economy" in Third World countries was deemed to be an obstacle to progress. The point to be made now is that the indigenous economy needs protection while the advanced economy in peripheral countries needs careful exposure to become fully competitive. A centre economy in a peripheral country which is fully competitive on world markets should be allowed to unfold its full potential, while the peripheral economy in the same country should be given the protected space to develop along its own lines. But under no circumstances should the centre economy be subsidised or given preference at the expense of the peripheral economy. It should also not be allowed to undermine the peripheral economy.

On the basis of these considerations we recommend the imposition of fairly hefty duties on imported energy and manufactures, especially luxuries. This will help to balance national accounts; make agriculture and transport with draught animals competitive;[18] restore relative prosperity to rural areas, with the result that migration to urban slums will slow down; make imported manufactures expensive and local trades competitive. By the same token a poor country should avoid debts and credits except in special circumstances and where amortisation can be guaranteed through definite contracts. As a rule of thumb imports should only be allowed as far as a sustainable compensatory export trade has been built up.

The third aspect: upgrading local technology

The fallacy of importing high technology, developed by the centre economy under centre conditions and for centre needs, into a peripheral economy has long been exposed.[19] The development of the peripheral economy should take local expertise and means of production as its point of departure. This does not mean that local technology cannot be upgraded and developed. The hoe can indeed be replaced by an ox-drawn plough,[20] the sledge can be replaced with a cart on wheels and the clay pot, used to carry water from the river, can be replaced with a handpump placed on a well.

There are those who believe that peripheral development is a wrong concept altogether. If you open up the peripheral hinterland to commerce and industry through a system of roads and rails, they argue, this invariably leads to the exploitation of the periphery by the centre economy and its subsequent deterioration. I do not believe this to be inevitable. I also do not believe that subsistence agriculture as such is worth preserving. The point is, rather, that the peripheral population must regain control over its life, just as in the case of the reduction of births and deaths. It must become capable of rejecting, with pride, the lure, not only of Coca Cola and television sets, but also of tractor driven machinery and artificial manure. The motivation for doing so should not be suspicion against technology and cultural conservatism, but a sovereignty which recognises its own long term interests and avoids the trap of shortcuts to wealth, pleasure and ease.

The important consideration here is that advances in technology should, as far as possible, not become dependent on imports of foreign materials, expertise or fossil fuels, but utilise local resources and initiatives. It is not wrong to learn from early and comparative stages in Western development. Spinning, weaving, sewing, cart building, water wheels and windmills are cases in point. One only has to go to an agricultural museum to be struck by the potentials of these seemingly obsolete gadgets for Third World development. But the liberation of potential local creativity is much more important. Unfortunately traditionalist assumptions and values tend to place obstacles in the way of the ingenuity of the population. We have discussed these issues in chapter 8. Once initiatives are set free, possibilities for innovations are endless: toilets based on microbic processes which turn faeces into compost; biogas for fuel; solar cookers, and so on.

Agricultural production, for instance, can be improved substantially without the risk of running into personal debts or burdening the national balance of payments. Animal husbandry and the cultivation of the soil should, once again, form a symbiotic relationship in which animals supply manure, traction for plowing and transport. They can be supplemented by pigs, fowls and horticulture. Monoculture has proved to be hazardous in economic and ecological terms. Other measures include contour plowing, planting of windbreaks and wood lots for fuel, consolidation of fields, crop rotation and rotational grazing, the improvement of the genetic stock of seeds and animals, building dams, constructing roads and storage tanks, and so forth.

The fourth aspect: a lean and clean state

The scourge of corruption in Second and Third World countries is proverbial.[21] Together with the abuse of military hardware by power seekers it certainly presents a major obstacle to economic advance.[22] When aid or development projects are hijacked by graft, the donors get disillusioned and the population despondent. Paradoxically, former communist countries turned capitalist seem to be particularly vulnerable.[23] This is not the place to go into the sociological and political reasons why poor societies often fail to establish institutional safeguards against corruption.[24] The point to be made is that it can be done.[25]

Apart from corruption it is imperative that the bureaucracy and the army should be reduced to bare essentials. In a country where these two sectors are the only ones which offer lucrative jobs, space for personal ambition, and possibilities for constructing a political power base of dependent clienteles,[26] this goal is not at all easy to achieve. Here we can only state the obvious: bloated bureaucracies and armies are parasitic sectors which drain the meagre resources of poor economies without contributing substantially to their productive capacities. The alternative to rationalisations and retrenchments would be to make government servants work for their income and account for their actions. In peace time, armies should be utilised for public works and government servants should be subjected to the criteria of economic achievement and public accountability.

Another aspect of this principle is the importance of keeping state interference in the economy to a minimum. To imagine that any economy, let alone a poor economy, could do without state guidance and state control, would be an illusion. However, the role of the state should not be to run the economy, but to make room for the free initiatives of the economically active population. The argument that development requires a strong government, often used to legitimate military rule, is also not convincing.[27]

The fifth aspect: avoiding violent conflict

To repair the devastations and heal the traumata caused by war in Third World countries takes decades - if they can be repaired at all.[28] While the

Swiss bank accounts of arrogant and corrupt leaders begin to bulge, the population pays the price: an endless stream of refugees, a destroyed infrastructure, abandoned farms and collapsing educational and medical facilities. There are currently almost 15 million refugees in the world. In Africa alone the ranks of half a million cross-border refugees in 1965 had swollen to 4.6 million by 1988.[29] We can hardly imagine the abysses of economic, social and psychological misery which hide behind these statistics. There are also historic costs which cannot be quantified. The fragile Ujamaa experiment in Tanzania, for instance, has collapsed at least partly because the war against Amin drained the resources of this desperately poor country.[30]

But Tanzania has not been the worst case. Entire societies, already impoverished because of underdevelopment, have fallen to pieces through armed conflict. Recent cases include Campuchea, Vietnam, Iran, Iraq, Kuwait, Curdistan, Israel, Nicaragua, the Lebanon, Afghanistan, Zimbabwe, Angola, Mozambique, Uganda, Somalia, Rwanda, the Southern Sudan, Liberia, Bosnia, Yugoslavia, Northern Ireland, and some former Soviet republics such as Armenia and Azerbaidjan.

One should not argue that armed conflict cannot be avoided; this is blatantly untrue. Due to responsible leadership on both sides South Africa has escaped the fate of becoming another case - at least for the time being. There are democratic procedures, legal codes which stipulate the criteria of justice, courts of law where animosities can be sorted out, institutions of arbitration and instruments of international conflict resolution. Research into the nature of ethnic consciousness and ethnic conflict is one of the most important tasks for the immediate future. The creation of general public awareness of its importance for cultural self-expression, as well as its abuse in the political power play, is a matter of life and death and should be placed on the national and international agendas.

All these ways to avoid violent conflict have worked countless times. The romantic ruins of fortresses scattered all over Europe give evidence of a time when each local lord had to defend himself and his subjects against his neighbours. Today the whole of Europe is under the rule of law. Switzerland, a country which accommodates four ethnic groups and is located in the middle of a continent which has seen numerous and devastating wars during the last millennium, has enjoyed peace for half a millennium. If the wellbeing of the public constitutes the first priority of the leaders, there is nothing to compel the human race to engage in self-destructive conflict. And if violent conflict could be overcome, a major obstacle to development would be removed.

If we mean business we have to go further than humble appeals to sovereign rulers. I consider the pacification of violent conflicts on local and regional level by the international community as imperative. Force must be used if the absence of such force leads to greater violence. It is the only legitimate use of the massive military arsenals which have been built up in the 20th Century by richer countries. Military pacification and subsequent democratisation is one of the important ways in which centre surpluses could be utilised to create the environment for peripheral development. What has worked in Germany and Japan after 1945 may work elsewhere in the world.

Alas, we are far from there. It is puzzling how the First World can prop up dictatorial and corrupt regimes, make huge profits out of trade with weapons of mass destruction, look on passively as warlords engage in the most atrocious adventures and then humbly send food aid to cater for the millions of victims and refugees. The warlords seem to have understood that rich donors will clean up the humanitarian mess they create and concentrate on their power struggles. But what are the interests guiding the extraordinary behaviour of the First World? Surely there is an alternative. We do not underestimate the immense difficulties involved, but the massacre in Rwanda should never have been allowed to happen!

The sixth aspect: reconstructing collective consciousness

Make no mistake: all this presupposes a change in collective mindsets. It is not possible to enjoy the material fruits of "development", generated by modernity, yet continue without the critique and reconstruction of traditional patterns of thought. The wisdom contained in traditional cultures can and should be retained, but the orientation to the past and the dynamistic-magical approach to reality, also found in such cultures, must be overcome. The loyalty found in traditional cultures can and should be retained, but authoritarian attitudes and paralysing role ascriptions must be dismantled. The tools and methods used in subsistence agriculture should not simply be discarded and replaced by imported technology, but they should be developed on the basis of modern insight and scientific research.

It is not prudent to overlook the fact that there are assumptions, social conventions and patterns of behaviour in traditional societies which are counterproductive in terms of economic development. Traditional communities have worked out sophisticated mechanisms which led to social cohesion, mutual protection and collective responsibility. These include the deference paid to fathers and chiefs in a patriarchal society. Under small-scale and pre-industrial conditions, such institutions were functional. In postcolonial situations, however, where Westernised elites have moved into vast vacuums of power, these attitudes can be abused by corrupt leaders.[31] The population must learn that under modern circumstances constructive critique, democratic accountability, institutionalised checks and balances and an outspoken opposition are important if the abuse of power and bureaucratic inefficiency are to be overcome.

Another example is the bride price. In many traditional societies cattle are used to balance out shifts in reproductive potential between extended families. The bride price compensates for the loss of a child-bearer through marriage and is again utilised to secure a bride for the brother of the bride. This practice has led to very stable marriages and a network of institutionalised relationships between extended families. But under modern urban-industrial conditions the practice has become problematic to say the least. The extended family has broken up. The bride price takes the form of money. It is paid not to the family but to the father of the bride. He spends it not on securing a bride for his son but on a car, alcohol or a prestigious marriage feast. Instead of supporting the couple, the elder generation exploits the younger.

Moreover, the practice helps to bring about promiscuity and social decay. In an urban situation many young men can no longer depend on the support of extended families. They have to secure exorbitant sums of money before they can hope to marry. Since most young men are unemployed or receive dismal wages, young couples enjoy sex without getting married. After the third or fourth child the mother may become unattractive, family obligations become burdensome and the young men find new sweethearts. Apart from the desolation of single-parent families, the social and economic costs for the society are considerable. Children are dumped in the laps of grannies who no longer understand the modern youth. The absence of male (and female) role models and parental discipline leads to cultural disorientation, gangsterism and crime

Obligatory family loyalty can also become a drain. I have observed the case of a man from the rural periphery who went to the city to earn money. His brother ostensibly took over the responsibility for the family of the migrant while he was away. This sounds reasonable. However, not only the wife of the migrant but also his brother's family ceased cultivating their fields. Moreover, the brother moved into the homestead of the migrant with his entire family. While the agricultural potential of both families lay barren, the two families all lived from the remittances sent home by the migrant worker. What a waste of potential!

In all three examples what used to be an important social safety net has become a racket. Obviously there is no way out of poverty where such practices persist. It is the communities who should begin to rethink their cultural heritage in group discussions and adapt it to the changed situations of their times. While the individual is not always able to withstand the pressures of social conventions and trends, a community may. The tribal leadership, the public media, educational institutions and the churches can all help to bring about a process of conscientisation and restructuring. Modern socio-economic processes depend on social and cultural accountability, flexibility and adaptability. Obsolete and rigid structures present obstacles to the progress of the periphery and must be tackled with determination.

It is ironic that changes are taking place in any case through the impact of the dominant urban-industrial civilisation, but not necessarily in the right direction. As we have seen in chapter 8, the social and mental constructs of the centre uproot the social and mental constructs of the periphery and render them impotent. This is where unredeemed entropy happens at the level of collective consciousness! It resembles the material destruction that occurs in war. The result is a breakdown of self-respect and self-confidence, rather than a surge forward into new possibilities.[32] The peripheral population is a victim of alien powers, not a sovereign decision-maker who consciously engages in its own reconstructive effort.

As a result the kind of development which occurred in the centre cannot simply recur in the periphery. Healthy economic development presupposes social and mental reconstruction. It is imperative that ways be found to overcome this paralysis.[33] As stated above, this would have to include the debunking of wrong expectations, the abandonment of claims to entitlements and the exposure of ideological rationalisations. The devastations caused by colonialism and slavery are sad facts of history. But the peripheral leadership should not allow itself the luxury of self-pity and scapegoating because it is not in their own interest.[34] The population too must be sufficiently conscientised to develop its own pride, refuse to live from handouts, take its destiny into its own hands and map out its own paths into the future. We shall continue with these reflections in chapter 18, section I.

Revision: Summarise the most important policy options suggested in this chapter.
Application: Think of a group of severely underprivileged people in your social environment. Would any of the policies suggested in this chapter benefit them if they were applied? How realistic would their implementation be in this concrete case?

> *Critique: (a) "Your Western middle class disrespect for the survival techniques, the hidden excellence and the cultural values of the poor is nauseating. I thank God that you are not a politician with clout to implement such inhuman ideas." (b) "None of the lessons to be learnt from the young tigers in South East Asia figure in your list of suggestions. You seem to have missed the boat somewhere." (c) "The cynicism with which you are able to equate abortion with warfare betrays your inexcusable ignorance of the issues involved." How would you react?*

Let us summarise

This chapter offered, by way of example, a few policy options for the periphery. From the outset we made it clear that *the goal* of the periphery cannot be to catch up or emulate the centre because that is neither ecologically feasible nor economically desirable. The aim should be to secure a *modest and sustainable* standard of living for the current and future generations. To forestal the long term deterioration of living standards, it is crucial that *population growth* be arrested. For that to happen the scientific manipulation of the birth rate must match that of the death rate. Those who oppose abortion must advocate contraception. Population control implies old age security, education, the availability of contraceptives, and female emancipation.

The second task, development, should be based on achieving a *balance between productive capacity and consumption*. The periphery must use local resources and develop its own productive potential to satisfy local needs. In most peripheral countries agricultural development should form the base on which trades and smaller industries can be built. Dependence on capital, fuel and fertiliser imports should be avoided. Where niches of genuine *comparative advantage* can be found, exports can be encouraged, but industrialisation should not be supported at the expense of agriculture, nor should imports be allowed to undermine local producers. *Traditional technology* and methodology should be taken as the point of departure, but they should progressively upgraded by scientific insight to suit local conditions. The underlying principle of all these measures is not an isolationist ideology, but the attempt of the peripheral population to regain dignity and sovereignty over its life.

Corruption and a self-serving bureaucracy should make way for a *lean and clean administration*. One of the most critical tasks is to overcome *violent conflict*. Social justice, conflict resolution through the legal system and democratic accountability are of the essence. All this presupposes considerable changes on the level of collective consciousness. A rigid *traditionalism* does not open up a viable future.

Notes

[1] There are hundreds of attempts to draw up specific guidelines for action, including my own (from "Affluence, poverty and the Word of God", Durban: Lutheran Publishing House 1978:261ff onwards). For a remarkably professional approach see Storkey 1986.

[2] At a recent visit to an agricultural museum I was struck by the ingenuity of the tools and the adequacy of life in commercial farming areas in the first quarter of the 20th Century. Compared to the current situation, energy consumption and technological sophistication were minimal, yet there was the excitement of constant innovation. Indeed those ox-driven pumps, horse-drawn carts and hand-held drills can be improved infinitely without an increase in fossil fuel consumption and pollution.

[3] "While it may be accurate to argue that most of the people added to the population will be too poor to make a consequental demand on the ecosystem, such logic condemns the majority of human society to perpetual poverty and runs contrary to the principle of justice" (Korten 1990:166).

[4] For the main positions in the discussion see Kieffer 1979:385ff.

[5] The UN Declaration on Population of 1967 affirms that "the opportunity to decide the number and the spacing of children is a basic human right ..." (Kieffer 1979:387).

[6] Cf Korten 1990:170f.

[7] For opposite ethical points of view see Fletcher J 1966: Situation ethics. London: SCM and MacFadden C J 1967: Medical Ethics. Philadelphia: Davis.

[8] Cf the classic argument of the inevitability of such a dualism by Reinhold Niebuhr 1932: *Moral man and immoral society: A study in Ethics and Politics*. New York: Scribner.

[9] For the just war, holy war, and just revolution theories see my essay on this theme in Nürnberger 1989:130.

[10] The dualism between private morality and public necessity implies that individuals do not have the right to kill. If abortion is accepted by the community, the community must lay down rules for its implementation. Female emancipation does not include the right to abort at will. The valid demand that a woman should be given "the right over her own body" cannot be used to legitimate abortion, because the fetus is not part of the body of the mother but another human being, whom she has no right to kill.

[11] For a detailed discussion see my article "Unmasking the horror" in Nürnberger 1989:156ff.

[12] For a list of policy preferences see Korten 1990:69ff and 78ff.

[13] See chapter 6 of Nürnberger 1998.

[14] A passionate appeal for "justice for the peasants" is made by Athanasius 1988.

[15] "African states tax producers through low prices in orer to shift resources from the agrarian sector to industry, to increase government revenue, and to mollify urban consumers who demand a low price for their food" (Herbst 1990:82). It needs to be added that the urban population has political clout, the disorganized peasantry has not.

[16] For a discussion of Zimbabwean agricultural policies see Herbst 1990:82ff.

[17] See chapter 6 of Nürnberger 1998.

[18] Cf the Kenya Network for Draught Animal Technology, Box 30197, Nairobi; also SANAT, Faculty of Agriculture, Univ of Fort Hare, P/Bag X1314, 57OO Alice, South Africa.

[19] See especially Schumacher 1973 and the work of his institute for intermediate technology.

[20] Cf the Kenya Network of Draught Animal Technology and SANAT mentioned above.

[21] Kotecha & Adams 1981; Andreski 1968.

[22] Michael Elliot says that "the universal language is graft, bribery and payoffs. Can the advocates of market reform and economic growth be heard above the din?" (*Newsweek* Nov 14, 1994, p.10ff).

[23] Elliot *op cit* p 12.
[24] For a discussion see Nürnberger, K: Democracy in Africa - the raped tradition. In: Nürnberger 1991:304-318. Further: Ekpo 1979, Kotecha & Adams 1981, Mowoe 1980, Mazrui 1980, Nyongo 1987.
[25] For the full argument see Klitgaard 1988 and 1991. Hong Kong became clean by instituting a powerful "Independent Commission against Corruption in 1974. Unfortunately it is now relapsing due to closer ties with the mainland (Eliott op cit 12).
[26] Kotecha 1981:67ff.
[27] Mowoe 1980.
[28] There is a difference between the destruction of a highly developed economy, such as those of Germany and Japan in 1945, and the destruction of an economy which is struggling to get off the ground. The former gets an enormous boost through the reconstruction effort, the latter may be crippled permanently.
[29] Broadley 1992:28ff.
[30] See chapter 6 of Nürnberger 1998.
[31] Nürnberger, K: Democracy in Africa - the raped tradition. In: Nürnberger 1991:304-318.
[32] "Examples are set, but they fail to spread. Taboos are fought, but obstinately they persist. A lack of flair for mechanics often looks ominously like a disbelief in mechanics. There is a lack of sense of maintenance ... Beyond occasional limited success, goodwill, and even transitory cooperation, there is the almost palpable presence of instinctive and deep suspicion. For the real, almost immutable obstacle is not economic but psychological. Somewhere in the cause of the encounter with industrial civilization ... there occurred a tragic breakdown in self-confidence. With the passage of time, it has turned into self-tormenting humiliation, into loss of self-respect and sullen introversion. Ultimately, and in economic terms, it has led to the present paralysis and to the almost universal inability to respond to the signals of the alien system" (Mende 1973:182).
[33] I have attempted a large scale investigation into the possibilities of the Christian faith to play a constructive role in this regard in Nürnberger 1982.
[34] In her book, *Et si l'Afrique refusait le developpment,* the economist Axelle Kabou from Cameroon is reported to castigate these attitudes severely, but the text is not accessible to me.

Policies for centres

What is the task of this chapter?

Our analyses have shown that global economic developments are moving in dangerous directions. In Part IV we indicate what would have to be done if we wanted to lead them into a more acceptable future. The previous chapter concentrated on economic peripheries. The current chapter offers some reflections on redirecting the industrial economy in centre regions. The emphasis lies on the compulsion of the modern economy to create artificial needs due to growing productive capacities, and on the entropic dangers of increasing throughput. The next chapter will deal with the relation between centres and peripheries.

General considerations

The industrial and commercial economy of the Triad, together with its peripheral offshoots, is becoming a juggernaut of unimaginable proportions. Though its present trajectory seems to lead into an abyss, the inherent dynamic of the centre cannot easily be derailed. Nor should this be attempted. We shall argue that, in spite of the wrong direction it has taken, it is still the *greatest potential asset* of humankind at the turn of the millennium. The corporation is, as Stackhouse has argued, the modern locus of production. The task is, rather, to channel it into more promising and useful directions.

What is the problem of the centre? The exponential growth of the industrial economy necessitates a simultaneous growth of consumption. This syndrome harms the affluent through unbalanced consumption patterns, the natural environment through pollution, the peripheral population through economic marginalisation, and future generations through the depletion of non-renewable resources. The task is to spread economic potential and to reduce economic activities to sustainable levels. To use a picture: limited and precious water supplies must be spread over the garden; it should not escape in gullies and cause erosion.

Much has been said on the topic of economic restraint for the sake of ecological sustainability, but there are many objections. Impoverished communities cannot afford environmental concerns when they lack the most basic essentials. The affluent do not want to risk a recession just to save the environment. Ecological dangers tend to be played down. In time, it is argued, technological innovations will cope with any crisis that might emerge, and industries are learning to clean up the mess they are creating.

The problem is that in global terms the crisis has already reached unmanageable proportions. In economic terms, millions of people are perishing or ailing even now. In ecological terms not even the emission of fluorocarbons and carbon dioxide nor the depletion of rain forests have been halted so far. We simply cannot afford to continue as if nothing was happening. So how do we respond to these objections?

It is true that the most basic needs of the poor must receive priority. This should be done at the expense of unnecessary and unhealthy consumption among the wealthy, not at the expense of the environment. But how are we to understand this? There have been economists who have advocated drastic measures to arrest the centre dynamic and create a "steady state" economy.[1] For them "growth" as such is the culprit.[2]

I cannot see how this verdict can be avoided. In *ecological* terms, we simply have to restrict further material growth, if we are not to destroy the earth for future generations, whether this is going to cause a recession or not. In *economic* terms, the reasonable consumption of the wealthy is bound to reach a ceiling sooner or later. Luxury demand cannot keep the economy going all on its own. And even if it could, this would be undesirable because luxury demand among the rich is no substitute for real need satisfaction among the poor. And again the earth cannot afford the ecological price. So a slowdown in the economy is inevitable. Unfortunately not even the prestigious Brundtland Report has had the courage to problematise the principle of economic growth.[3]

Yet the demand to arrest industrial growth needs to be qualified. In *economic* terms, the economy is a flow of resources, not a static deposit of wealth. A dynamic process can certainly be accelerated or slowed down, but to stop it may cause its collapse. The periphery would not gain from such an eventuality. On the contrary, the net result would be more generalised poverty, rather than a more equitable distribution of prosperity. In *ecological* terms reality is subject to evolution. Technology is part of that evolution. Technology, beginning with the utilisation of fire, the domestication of animals and the cultivation of crops, has altered social and natural reality to such an extent that there is no way back to a state of original innocence. We simply cannot reduce a world population of 5 billion to 5 million hunters and gatherers. Whether we like it or not, the centre dynamic is the most valuable asset the global society possesses and should be treated with respect, not with resentment. As I see it, therefore, the task is to redirect this dynamic to make it serve the society as a whole, including future generations, rather than pandering to the short term wants of an elite.[4]

At the very least, technology is needed to repair the ecological damage of industrialisation and population pressure. Much of the creative energy of the centre can be, and will have to be, channelled into ecological correctives, for instance anti-pollution industries. This is expensive, but in the long term it is more profitable than investment in the senseless waste caused by artificially induced luxury consumption. It is simply a redirection of productive activity from where it is not needed to where it is badly needed. Redirecting rather than braking the dynamic will also obviate much of the recessionary pressure. In practical terms, producers must be forced to bear the full social and ecological cost of their enterprises - or, in ecological jargon, they have to internalise external costs.

On the other hand these cleaning-up operations also represent growth in economic activity which demand energy and other natural resources. They increase rather than slow down throughput. We have to concede that, while they may alleviate the symptoms, they do not cure the disease. Ultimately we have to go to the roots. Human economic activity must change from quantity to quality. It must not rise above the thresholds of sustainability, that is, the capacity of nature to absorb the impact and the ability of humanity to find alternatives to depleted resources. Any other apparent way out is tantamount to drawing from our capital and calling it profit.[5] This is a gigantic task and piecemeal measures will not do. The following steps are meant to indicate the direction.

The first aspect: productivity and equity

The collapse of Marxism-Leninism in Eastern Europe has discredited the command economy beyond repair. If we do not wish to disregard the lessons of history we have to concede that in terms of innovation, progress and productivity the *free enterprise* system cannot be beaten.[6] Karl Marx was the first to recognise that.[7] But this does not inspire unqualified enthusiasm.[8] The first problem we have with this system is that unchecked freedom leads to growing discrepancies in economic potential so that even the principle of economic freedom is ultimately self-destructive.[9] Most grotesque are the income discrepancies in the United States, where chief executives of the biggest companies can earn $ 18 000 a day, while redundancies become more and more common due to "downsizing" and "rationalisation", the relative income of the lower sections of the population declines, and one seventh of the nation is classified as "poor".[10]

Social democracy has proved to be the most acceptable and viable system for countries of the centre so far.[11] Within a democratic state a reasonable level of equity and social security can be reached along these lines. In postwar Germany productivity related wage increases have led to mass purchasing power and long term growth.[12] Sweden has become one of the most affluent societies in the world. The success of Taiwan also proved that equity is not an obstacle in the way of economic progress but a powerful incentive.[13]

This is true even though social democracy presently faces a major crisis. One of the organisational problems of social democracy is that growing welfare expenditure must be covered by economic growth and high employment levels, otherwise state coffers will be overburdened.[14] In Sweden growth declined, unemployment increased, the number of old people increased relative to the active population, many people took advantage of the welfare system, retired early or absented themselves from work, housing was subsidised at unrealistic levels and the welfare system expanded regardless of the costs. The contribution of the state to the GNP grew from 40% in 1970 to 71% in 1994, taxation rose from 40% to 55% and the budget deficit grew to 13% of GNP.[15] These developments have discredited the system of social democracy to some extent, even in its classical home.[16]

However, this failure is not due to the principles of equity and social security, but to faulty allocation. These societies have simply lived beyond their means. For a variety of reasons, accumulating budget deficits have reached staggering proportions in many countries of the First World, notably the United States. It may seem ironic that precisely rich societies have to get used to the fact that an economy cannot spend more than it produces. This awareness must reach the average person on the street. It is important for the viability of social democracy that the state should cease being deemed able and obliged to provide unlimited benefits. Rather, the state should direct the private enterprise economy to bring about greater equity and security. The power of the productive sector must itself cater for social benefits.

Because pension, medical and unemployment contributions are nothing but deferred and translocated consumption, they should become part of the combined salary package, thus of factor costs. *Current* consumption (wages and salaries), *future* consumption (pensions) and *potential* consumption (medical schemes and unemployment benefits) must make up the factor cost of labour and the sum must not go beyond the level of sustainable productivity. If future and potential consumption are simply added to current consumption their combined weight can break the back of any economy. If the cost of welfare is borne by employers and employees, paid-out salaries and profits will not reach unrealistic levels, parasitism will be cut out and a bloated bureaucracy will not emerge.

Another problem connected with equity is unemployment and marginalisation. While unemployment has been growing in Third World countries for a long time, this is a fairly recent phenomenon in Triad countries. It has essentially two causes, technological advance and an imbalance in the costs of the factors of production. We shall deal with this problem below in the context of accelerating throughput.

Yet another aspect of equity is democratisation. It is imperative that the checks and balances typical of a free and democratic society in the political realm be applied to the economic realm as well. "The large private corporation fits oddly into democratic theory and vision. Indeed, it does not fit."[17] The corporation is the modern locus of production and as such indispensable.[18] There is no need to demonise it as such. The problem is, rather, that the greater the corporation, the greater the concentration of power and the greater the impact of the utilisation of this power for private interests.[19]

Bold steps have been taken in the direction of economic democratisation in countries such as Germany and Sweden. But social democracy in general is still a far cry from economic democracy. The idea of industrial democracy has run into problems recently. Capital has begun to float freely on the global market place under conditions of severe competition and new technological possibilities such as computerised automation. This trend has undermined efforts to democratise local enterprises, let alone multinationals.[20]

The problem is, as in other cases, the restricted horizons of the currently dominant mindset. If capital has become multinational, then democratic principles must be applied internationally.[21] The recent slump, beginning among the much praised Asian Tigers, has exposed the weaknesses of the globalised system. It is unacceptable that predatory capital breaks lose from its national fetters to roam freely on the international market place and savage the most vulnerable. It is intolerable that a bunch of currency speculators can send a whole national or regional economy into a tail spin.

But the imperative to go beyond the nation state is not restricted to the control of capital. It has often been observed that, while social democracy is fine for a centre economy, it cannot be applied on a global scale. Alleviating discrepancies within a rich society can actually exacerbate discrepancies between rich and poor societies. The development of international instruments to regulate the flow of economic assets in the global economy is one of the most pressing tasks of the future. We come back to that in the last chapter.

The second aspect: market correction through the tax system

We have opted for the retention of a reconstructed free enterprise system mainly because of its innovative and productive dynamic. But this confronts us with a seemingly paradoxical situation. The current centre dynamic is based on the acceleration of the throughput of material from resource base to waste and it is not clear whether it could survive a drastic slow-down of this process. We have also argued in previous chapters that such a slow-down is absolutely imperative. There are essentially three kinds of reasons for that:

(a) There is a correlation between overdevelopment in the centre and underdevelopment in the periphery. Entropy caused by the centre is off-loaded in the periphery. This includes the marginalisation of labour due to the deployment of advanced technology nationally and internationally.
(b) There is the depletion of non-renewable resources, the over-exploitation of renewable resources, the generation of waste and the pollution of the environment.
(c) There are the side-effects of overconsumption and unbalanced need satisfaction.

As already stated, our answer to this impasse is that the centre dynamic must not be arrested but *redirected*. There can be growth in quality and reduction in quantity. Wider horizons and deeper insight, cultural excellence and enriching relationships are ecologically not harmful but beneficial. The electronic media can bring fulfilment to millions if they are not abused to raise material wants through deceptive offers of happiness. The state must institutionalise incentives and disincentives both on the supply side and the demand side of the economic equation to channel economic activity in more desirable directions. At the institutional level this can best be done by means of the tax system. At the level of collective consciousness awareness can be created by appropriate cost-benefit analyses. Let us begin with the institutional dimension.

Taxation and tariffs are among the most powerful instruments a state or a community of states can muster. Not surprisingly, the legitimating ideology of producer lobbies strongly decries taxes as counterproductive. But "the contention of the supply-side economists that low taxes by themselves guarantee economic health and growth has not been proven. Their contention that high taxes inevitably mean economic stagnation has been decisively disproven."[22] The question is, rather, what kind of taxes one imposes and what one does with the revenue.

The particular form and quantity of taxation strongly determines market forces. Because productivity depends on capital growth and technological advance the state often grants tax concessions for capital investments while raising income tax. The result of such a policy is that capital becomes cheaper and labour becomes more expensive. Obviously injudicious trade union action can considerably exacerbate the problem. As a consequence firms retrench workers and mechanise production. This again leads to a concentration of wealth in an elite of owners of capital and skills. The same can be said of support given by the state and the private sector to research and development aimed at labour-saving technologies.

This long term distortion of the factor market flies into the face of free enterprise principles. It is also at least partially responsible for the economic imbalances discussed in this book. *The rationale of capital investment and technological advance is to help people produce, not to throw them out of production.* Meaningful and dignifying work for the masses is surely a more profound social goal than growth for the sake of growth and profit for an elite. Until a satisfactory level of employment is reached, therefore, it is labour-intensive production which should get tax relief, while capital- and technology-intensive production should be taxed more heavily. The state and the private sector should also support research and training aimed at making labour-intensive means of production more competitive.

Taxation can also be used to address the issue of resource depletion and pollution. Ecologists have recently exposed the fact that present prices do not reflect the true costs of industrial throughput to the society and the environment. Chemical industries, for instance, can simply dump their effluent into a river to the detriment of households, fishermen or farmers downstream. The cost of air pollution has to be paid, among others, by the health system of a society and by countless individuals suffering from respiratory diseases.

Ecologists insist that all external costs of a productive process, that is costs which are not borne directly by the producer, be "internalised" by the firm.[23]

Such a policy would immediately put an end to the extravagant way in which scarce resources are currently being depleted. This is of particular importance for fossil fuels (crude oil, natural gas and coal), because these sources of energy are both irreplaceable and highly polluting. The pollution caused by nuclear energy is potentially much worse. To an occasional visitor of the Triad it is mind-boggling how pampered its inhabitants have become during the last few decades. Tropical temperatures in homes, shops, factories and trains in the midst of winter are neither necessary nor healthy. Why should a giant building, such as the Frankfurt airport, be heated to such a degree in winter that one has to take off not only one's coat but also one's jacket and pullover to avoid sweating? Equally problematic is the large scale use of fossil fuels for private transport. All this would change immediately if the prices of these resources would be raised dramatically. Incremental taxation is now widely discussed as a means to bring this about:

> A socially and economically feasible strategy would consist in the raising of fossil and nuclear energy prices in annual and predictable increments of 10% per annum for about 15 years which would amount to a nominally fourfold, in real terms perhaps threefold, rise in prices. And all the money siphoned off should be plowed back into the economy through other tax reliefs and rise in welfare grants to the socially weak.[24]

This policy would give clear advance signals to firms and they could react and adjust. Production would become not only less energy-intensive, thus less polluting, but also more labour intensive, thus helping to solve the unemployment issue. The singularly precious and non-renewal assets of crude oil, natural gas and coal would last longer as raw materials for industrial use rather than as sources of energy. Oil producing desert countries would be given a longer lease of life. The policy would also generate strong incentives to explore alternative sources of energy, develop energy saving technologies and make such innovations more cost-competitive.

A similar policy can be based on progressive taxation of pollution and the generation of waste. Whether direct regulation of pollution is preferable to penalty taxes or effluent charges, as some authors suggest,[25] cannot be investigated here; the ecological effects would probably be largely similar. Obviously such measures have to be taken on the basis of international agreements, otherwise economies subjected to energy taxes could become uncompetitive on world markets. The same is true for all ecological measures. The imports of countries which do not become part of these agreements should be subjected to tariffs which correspond with domestic taxation to cut out unfair competition.[26]

The revenue generated by these levies can be plowed back into the economy in many forms: public transport, the development of alternative sources of energy, research and development directed at energy saving means of production and patterns of consumption, and so forth. Support for the generation of non-material sources of satisfaction, such as adult education, the retrieval and enrichment of a family culture, arts and crafts, outdoor activities, sport, and so on, is important for the consumer side of the economic equation. Once incentives are in place to switch from quantity to quality in consumption, supply will follow the demand and producers will improve the quality and longevity of their output.

The third aspect: declining marginal utility

The accumulation of unproductive economic potential in rich societies is matched by the accumulation of goods and services with minimal utility in private households. This brings us to the demand side of the economic equation. The amounts of perfectly usable furniture, crockery, television sets, clothing, and so on, which spill over from oversaturated homes in middle class suburbs into garbage vans is mind boggling. But they are only the tip of the iceberg. Much greater amounts of underutilised artifacts remain in the homes. Examples are bulging wardrobes; television sets in virtually every room; hifi sets; hand guns; mechanised kitchens; toys of such quantities that children sit among them bored stiff; superfluous luxury cars; yachts moored in distant harbours for most of the year. All these items constitute masses of wasted resources and represent vast economic potential turned into stagnant material with minimal practical utility.

Yet the spending of affluent societies tends to grow relentlessly. There is a subjective and an objective reason for this. Subjectively the law of declining marginal utility is in operation. This law says that, as we approach saturation point, ever greater inputs are necessary to produce ever smaller increments in satisfaction. The acquisition of the first little transistor radio can make an enormous difference to one's life, the addition of a fourth television set to the existing arsenal does not.

The fact that the level of marginal satisfaction declines with greater affluence is an uncanny yet economically highly significant phenomenon. It is puzzling why economists have not paid more attention to the *economics of consumption*.[27] Why has there been such an enormous emphasis on efficiency of production and such a glaring neglect of the efficiency of consumption to bring about real utility? Why has capitalism generated rational procedures on the factory floor and in the balance sheets of commercial undertakings and, at the same time, promulgated the most irrational behaviour among consumers? The only explanation I can find is that economics is the legitimating ideology not of consumer, but of producer and distributor interests. More efficient consumption patterns would cut down on demand and threaten profit margins.

It is, of course, not in the interest of consumers that the most meticulous cost-benefit analyses are conducted on the supply side, while nothing comparable is done on the demand side. The overconsumption of material goods and services leads to serious imbalances in the satisfaction of needs. Endless hours spent in front of television sets, for instance, impact on one's physical health, impoverish personal relationships within the family, stunt the budding creativity of the young generation, undermine the capacity of ordinary people to think for themselves and expose the subconscious of the population to sophisticated manipulations by profit seekers, film stars, football idols and mavericks. The potential beauties of a family hike in the mountains are overrun by the compulsion to cover kilometres in luxury cars. The spiritual depths of silent meditation are crowded out by the incessant thumping of drums pouring from powerful loud speakers.

More serious is the fact that humankind can really not afford this waste, both for social and for ecological reasons. In the first place the resources ploughed into overprocessed but underutilised gadgets in the centres are dearly needed by the impoverished masses in the peripheries. The standard argument of economists that utility cannot be measured is ridiculous in view of the stark facts of the situation. The marginal utility of the proverbial piece of bread, thrown into the dustbin of an affluent home, would rise dramatically if it could reach one of the millions of starving people all over the world.[28] The change from a manual to an automatic luxury sedan adds nothing to satisfaction of the owner except the convenience of not having to change gears. Yet the additional cost of this insignificant increment could be sufficient to provide an entire homeless family with adequate accommodation. In short, marginal utility calculations should be high on the economic agenda and deeply ingrained in the consciousness of the population.

Even more important are the ecological reasons. As we have emphasised again and again humankind simply cannot continue to accelerate the speed of throughput from resource base to waste much longer without doing irreparable harm to the future prospects of the species. Those enormous quantities of fuel senselessly burnt up in our cars and heaters will be sorely needed by our progeny only a few generations from now. We have made the point in chapters 4 and 13 and do not need to go into further detail.

The fourth aspect: advertising and marketing

Declining marginal utility indicates the subjective reason for rising levels of consumption. The objective side is represented by advertising and aggressive marketing. We have argued in chapter 12 that the situation of overdevelopment in the centre leads to the artificial creation of "needs". The word is used deliberately because the "wants" of consumers are the needs of the producers. A growing economy has to dispose of its exponentially increasing output. The oversupply cannot be made available to peripheral population groups, where there are abysses of unfulfilled needs, because they have no purchasing power to generate substantial market demand. On the other hand the basic needs of centre populations, where the purchasing power is located, are largely fulfilled. As a result, the growth of productive capacity necessitates the investment of ever greater resources in efforts to stir up demand in a largely saturated market.

The first step in addressing this issue is to get clarity in one's mind on a fundamental question: is the artificial creation of needs in the centre, which spills over into the periphery, indeed at the root of the syndrome, as we have argued in chapter 12? If one has been convinced of that, the second step is to realise that a solution of the problem demands some innovative and revolutionary thinking. Certainly the idea of a drastic curtailment of advertising implies slaughtering one of the holiest cows in the capitalist economy. Even the thought seems to be ridiculous. But is it really?

It is crucial to understand that what is in the interest of those seeking profit is not necessarily in the interest of those seeking utility, that is the consumers, nor in the interest of the national, or even the global economy. If the aim was to fulfil the basic needs for healthy survival and reasonable satisfaction - which should be the economic goal both of individuals and the state - people would not have to be instigated to buy anything at all. They would seek and find the products of their choice as they needed them. Aggressive marketing serves the interests of profit, not utility. And the sooner the idea that profit is the prime value which has the right to dominate every other consideration is debunked the better.

With this we do not mean to apportion blame. To assume that an enterprise can operate without making a healthy profit would be ludicrous. Moreover, the present system forces producers and distributors to advertise, otherwise they could not survive. In fact, we shall argue that the current system is not even in the interest of the supply side. The point to be made here is that we need to build up the collective insight and the political will to change the system. This includes the consumers, the authorities and the suppliers.

The only utility the avalanche of adverts has for the consumers - who, after all, are supposed to be the ultimate beneficiaries of the economic process according to economic theory - is the information it provides about different products on the market. But that is largely a theoretical advantage. Adverts deliberately exaggerate the merits and conceal the weaknesses of the products offered. At worst they contain vicious disinformation. When cigarette smoking is presented as the peak of an outdoor experience in clear mountain air, or when ample consumption of sugar is presented as a necessity for a healthy child, we have to do with outright deception. It is not clear why the state should allow such immoral and detrimental activities to continue, if it does not allow robbery or rape.

Moreover, the great majority of people who are forced to consume this "free" delivery of adverts has no desire to be informed. If I already possess a car with which I am reasonably satisfied, why should I be forced to endure the praises of a luxury sedan day in and day out in the middle of informative television programmes? Why should I consume a barrage of nerve-wrecking images and music, which violate my aesthetic and moral instincts, when I have tuned in for the enjoyment of classical music? It is an encroachment on the freedom of the individual which should have no place in a liberal society. In ordinary life it would be highly improper to simply enter a home, make oneself comfortable on the sofa and waste the time of the family when it is about to enjoy a programme. Why should a powerful company be given the right of wasting the time of, and imposing its presence on, millions of helpless consumers?

It is puzzling why the populace does not react more strongly against this malpractice. It is the consumers who should most readily come to the conclusion that they are being manipulated and that it is not in their own interest to submit to that manipulation. In practice consumers are too isolated, disorganised and brainwashed to put up their defences. Advertising, developed with psychological and sociological expertise, packed in suggestive imagery and propped up by seductive music, has become addictive. We have dealt with the sophisticated onslaught on the convictions and inhibitions of the consumer by the modern advertising industry in chapter 11.[29] Consumer conscientisation and solidarity are imperative to match the sophisticated techniques of the marketeers. That again presupposes deep reflections on the meaning and purpose of life. Educational institutions, churches, cultural organisations, the academic community, the mass media and civic action groups have an indispensable role to play in bringing about public awareness on grass roots level.[30]

But marketing is only the means; the goal is to create higher need levels to prop up demand.[31] This is where the crux of the matter really lies. Industrial production has led to such devastating consequences for the natural environment that a further rise in material living standards can only be attained at the cost of a rapidly deteriorating quality of life. Affluent societies have to cope with unacceptable stress levels, noise pollution, mechanised processes of life (most frightening, perhaps, in hospitals), polluted air, dying forests, foul rivers and lakes, traffic congestion, overexposure to stimulants, including some beverages, drugs, rock music, reckless films, and so on.

Overconsumption in the material dimensions of life also leads to deficiencies in the non-material dimensions of life. Interhuman relationships, creative activities, cultural achievements, reflections on the deeper meaning of life all suffer. Feelings of futility, impotence, loneliness, boredom and aimlessness all undermine the human being from within. Where there is no challenge to invest one's life in a worthy cause, external freedom and affluence cannot satisfy. The deep-seated frustration of the youth in affluent societies, often leading to drug addiction, delinquency, riots and suicide, are symptomatic of this need. A reduction of material consumption to more healthy levels will not harm the balanced satisfaction of needs of the centre population but serve it.

Moreover, our analyses in chapter 12 has shown that sophisticated advertising and aggressive marketing are highly detrimental to the economy as a whole, and especially to peripheral communities. Artificially raised standards of living in the centre have a demonstration effect across the board. The impression gains ground that one misses out on life if one does not possess and enjoy certain gadgets and pleasures. Simple people are lured into purchases which are not in their overall interest. Their needs soar above their incomes and the result is overall frustration and unhappiness. Such a population becomes restive. It may resort to crime if there are no avenues to attain higher incomes. The society has to bear the financial burden and suffer the social consequences.

To achieve a reduction in consumption the rank and file of the population must wake up, realise what happens to them and take control of their lives. Liberal economics explicitly posits the sovereignty of the consumer. While this is, to a large extent, nothing but eye-wash, consumers could at least claim what is theirs by right. Are they really consumers or are they being consumed? Consumers can demand quality. Consumers can refuse to consume. Consumers can object to being pestered. Heightened awareness and indignation must then translate into democratic action by citizen advocacy groups.[32] And this is where the state comes in. In practice consumers are so heavily bombarded by advertising that they are hardly able to defend themselves against the psychological onslaught.[33] It is the responsibility of the state to serve the common good and it is the responsibility of a mature citizenry to ensure that the state does just that. The profit seekers are, after all, a minority in a democratic state.

We have argued that the freedom of a free society cannot be taken for granted because it is precisely the free market which leads to cumulative discrepancies in economic power. Freedom has to be achieved and maintained by the state as the instrument of a free and responsible populace.[34] If it is unacceptable that a powerful political party dominates and manipulates the mass media in its own interest and at the expense of other such parties, on which grounds is it acceptable that economic interest groups do exactly the same?

In this particular case the least that should be done is to level the playing fields between two messages - those of the advertisers and those of reason. This is similar to the policy of balancing out a distorted factor market which we have discussed above. As in the case of energy, capital and technology, advertising must become much more expensive in relation to the alternative message. If producers are allowed the freedom to exercise their clout in the mass media, then consumers must at the very least be empowered to withstand their impact in the form of education, religion, popularisation of research, conscientisation programmes, and so on. This would be the first option: the legal requirement that the prime time and sophistication invested in advertising may not surpass the prime time and sophistication invested in conscientisation. The imposition of a warning that smoking is a health hazard forms a useful precedent in this regard.

Before we continue with more drastic proposals let us look at the perpetrators. Deeper reflection reveals that the current system is also not in the interest of private enterprise. As mentioned above, they are forced into the practice by the system. In fact, advertising is extremely wasteful in terms of cost-effectiveness. For millions of viewers or readers the advert is irrelevant. The greater part of a newspaper is never read. Nevertheless both advertisers and subscribers have to pay for the cost.

Moreover, the law of diminishing marginal productivity comes into operation. Each firm has to build up its advertising to a level comparable to that of its major competitors before its investment in advertising begins to have any impact at all. Beyond that level, each firm has to try and gain an edge over its competitors. It is only the small increment in the psychological impact which lures potential buyers away from other firms. Yet only one of the competitors can gain such an edge over others at a time, forcing others to devise means to neutralise and surpass the latter.

Is this really the most rational way of using one's financial resources? All this money could be invested much more profitably in quality improvements which would benefit the consumers and relieve ecological pressures. Cut out fifty percent of the adverts in giant newspapers such as the New York Times and thousands of hectares of forests will be saved annually. Yet, while the present system is in force no enterprise can afford to abandon the senseless practice if it does not want to be outcompeted. Advertising is similar to drug addiction: after a time a drug does not offer much satisfaction to the addict, but to abandon it causes severe withdrawal symptoms. For this reason it cannot be left to voluntary initiatives of the enterprises concerned. There is no substitute for state intervention when free actors endanger each other's wellbeing. The state can at least draw up and enforce norms of what is acceptable in advertising. Quality controls, impositions on tobacco adverts, requirements to reveal the contents of processed foods and other measures are already in force in many countries.

But these instruments are still too feeble. The state should take at least one further step. New products could be allowed a strictly limited time of public advertising. Subsequent adverts could be subjected to steeply progressive taxes. Advertising must at least reflect its social costs. The problem with progressive taxation of adverts is that this would give powerful firms increasing advantages over smaller competitors, which would violate the norm of the equality of opportunity. Moreover, because of the social harm done by the creation of artificial needs, both in the centre and the periphery, we should become even more radical and cut out advertising altogether.

In the age of electronics there are far more rational ways of informing potential customers. Say products were categorised and their specifications were entered into a data bank which could be called up through cable television in homes or computer modems in shops.[35] The accuracy and comparability of the descriptions could be checked by the competitors themselves under the guidance of an arbitrating body. When one needed an article one would compare the specifications and ask for a demonstration. But when one did not, one would be spared the annoying or tempting brain massage.

Meanwhile producers would be relieved of a drain on their resources which could be channelled into the much more productive areas of research and development. With competition in deception no longer operative, competition in quality enhancement could take its place. It is in the area of quality and service where competition in a genuinely free and productive economy should be located. In short, this approach would be to the advantage of the consumer, the producer, the society as a whole and the natural environment.

It will be argued that newspapers, radio and television stations could not survive financially if they had to do without advertising. But this argument is ideological. It assumes that in a liberal society everything must of necessity be left to private enterprise. In fact, a liberal society can only remain liberal if it distinguishes between private interests and the common good. Radio and television, especially, are critical means of public communication in a modern state. Democratic principles demand that they should not serve sectional interests. The cost of such a service should, therefore, be borne by the population as a whole. They should, ideally, also be accessible to all citizens.[36]

The point to be made is that dependence of the mass media on advertising is not an inescapable fate but the result of a particular way of organising the economy, and a particularly irrational one to boot. At present producers *and* consumers *and* third parties, who are neither producers nor consumers, have to bear the cost of a wasteful and inefficient system of communication. And ultimately it is the society as a whole which loses out. The society as a whole could easily maintain the mass media with the same amounts of money which are now sunk into advertising through other and more acceptable channels.

Before we leave the topic of marketing let us add one more item, namely credit cards and hire purchase schemes. They too are designed to boost consumption. They offer a deceptive way of satisfying needs which have been raised artificially in the first place. As such they cause immense hardship. The affluent can pay cash and are not affected. The poor are lured into consumption patterns they really cannot afford. They pay much more for the same article because of cumulative and often extravagant interest rates. Once they cannot meet their obligations the goods are retrieved and resold elsewhere. The customer loses everything. Apart from its social effects this practice is a blatant form of exploitation in the classical sense of the word. As such it is not only harmful to the individual but also to the national economy. It prevents saving, thus capital formation, among the masses of the population. While the majority is increasingly indebted, a minority monopolises capital accumulation. Just as gambling, it should be forbidden rather than encouraged by a responsible state.

The fifth aspect: curtailing waste production and pollution

We have argued that the thrust of the centre dynamic must be redirected from quantity to quality. There is no limit to growth in quality. Let us begin with small and easy things. Cars that last for fifteen years rather than five, and which are more fuel efficient than current models, will also cause less strain on the environment. As some outstanding manufacturers have shown, the prestige of a well constructed car, of excellent garage services, of reliability, of consumer friendliness, of fuel efficiency and emission control can determine market demand as much as blown-up advertising and finicky features. Does one really need an electric motor to close a car window? These "comforts" are ridiculous. Fortunately computer hardware has consistently become less bulky and more efficient - but only to make room for the cancer growth of overdeveloped software.

Existing resources must be recycled as far as possible and the amount of new resources utilised must be kept to a minimum. This implies that we cut out all unnecessary production, thus all unnecessary generation of waste. Unnecessary production is the output of goods and services which add only marginally to the satisfaction of real needs and which are, in this sense, not cost-efficient, even though they may bring in profits for producers. Unnecessary production happens on many fronts. Many articles, such as toys and kitchen tools, are unduly sophisticated and bulky. Instead of arousing the creativity of children, the toy market is swamped by technological gimmicks. My generation grew up during the war. We had no choice but construct our own toys. From watching children at play, it appears to me that our self-made dolls, model cars and kites yielded more fun and were valued more highly than the glossy throw-aways which clutter up the rooms of affluent youngsters today. The thrill of novelty evaporates almost overnight, while the pride of achievement can last for years.

Many other products from washing machines to computer programmes to medical supplies are overladen with unnecessary complexities. Why not return to some of the wholesome and inexpensive home cures of a generation ago, rather than exposing our bodies to ever more sophisticated and ever more expensive gadgets and chemicals? Why not raise taxes on health-impairing consumer items, such as household detergents and poisons, fatty meats, refined cereals and preservatives, rather than on products conducive to health, such as fresh fruit and milk? Why not grant tax concessions for organic gardening, rather than artificial fertilisers? Why is the noise level of loud speakers not subjected to controls?

Another example of artificial wastage generation is packaging. Why should coated pills be packed individually in plastic foil, when a decade or two ago a simple container worked perfectly well? Without doubt additional packaging benefits the chemical industry; but does it benefit the consumer and the environment? Why is it no longer possible to bring one's own basket and milk can to the grocery store? In Germany firms have been obliged by law to take back the packaging material of their household gadgets. When factory floors and commercial outlets, rather than homes and rubbish dumps, were in danger of being swamped with waste, the output of packaging declined dramatically. A similar result was achieved by requiring shoppers to pay for plastic bags or glass jars in supermarkets: they changed their habits almost immediately and began carrying their own containers.

Cutting down on bulk implies new methods of construction. At present, for instance, one often has to discard a gadget (say a lamp setting) because a single component (say the globe) is in disrepair. This is a horrendous waste. Cars, computers, washing machines and all kinds of other products can be put together from single, standardised and replaceable components which can easily be exchanged when they wear out or when technological upgrades become available. In practical terms this would happen automatically if suppliers were forced to trade in old models for new, or when producers retained the ownership of their products and consumers would lease them. In these cases it would be in the producers' own interest to increase the life spans of products and components and make them suitable for multiple recycling.[37]

Cars themselves, of course, are a wasteful and highly pollutant means of transport. In Germany, 70% of all carbon dioxide emissions are due to traffic. The pollution caused by road transport per person and kilometre is 9 times higher than that caused by railroads.[38] The trouble is that the market price of private transport does not reflect this cost. In fact, it is cheaper for the traveller to go by car than by rail. It would be in the public interest if the state rectified this distortion of the market, make convenient and affordable public transport available and obtain the money for doing so from taxing motor transport more heavily.

Another instance is the practice of planned obsolescence. This means that products are deliberately made to last only a limited period of time. When nylon was first invented, stockings made of this "miraculous material" lasted and lasted. I still own a pair dating from 1952 which has never been mended. Today the slightest brush with a rough chair ruins a lady's nylon stockings. Of course this lack of resilience is deliberately programmed into the product. It leads to a higher turnover for the firms concerned, but is this also in the interest of the consumer and the environment? We could profitably return to the manufacture of quality products and the old fashioned repair, restoration and maintenance of used articles by skilled artisans. It would create jobs, self-esteem, satisfaction and a more healthy environment.

Such examples can be multiplied indefinitely. Countless people are busy designing and testing alternatives to present structures, processes, goods and services. The problem is how ingrained patterns, upheld by powerful and entrenched interests, can be broken. A democratic state depends on informed initiatives of mature citizens, in this case the consumers. Committed groups should demonstrate that a different lifestyle is as fulfilling and as aesthetically acceptable as following the latest artificially induced fads, while also being much more responsible. Consumers must become self-confident, get organised and conscientise each other. They must build up collective resistance against waste, against products which are so much rubbish, against unnecessary complexity and against senseless throughput. Market supply will adjust to a changed market demand very rapidly.

Economic restraint and rationalisation are also possible in the public arena. It is often overlooked that vast amounts of resources and economic potential have been built into stagnant structures of highly developed countries. Their productivity or utility is questionable in many cases, to say the least. This too is a symptom of overdevelopment. Most of the Gross National Product per capita, which is found in the statistics of rich countries, does not reach the homes of the average citizen but is embedded in elaborate constructs and overcomplex processes. Thus enormous economic potential is devoured by the military-industrial complex; the army;[39] space research; unwieldy, unnecessary and often counterproductive bureaucratic complexities;[40] costly systems of legal procedures;[41] super-highways and luxury airports; exorbitantly posh foreign missions, and so on. A conscious effort to clean up, streamline and rationalise both the private and the public sectors could save trillions of dollars worth of resources and wastage.

The sixth aspect: arms production

The arms production is perhaps the single most prolific and abhorrent form of wasteful production.[42] This is an international as well as a national issue. During the times of the cold war, the arsenals of both superpowers were sufficient to destroy the planet a few times over, yet the senseless competition for an edge over one's rival had to continue.[43] Fortunately the cold war between the superpowers has ceased and a few tentative initiatives have emerged in connection with armaments reduction, especially concerning the control of the nuclear arsenals of the great powers.[44] Yet most of these weapons are still deployed.

Currently arms reduction efforts by the great powers are overshadowed by the incredible misery created by irresponsible leaders and their followers in many parts of the Third World. They are being supplied with sophisticated killing machines on a vast scale. Unfortunately the arms race has completely engulfed many of the poorer nations. There is also an arms race within nations, especially in situations of latent or acute civil war. We have already discussed the social, economic and ecological costs of warfare above. We have not even begun to build up a responsible international community in this regard. But there are developments, for instance the transformation of NATO into a regional security arrangement, which point in the right direction.

Weapons are wasted when not used, and economically destructive when used.[45] Part of the problem is that arms production is still one of the most lucrative industries of the richer nations, and they will always find ways of creating a need for their products. The tragedy is that the poor and powerless are the victims. This is true both within a country and internationally. The elites make the policies and give the orders and the most miserable among the poor suffer the consequences. Instead of life-enhancing aid, death-dealing weapons are poured into poor countries in large quantities. There they do not only devour resources badly needed for feeding the hungry, but also make oppression, corruption, rebellion, civil war, international war and all their consequences more devastating and lethal.

Seen against our goal of comprehensive wellbeing, this sort of "economic activity" is atrocious and should be stopped with all national and international means available. The funds invested by governments in arms should be invested in socio-economic upliftment, justice, education and democratic institutions designed to overcome rather than fuel social conflict in the first place. It is also not true that a curtailment of the weapons industry would lead to recession, because the same economic activity can be channelled into more profitable directions, for instance into the remission of debts, infrastructure and ecological concerns. If parts of the highly capital-intensive arms industry could be abandoned, the liberated capital would suffice to cater for a lot of labour-intensive industries.

In short, the whole area of arms production and trade is one of the most vicious sources of underdevelopment, inequality, wastage of scarce resources and ecological destruction. It must be recognised as such by the international community and subjected to strict control. An independent judicial authority similar to the International Court of Justice must be established to determine the legitimate security needs of states, excluding the stabilisation of authoritarian regimes, and given the powers to veto any arms deals going beyond those needs.

Chapter 17 - Policies for centres | 435

> *Revision: Can you summarise the basic rationale of the suggestions made in this chapter?*
> *Application: Take an income tax form and figure out how it should be organised to implement some of the suggestions made in this chapter.*
> *Critique: (a) "When do-gooders and world transformers begin to dabble in economic matters they can only make things worse. Your suggestions are entirely out of step with economic realities and no economist, politician, businessman or union leader can take them seriously." (b) "I presume that you have used electricity, computer technology, a car and a healthy diet when compiling this book. If you had taken your own suggestions seriously, you should have gone into the desert, lived on locusts and carved your dreams into stone slabs." (c) "If the price of fuel was raised, you would get a revolution led not by the rich, who can afford the higher prices, but by the poor who depend on public transport and affordable traction." Do you think that there is at least some truth in these contentions? Explain.*

Let us summarise

The dynamic of the centre economy is the greatest economic asset of humankind and should be *redirected rather than arrested*. Economic equity can be enhanced by the development and wider application of *social-democratic* principles. Exponential growth of the industrial economy has to make way for growth in *quality*. Scientific insight and technological advance must be harnessed to reduce and ultimately forestal the destructive impact of industrial throughput on the natural environment and the resource base. The most potent instrument to balance out distorted factor prices is the *tax system*. Energy, capital, technology, pollution and waste should be progressively taxed. The proceeds should be plowed back into the economy where the greatest needs occur, notably in the ecological realm.

An important aspect is the development and popularisation of an *economics of consumption,* based on the law of declining marginal utility and the balanced satisfaction of needs. Because the artificial creation of needs is socially detrimental and economically irrational, *advertising* should be curtailed, if not replaced by a computerised product information system. This is a case where the state must intervene to overcome abuses of power and secure democratic freedoms. Enterprises must be forced to *internalise external* (social and ecological) *costs* and prices must reflect these costs. Wasteful production and the production of waste must be curtailed. Private and public institutions must be *rationalised* by means of cost-utility analyses. Unnecessary and counterproductive economic processes must be eliminated. *Arms production* and the arms trade are the greatest culprits in this regard. All this presupposes the development of international institutional arrangements.

Notes

[1] E.g. Daly 1991 (1973).
[2] Mishan 1993 (1967).
[3] Korten 1990:166.
[4] Cf Korten 1990:133.
[5] As far as I can see this fact has first been highlighted by Schumacher 1974:10ff.
[6] For a useful discussion of current alternative macro-economic systems see Küng 1997:184ff.
[7] The *Communist Manifesto* of 1848.
[8] The free enterprise system, as practiced today, is definitely not the "end of history" in economic terms as Fukuyama 1993 suggests.
[9] See chapter 7 of the companion volume, *Beyond Marx and Market,* Nürnberger 1998. Neo-liberal policies in the United States under the Reagan and Bush administrations have seen an unprecedented concentration of wealth and the economic stagnation of the lower classes. In the last two decades the share of the top 1% of the population increased its share of the wealth from 18% to 40%, while three-fifth of the population suffered a real drop in wealth (Küng 1997:176).
[10] Küng 1997:176f.
[11] For an elaborate treatment of these issues see the companion volume, Nürnberger 1998
[12] For detail see Küng 1997:196ff.
[13] Fields 1980; Vestal 1993.
[14] For the following see Küng 1997:170ff.
[15] In 1993 the budget deficit in Germany was 3.3% of GDP, 3.4% in the US, 5.8% in France, 7.7% in the UK and 9.6% in Italy (Froyen 1996:447).
[16] Cf Michael Elliott: What's left? *Newsweek* Oct 10, 1994, 12ff.
[17] Lindblom 1977:356.
[18] Stackhouse 1987:118.
[19] See Korten 1995 for an analysis.
[20] Junne in Caporaso 1987;71ff.
[21] See Küng's attempt to generate an international ethics (1997).
[22] Drucker 1993:150.
[23] For pollution taxes see Turner, Pearce and Bateman 1994:166ff.
[24] Von Weizsäcker 1990:77.
[25] E.g. Buchanan & Tullock in Congleton 1996:31ff.
[26] For the link between international trade and environmental policies see Leidy & Hoekman in Congleton 1996:43ff. For the consequences of non-cooperation see T Sandler in the same volume, pp 251ff.
[27] Cf Leiss 1988.
[28] At the time of writing there are close to 20 000 000 people at risk of starvation in the horn of Africa alone according to a 1994 *Newsweek* report based on statistics supplied by the US Agency for International Development and the UN World Food Programme.
[29] Note again the quotation from Swimme 1996:16f on page 279.
[30] For action groups see Korten 1990:167.

[31] Cf Korten 1990:165.
[32] See Korten 1990:167.
[33] For a critical analysis of advertising see Schudson 1984.
[34] Korten argues that the idea of a "benevolent state" is deceptive. One should take the sovereignty of the people as point of departure and interpret the state as one of their instruments. The state should not replace but complement private and communal initiatives (1990:159).
[35] As Bill Yates has predicted in his book *The road ahead* (Viking 1995), the internet is going to revolutionise advertising in any case.
[36] We have suggested above that television sets could be installed in schools and community centres in less privileged neighbourhoods and controlled by the local staff. But we are at present busy with the affluent centre population where accessibility presents no problem.
[37] I owe these these practical suggestions to Von Weizsäcker 1990, 1994.
[38] Von Weizsäcker 1990:87.
[39] We mentioned above that, in spite of the end of the Cold War, Reagan's "Star Wars" programme still devours $ 3 billion in Clinton's defence budget (*Mail & Guardian* Dec 9, 1994, p. 13.)
[40] Consider a humble matter such as the petty complexity of postage rates, or more seriously, the enormous and unnecessary complexity of income tax returns, which is serviced by an army of bureaucrats, tax advisers, lawyers, and so on.
[41] Robert Samuelson speaks of the US legal system as a "regulatory juggernaut": "There is more regulation with fewer benefits, and the whole process grows increasingly arbitrary and murky. The totality of federal regulations now comes to 202 volumes numbering 131,803 pages. This is 14 times greater than in 1950 ..." (*Newsweek* November 7, 1994, p 41).
[42] See Drucker 1993:145ff.
[43] Prinx 1984. Carlton & Schaerf 1989.
[44] Nuclear warheads will be cut by two thirds by the year 2003 according to recent US-Russia agreements and two million people have been demobilised from the armed forces (*United Nations Development Report* 1993:2).
[45] Korten 1990:138.

Policies for centre-periphery interaction

What is the task of this chapter?

The model we designed in chapter 15 distinguishes between factors located in the centre, factors located in the periphery and factors located in the relation between the two. The two previous chapters dealt with policy options for the periphery and for the centre respectively. The present chapter is devoted to policies concerning the relation between centre and periphery.

It is convenient to distinguish between the *national* and the *international* scenes and to begin with the former. The discrepancies between centre and periphery at the national level are particularly visible in semi-industrialised countries and we shall take this situation as background for our reflections.

Section I: Centre-periphery interaction at the national level

The greatest problem of the periphery is the existing *concentration of power* in the centre. A greater spread of economic potential in the population is essential. A society has to institutionalise countervailing processes against the natural tendency of any asymmetrical interaction to benefit the powerful at the expense of the weak and vulnerable. The private and public sectors should cooperate in that. Examples are investments in training, housing, transport, health services, labour intensive industries, supporting small enterprises and the informal sector, building up the purchasing power of the masses through productivity-related wage increases, opening up avenues for shared decision making on the boards of enterprises, and so on.

To secure widespread consensus and cooperation a *social contract* should be agreed upon by all the major actors in the society. This is difficult but not impossible to achieve.[1] It is natural that a privileged elite feels threatened by the socio-economic advance of impoverished masses. This fear must consciously be addressed and allayed. The experience of social democracies has shown that high income groups have much to gain from growing peripheral potency, greater participation in the economy and the democratisation of economic decision making. A more equal distribution of the national income not only reduces the conflict potential in society, but also opens up greater markets without causing depression or inflation. Both are to the advantage of the elite and the society as a whole.

Where power has accumulated in one section of the population it can be abused at the expense of other sections. Building up the moral fabric of a society is important but not sufficient. Abuse of power must be prevented by means of statutory measures, that is, by laws administered by appropriate courts, including constitutional courts. It is important that democratic checks and balances be extended to the economic sphere.[2] Powerful lobbies must be neutralised and corrupt governments must be held accountable.

Abuse of power often involves international investment and trade. International financial agreements must include provisions which prohibit actions detrimental to poor communities and such provisions must be enforced by the courts. As far as legal resorts are non-existent or ineffectual, blatant instances of corruption must be ruthlessly exposed and pilloried in public by mass media, churches and activist groups.

The rural periphery

We have said in chapter 16 that the peripheral population must be empowered through development. For that to happen it must gain access to resources. Equal access to resources and equality of opportunity are the economic corollaries of equal access to political power in a democratic state. The primary peripheral resource is agricultural land. It is imperative in most semi-peripheral countries that property rights be redefined and *land reforms* be carried out. The principle to be applied here is not that the property rights of the haves be protected against the claims of the have-nots but the other way round.

It is obvious that powerful landowners will resist land reform. Their objections are not necessarily without substance. To expropriate established commercial farms, who have fed the nation, and resettle thousands of squatters there who have no farming experience nor the desire to farm, is not very prudent. In contrast, the policy of breaking up great estates and transferring ownership to peasants, who have already been working the land in the service of their landlords, has yielded much more acceptable results.[3] Another way to democratise agriculture is to establish means of joint ownership, decision-making and profit sharing between former land owners and former workers. The problem with both methods is that they may not cater for thousands of people with insufficient land living in overcrowded reservations. In such cases a double-pronged policy is essential: the acquisition of land for small scale farming and the creation of jobs outside the agricultural sector.

Commerce and industry

To counteract concentration of economic potential in commerce and industry is equally difficult. One way is to enforce strict *anti-monopolistic legislation*. But then the relevant laws must be given sufficient clout to make a difference. It must be emphasised that this is in line with the liberal creed, because monopolies destroy the smooth functioning of the market mechanism. Another way to ensure a greater spread is to limit the share-holdings of single or corporate owners to, say, 10% of the stock. Another way is to grant shares of enterprises to workers as part of the remuneration package. Some of these measures have been tried out in various contexts with varying success. Such experiences must be utilised and adapted pragmatically to local conditions to find workable approaches.

Membership of pension funds, life insurance agencies and medical schemes also spreads potential in the population. One could make membership in such schemes compulsory and create tax incentives for the latter to invest in smaller, labour-intensive companies. A popular measure in Third World countries is to make the state a major shareholder in large enterprises. The benefits of such a policy are not always apparent. To benefit the majority, the proceeds of such share holdings should at least be channelled into the social infrastructure of the country rather than into an inflated bureaucracy. Centre firms can contribute to the spread of economic potential by giving subcontracts to smaller companies rather than keeping the whole production process under their control. Such subcontracting firms are also known to be successful generators of employment.

A way to spread power more equally is *industrial democracy*. There are a number of alternatives here: (a) Workers are free to organise trade unions who then negotiate on equal terms with employer organisations (the capitalist model). (b) Workers are represented on the boards of larger enterprises (the

Chapter 18 - Policies for centre-periphery interaction | 441

social-democratic model). (c) Workers control the enterprises in which they are working and appoint the directors (the former model of democratic socialism). (d) All enterprises are owned by the state and the workers are employed by the state (the former Marxist-Leninist model). The second model seems to combine responsibility with efficiency in the most optimal way. The experience of Marxist-Leninist countries in Eastern Europe suggest that the last model should be avoided.[4]

Going beyond the industrial scene, *small businesses* and the so-called informal sector should be supported. This cannot mean that undertakings which are not viable should be propped up artificially. The idea is, rather, to create the space for smaller initiatives to get off the ground. Statutory and administrative obstacles in the way of their development should be removed.[5] It is counterproductive, for instance, to insist on First World building standards in slum upgrading operations. Positive measures include credit facilities for firms without fixed assets, together with safeguards against early bankruptcy, the provision of training, marketing channels, and so on. Companies can be taxed progressively according to size, so that the scale advantages are neutralised and the competitiveness of smaller companies is enhanced.

This measure would also induce big companies to break up into smaller units. This is indeed desirable, except in the case of industries which need a minimum scale of operations to become internationally competitive. But even in these cases capital ownership and control can be spread more widely. It is not in the national interest if a handful of corporations control most of the formal economy.

The effects of the policies of *multinational corporations,* in particular, should be kept under close surveillance.[6] This is not because they are necessarily evil as such, as radicals have believed until recently,[7] but because the scale of their operations makes the impact of any harmful move so much greater; because they are controlled from outside the realm of local interests and have no local stakes, and because their international connections enable them to pull out where problems arise and come in where a killing can be made. More harmful is the volatility of financial capital and the scourge of currency speculators. Recently the celebrated Pacific Rim countries, Russia and other "emerging economies" such as South Africa have become victims of their ravages. At a time when roving financial capital is behaving on the world scene like a pack of wolves in the steppes, we desperately need judicially potent international controls.[8]

Once again, experience has shown that the development of new economic initiatives should be based on the principle of free enterprise.[9] A rather problematic way of spreading the national cake more equally, therefore, is the extension of the bureaucracy. This method has been used widely, especially where previously disadvantaged groups have come to power. But the policy can backfire severely. Most bureaucracies are not engaged in material production. Those that deliver services are notorious for their inefficiency. When people receive a salary for making no noticeable contribution, there is some distribution of consumption, but there is no distribution of production. We have emphasised that this is the wrong way to reach greater equity.

Moreover, such consumption is parasitic. In affluent societies the growth of the bureaucracy may be a means to channel financial resources derived from a capital-intensive industrial sector to a labour-intensive services sector, thus creating jobs. But if the bureaucracy is not living off a powerful industrial centre but, instead, sucks resources out of a weak rural periphery, the latter may ail and perish. It is worse when bureaucracies keep themselves busy with creating administrative obstacles in the way of private producers. The first symptom that there is something wrong is a high rate of inflation. The practice also lends itself to the protection and enhancement of some groups at the expense of others, thus to discrimination, nepotism and corruption.

But the principle of free enterprise implies that the state must make economic freedom accessible to all. While the state must not run the economy, it must indeed level the playing fields, empower would-be competitors, and police the economic highways.

Employment creation

In most semi-industrialised countries the emphasis has to fall on employment creation.[10] Research in industrialised countries has shown that governments which have made unemployment a definite policy priority have usually succeeded in overcoming the problem to an appreciable degree.[11] In industrialised countries the policy of making the work force versatile and adaptable to the changing demands of the labour market through training and retraining is indeed of critical importance. However, it is no solution for the problem of large scale marginalisation in semi-industrialised and peripheral countries.

Current wisdom has it that *capital investments* create further employment opportunities. Our analyses, especially in chapter 12, have shown that this is a fallacious argument. To a large extent capital consists of labour saving devices. They raise the productivity of labour but lower the rate of employment. While demanding higher levels of expertise, capital displaces labour, if not in the industry concerned, then elsewhere in the system. For capital investments to secure higher levels of employment, the growth of the economy must outperform the rate at which labour is displaced by capital. For this to happen markets must be found either in the centre, and that means in the realm of luxuries, or in the periphery, where it will destroy jobs.

There is also the possibility of export promotion, this possibility is much more limited than commonly assumed and it only transfers the problem to the international scene. It was one of our fundamental insights that unemployment can be exported. Moreover, if employment creation is to remain ahead of labour displacement, market growth must constantly accelerate. This cannot continue indefinitely without undermining the system as a whole.

It could be argued that industrial growth raises the overall wealth of the nation which will benefit all its members at least indirectly. A stronger industrial sector means that the state can gather more taxes which it can use to build schools, roads and so on. This argument is valid only under certain conditions. In the first place the destruction of productivity caused by the marginalisation of a great proportion of the work force must not be greater than the gains in productivity through mechanisation, otherwise the wealth of the nation as a whole actually deteriorates.

Secondly, the gains achieved by such growth must indeed be used to satisfy the needs of the poor and not lead to even greater discrepancies between the rich and the poor. Highways are fine for the elite but the elite is not the nation. These mechanisms must also be seen in their international contexts. The growth of local offshoots of centre power in peripheral countries can indeed cripple local production in these countries. The experiences of the South East Asian "tigers" suggest that it takes some shrewdness and determination to avoid such consequences.

It is the wellbeing of people which should receive priority. The owners of capital are a wealthy elite while the workers have only their labour to feed their families. Under modern conditions, capital is more prone to outcompete labour than the other way round. So it is labour which needs to be protected, not capital. The effects of technological innovations on the labour market should be monitored carefully. Obviously this does not mean that powerful trade unions should be allowed to cripple the economy.

As mentioned above, the *tax system* should support employment creation rather than mechanisation, especially under conditions of high unemployment. Moreover, the state should sponsor and subsidise research aimed at the competitiveness of labour-intensive methods of production, rather than labour saving devices. Labour saving devices make sense when people are subjected to long hours of exhausting work under unhealthy conditions, or when labour is in short supply. But they do not make sense when a large percentage of the work force lies idle and has to be supported by the national economy.

The easiest way of achieving these goals is not by trying to regulate the economy through the legal system but by balancing out distorted factor prices through the tax system. As we have seen in the previous chapter, progressive taxation of energy, capital, imported technologies, pollution and waste generation will automatically make labour and local technologies more competitive.

As the economy grows, jobs can be created in schemes to protect and reclaim the natural environment, in upgrading neglected amenities and services such as housing, transport, street cleaning and recreational facilities, especially in poor communities squatting on the verges of industrial areas. Why should bulldozers be used to construct rural roads when thousands of people would be willing to work with picks and spades for low wages! Why should mechanical lawn mowers be used on public lawns! The mechanisation of agriculture and the automation of the motor industry are further examples.

The facilitation of development among the marginalised

What can be done about the marginalised in a situation where the formal sector is too small to integrate the entire population? We have argued above that a provisional second economy must be developed alongside the formal economy if the poor are to get out of their misery. In the long term these two economies must, of course, be integrated. For the development of a second economy, agency is of the essence. On the one hand it is the poor who have to analyse their situation and take the initiative to improve it. Outsiders, however well intentioned, can at best be facilitators. They have to respect and learn from the wisdom and the survival strategies of the poor.

On the other hand, the skills and means of the peripheral population can be enhanced deliberately. Here the state, the private sector, the church, the scientific community and the NGOs can all cooperate to overcome poverty. The goals should be to level the playing field; release local initiatives; provide access to resources; build up skills; facilitate networking; create the environment for micro and small businesses, and nurture the disempowered until they are competitive.

The marginalised must have the constitutional right to be represented on all decision making bodies, including larger enterprises, unions, government departments, and local authorities. Municipal or state services can be transferred to community responsibility. Potential initiators and leaders among the population can be identified, trained and given opportunities to gain experience. To overcome the administrative maze and nightmare one can set up one-stop centres where all official documents can be obtained and processed. Tenders can target the poor, rather than the lowest bidder, and develop the potentials of the informal sector. Contracts can include provisions to benefit the community in terms of job creation and capacity building. To overcome unemployment, the youth can be involved in national service schemes and military facilities and resources can be redeployed for development efforts.

Formal education should include economic and ecological literacy; policy and advocacy; enterprise, management and leadership skills; rudimentary health and healing; basic artisan skills such as carpentry, brick laying or sewing. The scientific community can do its part: a lot of research has been done which must be disseminated and activated. But research may not become a substitute for action.

The private sector can formulate clear policies for community upliftment and subject it to accountability. Companies can set up ethical codes for their operations in poor communities and demonstrate social and environmental responsibility. Low key credit facilities can be established. Low capital-intensity components of the productive process can be outsourced. Enterprises can recruit and fascinate talented young people and help them to create jobs for themselves in small undertakings rather than hunting for a job in great organisations.

The state and the private sector should support civil society and its initiatives. Churches and NGOs can become the voices of the poor, expose corruption and inequitable structures, and challenge the state, the private and sector and even the beneficiaries of development to their moral obligations. They can facilitate the emancipation and empowerment of women. They can help to rebuild the family and castigate unhelpful attitudes, such as the culture of violence, cheating, and entitlement. They can inculcate proven values, such as diligence, efficiency, frugality and integrity. They are capable of international networking to share experiences of projects and policies that have worked elsewhere. They can re-activate retired experts as facilitators of development. They can become involved in HIV prevention and the care of AIDS orphans. They can take young people out of their environment for a year to serve the community.

The media are powerful agents of development. They can begin to concentrate on success stories, rather than sensational calamities. They can convey a sense that poverty can be beaten; include match making sites; give publicity to public dangers such as HIV and pollution; avoid the creation of unattainable expectations.

Revision: Explain how the peripheral population can be empowered and what role the centre population can play in this task.

> *Application:* Which of the measures suggested above are actually applied in your country? In which instances are opposite policies followed? What is the effect of either?
> *Critique:* (a) "You simply assume that justice means that the rich must cater for the needs of the poor. In fact it is the crown of injustice if you subsidise incompetence, inefficiency and carelessnes while defrauding intelligent, hardworking and responsible people of the fruits of their labours." (b) "The problem is that the rich do not want to share. The solution is to transfer their assets to the state and see to it that all have an equal access to the wealth of the nation." Comment.

Section II: Centre-periphery interaction at the international level

Classical economics has originated at a time when the problems of the nation state were paramount. Its very methodology tends to be ethnocentric in the sense that it tries to serve the particular interests of the nation state.[12] In a globalised economy this is no longer appropriate. Globalisation may be unavoidable,[13] but our analyses have also shown that international integration leads to international concentration and international marginalisation.[14]

The situation is not hopeless.[15] Although there is no world government which could enforce "social democratic" policies on a global scale, many statutory bodies with an economic agenda have emerged on the international scene.[16] The agonies of two world wars, countless regional conflicts, the costs of the cold war, the prospects of a nuclear holocaust, the need for arms control, the rise of international terrorism, the misery in the Third World and the increasing threat of ecological disasters have led to the growth of international accountability and responsibility since the beginning of the century.[17]

The dominance of the nation state as such is being undermined by the development of the global economy, especially concerning information and financial capital, and this calls for new structures.[18] Regional formations with supranational institutions have also sprung up in various parts of the world, the most prominent being the European Union. The nation state is also being undermined from another end by the rise of ethnic tensions, often with transnational connections.[19] It is against this background that we should be bold enough to formulate policy directives for the transnational scene.[20]

Non-statutory agencies have also proliferated.[21] Religious organisations, activist groups and non-governmental organisations play a critical role in the global reconstruction of the social fabric. Most of them have aid programmes and multiple local constituencies which sense the heart beat of grassroots agonies and struggles. Their extended networks facilitate cross cultural contact and communication. Some have enough clout to lobby in the corridors of power. Insights gained by such international bodies should filter down to local constituencies, while insights gained on the ground should be channelled upward to decision making bodies. Information and advocacy should also be fed into the mass media.

Again, curtailing the abuse of power by governments, multinational corporations and international financial institutions is the first task. International law should detail criteria for lawful economic interaction. It should protect the interests of the weak and vulnerable against blackmail; undue political and military pressure; exploitation of ignorance and corruption among local administrators; covert transfers of capital; unilateral determination of the terms of trade; credits granted with unfavourable strings attached; tariffs which protect centre industries against competition from the periphery; the manipulation of interest rates at the expense of Third World debtors; bribing governments to accept toxic waste; dumping of cheap products to destroy local industries; protection of endangered species; large scale deforestation; the pollution of the environment; and so on.

There should be an *International Court of Economic Justice* with teeth to pillory and punish transgressors. If sanctions could be applied against certain countries because of their abuses of political power, why not against abuses of economic power! The policies of institutions like the World Bank and the International Monetary Fund should also be subject to scrutiny and audit by such a court of law. The structural adjustment programmes of the World Bank and the International Monetary Fund, while imposing some inescapable and beneficial reforms, have also led to a dramatic deterioration of welfare achievements, such as education and health services, in many Third World countries. These losses undermine the future prospects of poor countries.[22] The whole idea that local resources should be channelled into servicing remote creditors and affluent markets leads priorities in the wrong direction and creates resentment and despondency in the population.

Because international controls are not yet very effective, such legal instruments must also be institutionalised on regional and national levels. The European Community, for instance, has ambitious programmes to balance out inequalities within its sphere and to address ecological deterioration. The danger is, of course, that such a region will develop a provincial selfishness at the expense of other members of the international community.

The debt crisis

So the first task is to curtail the abuse of power. The second task is to neutralise structural mechanisms which operate to the benefit of the centre at the expense of the periphery by institutionalising countervailing process. By far the most important measure to be taken is to write off *untenable debts which cripple Third World economies*. For people with a global consciousness, the maintenance of the current debt burden of the poorest countries has become totally irresponsible. Third World countries overexploit their scarce resources, such as rain forests, to service their debts. Their agricultural potential is utilised for export crops, such as soya beans to feed cattle in rich countries, at the expense of food for their own populations. Foreign exchange earned is not used to import capital goods for production, but thrown into the abysses of debt. Under such circumstances, development becomes illusory.

Peripheral countries should not be forced to push exports at the expense of local supply. They should also not be forced to push exports of goods for which there is a declining market, such as coffee or sisal. While it is certainly true that Third World governments have been irresponsible both in borrowing and spending such funds, it is the rich governments who are responsible for the dramatic rise in interest rates, which made debt servicing by poor countries such a nightmare. If such hikes have really been necessary, which can be disputed,[23] the victims should have been compensated. This at least would have been the natural attitude of Western government towards their own subjects, so why could the same not be done for those who are much less able to cope with the impact?

Moreover, there are not only irresponsible borrowers but also irresponsible lenders.[24] A responsible bank manager does not grant an overdraft facility without checking the creditworthiness of the client. Why has this principle not been applied in the case of international loans where much greater sums were involved? Are military dictators ever creditworthy in a society based on liberal principles? If a contract made with a lunatic is null and void in private law, should the same not apply to a corrupt dictator in international law?

Without doubt many credits have been advanced overeagerly to boost sales or to invest petro-dollars. But even after the oil crisis and the subsequent raising of interest rates, the lending of US banks to Third World countries continued. In fact, it increased fourfold from 110 to 450 billion between 1978 and 1982. Bankers seemingly disregarded information on creditworthiness, speculating that governments would bail them out, thus making citizens unwilling insurers. But can the population of a poor country, itself a victim rather than an accomplice, be held infinitely responsible for the debts of irresponsible rulers and capitalists, even long after their demise?

One of the oddities of the modern world is that there is no equivalent in international law to the bankruptcy provisions found in private law, which protect the debtor from total privation.[25] Because bankruptcy does not apply to a sovereign government, it can use its power "for the nearly unlimited extraction of repayment from its people."[26] Surely this is an anomaly which should be overcome. If defaulting or the remission of debts would lead to the collapse of the world financial system, as it is often argued, what about the collapse of peripheral economies? It is famine-stricken people who should be pitied, not rich financiers and gullible governments.

Recently the idea of a *nature-for-debt swap* has been mooted.[27] According to this scheme debt releases are to be offered, for instance, in exchange for a cessation of deforestation in the Amazon basin. While the idea is interesting, the underlying motives are questionable. Accusations, bickering and blackmail on both sides provide the evidence. The rich say that the poor are not willing to make their contribution to the health of the global environment by protecting common assets of humankind. The poor say that the industrial countries have chopped down their own forests long ago and that it is now their time to pollute, deplete and get rich.

There is also a lot of hypocrisy in the idea. If the rich suddenly have enough money for a debt-for-nature swap, why not write off the debts in the first place? It is clear that deforestation bothers Westerners, not because of the consequences this has for Third World populations, but because it affects the capacity of nature to clean up the atmosphere which they pollute and which is beginning to revenge itself on the polluters.

Third World governments, therefore, react angrily, saying that the forests are by no means the common property of humankind, but national assets which they have a right to exploit. If the rich want them to conserve their forests, therefore, they must pay full compensation for what the poor forfeit by not utilising these assets. It can also be argued that deforestation is largely conducted by commercial interests based in the centre and could be stopped at its roots by centre governments if they were serious. On the other hand these protestations cannot remove the fact that it is the population of poor countries whose assets are being destroyed.

International trade

After addressing the debt crisis, balanced trade relations may be the most important consideration. The "New International Economic Order" was a relatively tame initiative to establish a more balanced interaction between the rich and the poor nations, but even that proved to be unacceptable to the rich countries. This shows that the latter are still determined by very narrowly defined interests and have not developed global horizons yet.[28]

If raw materials from Third World countries constantly drop in price and manufactures from industrial countries constantly rise in price, the rich countries get more than their share out of the deal. Rising prices of manufactured goods are largely due to rising wages and other privileges accruing to workers in the manufacturing country. So the periphery loses out to the centre, even to the periphery in the centre. The same is true for loans which are tied to the provision that the funds must be spent on the products of the creditor country and at the latter's price. The same is true for import restrictions imposed by rich countries on agricultural products from poor countries, while the latter have no means to retaliate. The same is true for debts incurred by poor nations at low interest rates which are subsequently raised.

The combined effect of all these practices is a net flow of financial resources from poor to rich countries, which is not only grossly unfair but also dangerous for world peace and ecological health. To offset these imbalances, countervailing measures have to be institutionalised. There should be regular checks on actual cash flows between rich and poor countries and these should be balanced out by means of trade agreements or through direct transfers through investments in the peripheral infrastructure.

All this cannot materialise without a change in collective consciousness. What is needed is a deliberate decision by the rich not to utilise their positions of power to their own advantage. In a world of devastating misery and unheard-of affluence, the rich should not allow themselves the luxury of making profits at the expense of the poor. The demands of justice are not economically irrelevant. The rich minority must also begin to realise that an impoverished majority is not in their own long term interest. An avalanche of poverty-stricken people in search of means of survival may soon begin to swamp even the protected spheres which centre elites have carved out for themselves. The environmental impact of poverty will also not stop before their door steps. Moreover, much "ambulance" service in the form of crisis aid can be avoided by timely measures to compensate for imbalances in the interaction between rich and poor countries and communities.

International aid

Only after protection against the abuse of power and a greater balance in trade relations have been placed forcefully onto the agenda should international aid also be considered.[29] As we have seen in chapter 12, centre developments lead to the marginalisation of the peripheral population. It is only reasonable to expect the centre to help strengthen the periphery to withstand the impact. There is no end to the possibilities of facilitating home grown development. They include the financing of infrastructure such as telecommunications and transport networks, training in technical and administrative skills, locally based research, development of appropriate technology, compensation for ecological discipline, such as the non-utilisation of areas under tropical rain forests, and so on.[30]

As mentioned above, however, the centre cannot develop the periphery; the latter must develop its own potential. Therefore any assistance only makes sense if it liberates and empowers the initiatives in the periphery. Development projects may not be the most appropriate instrument because they tend to be artificial creations which collapse when foreign leadership and support is withdrawn. What is true for projects is true for expatriate personnel. Short term lucrative contracts of highly specialised experts sent to poor countries can severely damage the cause of the latter. It takes about 5 years simply to become fluent in the vernacular, adjust to the cultural assumptions and understand what happens in the society. Before presuming to be of assistance, expatriate experts and professionals would have to be motivated to identify with the local population, experience its predicament, study local conditions and needs, facilitate local initiatives rather than taking over control, subject themselves to local leadership and be prepared for long term service.

Presupposing such a change at the level of collective consciousness, one could indeed consider the possibility of offsetting the brain drain with an injection of expertise from the centre. Experience with the Berlin Wall suggests that one should not try to force academics and professionals to stay within the boundaries of their country of origin. Restrictions do not generate motivations. Rather, the loss to the periphery should be balanced out by volunteers from the centre who are capable and eager to live and serve under peripheral conditions. Hundreds of thousands of traders, missionaries, soldiers and administrators have once been lured to far and dangerous places by the spell of nationalist and imperialist visions. Perhaps the same can also happen on the basis of a global and more appropriate vision?

At this juncture a few reflections on *food aid* may not be out of place. It is often argued that there is sufficient food in the world, the only problem being unequal distribution. To the extent that wheat or maize or soya from the periphery is fed to pigs which end up on the tables of wealthy protein consumers in the centre, this may be true. However, we should be wary of this argument for at least three reasons:

(a) In most industrialised countries agriculture cannot compete with manufacturing. The "food mountains" found there are the result of state subsidies. In other words, high yield agriculture can only survive as a parasite of industrial production. At the same time it cannot compete on world markets in this form. That is why it is protected against agricultural imports. Agricultural overproduction is a symptom of the disease, rather than a cure for the nutritional impasse. Western high yield agriculture is also highly artificial and therefore wrought with dangers. Ecological, technological, economic or political pressures may lead to gross distortions, even to the collapse of this system of production. I just remind you that modern agriculture is firmly based on fossil fuels which are bound to run out within the foreseeable future. It would be shortsighted to base the food requirements of Third World populations on such a precarious foundation.

(b) A mere transfer of food from the centre to the periphery without reciprocal trade leads to increasing dependency of the periphery upon the centre. As a result the periphery becomes more vulnerable to blackmail and manipulations. For balanced trade, the periphery must have something to offer. Food imports also destroy the market of local producers, the will to produce may collapse and the dignity and self-respect of the peripheral population may be seriously impaired.

(c) Food aid cannot solve the problem of an imbalance between production and consumption. The law of declining marginal productivity is still in operation. When the graph of accelerating population growth intersects with the graph of declining agricultural production, famine on a vast scale can no longer be avoided. While there is uncertainty about the time scales of futurological calculations, there can be no doubt that unlimited growth is impossible on a limited planet. Either one balances out the system now or it will "balance" itself out in the future through mass starvation.

So the answer to the problem is not the transfer of agricultural products from the centre to the periphery, but the *enhancement of peripheral production.* One could consider whether in the place of food aid the subsidies now granted to centre agriculture should not be transferred to the periphery to build up productive capacity and purchasing power there. Such a policy would drive many centre farmers out of business and the centre would be forced to address structural imbalances now concealed by agricultural subsidies. There are already some schemes in central Europe which compensate farmers for not using their land. Is this a utopian dream, or is it part of a vision which could one day materialise?

Trade liberalisation in agriculture as such would not automatically benefit the peripheral population. Agricultural protectionism in centre nations is widely believed to deprive poor countries of the opportunity to sell their produce in return for manufactured goods. If subsidies and protective tariffs on behalf of centre farmers were lifted, and food surpluses could be produced in the periphery, genuine comparative advantages would emerge on both sides and agrarian countries would be able to offer a much needed product on international markets.

But as we have argued in chapters 5 and 12, free trade causes massive problems both in central and peripheral countries. For the periphery it conjures up the spectres of asymmetric interaction. Food production in the periphery would be geared not to the needs of the poor in the periphery, but to affluent centre markets in return for manufactured goods consumed by peripheral elites.[31] These goods include not only urban luxuries but also weapons used to oppress restive populations. Commercial agriculture in peripheral areas would be encouraged to mechanise its operations at the expense of workers and draught animals. We have argued, therefore, that peripheral agriculture should be developed to enhance food production for local use. Surpluses could be exported in case they materialised.

Free trade would also affect the rural periphery in centre countries. While urban consumers would benefit, farmers would go out of business and rural areas would be depopulated. The social and political repercussions would be considerable. In view of the possibility that the present productive clout of the industrial system may have a rather short life span due to dwindling resources and growing pollution, the destruction of centre agriculture would not be prudent at all.

Isolation or integration into the world market?

It was believed for some time, particularly among dependency theorists, that an underdeveloped economy which cannot compete on the world market, can only get off the ground if it is detached from this market.[32] Japan, the Soviet Union, China and Tanzania have followed this policy to some extent, some with more success than others. The problem is that, while isolation offers protection against overwhelming international competition, it also reduces challenges to become more efficient. If protected indefinitely, infant industries never grow up. Eastern socialist countries, for instance, lost the technological race with the West partly because they were not sufficiently exposed to international competition. The same can be said of Tanzania if compared with Taiwan.[33] Moreover, local markets are often too limited to allow for economies of scale and thus make industrialisation for import-substitution a viable proposition.

Current economic wisdom has it, therefore, that there is no substitute for *export-led industrialisation.* However, the theory of comparative advantage is misleading in the case of vast discrepancies in economic power between the trading partners.[34] The value of potential export products generated by a typical Third World country can often not pay for the sophisticated consumer goods and services desired by its population. The demand for, and thus the prices of, many raw materials have been falling while the prices of manufactures have been rising during the last few decades. Agricultural exports cannot provide the cash because most poor countries have no surpluses; agricultural production in centre nations surpasses demand and agricultural protectionism is not likely to disappear soon.

Cheap imports also undermine the productive capacity of the periphery. It is true that a local industry does not become efficient if it is not exposed to foreign competition, but this argument does not hold water in the case where traditional potters have to compete with automated plastic factories and are simply ousted from the market. Poor countries should consider such imports as "concealed dumping". Ecological economists have insisted that market prices should reflect the true costs of the product. This idea can also be applied to the social costs of cheap imports. The price of an imported gadget for the nation is not necessarily what the importer or the consumer pays for it, especially if the import destroys local production. The state should impose import tariffs accordingly.[35]

To stay within their means, peripheral populations have to become less dependent on foreign exchange and imports. Even the import of energy and capital goods must be restricted to the absolutely essential. The fact that existing industrial capacities in impoverished Third World countries are largely underutilised because of a lack of fuel, spare parts or expertise, demonstrates that the First World type of development is not the answer to their current economic problems.

Centre trade negotiators are also not necessarily sincere in their advocacy of free trade. The measure of free trade allowed by centres depends not on the beauty of the theory but on the economic interests of the most powerful players. As mentioned above, "development aid" is often granted with strings attached and peripheral economies are not always allowed to shop around on the open market. The European Union still protects its own agriculture. The ideological agenda of the free trade lobby is captured rather well in Krugman's cynical remark: "Then you might very plausibly advocate a temporary tariff to ensure that you get the core. Once you have established a decisive lead in manufacturing, you can remove the tariff - and lecture to the other country, which has effectively become your economic colony, on the virtues of free trade."[36]

The Marxist picture of free foxes in a free fowl run is applicable here. First world lobbies for free trade negotiate around GATT tables from positions of strength. If they were vulnerable to destructive competition, they would be much less enthusiastic about free trade. First World self-interest is so ruthless that Third World governments should not have any qualms in protecting their own economies against being swamped by cheap manufactures imported on the basis of deceptive credits. The problem is that in many cases they do not have sufficient clout to defend themselves against blackmail. Nor do Third World elites necessarily have the will to restrain their own consumption patterns and do what is in the interests of their countries.

Assuming that the trend towards concentration is undesirable, more regulated trade patterns would also be in the interests of the world economy as a whole. For all its undesirable side effects, tariffs are a powerful instrument to avoid, or break up, economic concentrations. Economically speaking, the cost of transportation as a factor of production, including tariff barriers, is an important determinant of competitiveness. The reduction of transportation costs (or tariffs) gives highly productive capacity from the centre greater access to the periphery, thus undermining less advanced peripheral enterprises and the informal subsistence economy. It also allows disintegrated peripheral factors of production to move more easily to the centre where economies of scale dominate, thus depriving the periphery of its productive assets.

The higher the cost of transportation (including tariffs), therefore, the lower the conglomeration effect, the greater the chance of localised industries to compete and the greater the spread between regions.[37] "If transport costs are low, economies of scale large, and the share of footloose industry in national income large, the result will be a single core; if the reverse is true, there may be no core at all; intermediate levels will support a multiple-core structure."[38] That is what is needed on a global scale in the longer term!

How much more would this not apply to restrictions imposed on inside trading of multinational corporations and, especially, the movements of highly volatile international financial capital! Note that the alternatives of free trade policies versus tariffs and other restrictions, are an example of the manipulation of factor costs, just like the taxation or subsidisation of fuel rather than labour, which we discussed in chapter 17.

Free trade is neither good nor bad in principle. It depends on what it does to the economy in each case. The *modern* sector in a poor country should concentrate on lucrative niche markets where it can be shown that it has substantial comparative advantages. It should not try to compete internationally where it has little chance of success. In many cases a policy of selective import substitution can still be applied, where infant industries stand a fair chance of becoming competitive when protected for some time. But they should be gradually exposed to international competition so as to mature in terms of efficiency and productivity.

As far as the *peripheral* economy in poor countries is concerned, a policy of self-sufficiency should be followed. Internal food, clothing and utensils markets, for instance, should not be allowed to be swamped by cheap imports. Peripheral production should also not primarily be geared to export markets, but serve the needs of the local population. This presupposes financial discipline and minimal dependency on foreign currency. Peripheral elites must learn to adapt to the limitations of a Third World economy and not try to emulate Triad living standards.

In contrast with the two extremes of total isolation and full integration into the world market, therefore, we opt for economic *sovereignty* and *self-sufficiency* on the one hand, and *selective trade* on the other. In the place of a "Wild West" type of free-for-all, we should strive for a dynamic, flexible and balanced trading system, which is negotiated on the basis of equal dignity and which heeds the interests of the more vulnerable partners.

Revision: What can be done to level the playing fields between industrial and developing nations?

Application: What consequences would an implementation of the measures suggested above have for your country, and for which population groups in particular?

Critique: (a) "Experience shows that the industrial countries will never part with their ill-begotten wealth and the developing countries cannot force them to do so. The only way out for the Third World is to go its own way and leave the West alone." (b) "All the billions of dollars of development aid from the North have not made the slightest impact on the poverty of the South. On the contrary, all that money has been squandered by irresponsible leaders. If the South chooses to live that way, why should the North be held responsible?" (c) "Your idea that debts could simply be written off is ludicrous. In the first place it would lead to the collapse of the global financial system. In the second place the leaders of poor countries would simply take up more loans and squander the money." What do these statements have in common? How would you respond?

Let us summarise

In this chapter we concluded our discussion on policy options with a discussion of the interaction between centres and peripheries. Section I was devoted to policies designed to balance out the interaction between centre and periphery *within a country*. Such discrepancies are particularly marked in semi-industrialised countries. Large differences in power must be overcome and the abuse of power must be curtailed by statutory provisions. A social contract must be drawn up to secure widespread support and cooperation for a balanced economy. The periphery must gain access to resources, especially agricultural land. Commerce and industries must be democratised and the assets must be distributed more equitably. The state should not run the economy but make free enterprise accessible to all citizens. The first priority should be accorded to employment creation; capital investment and technological innovation should be subject to this criterion.

In section II we dealt with *international* relations. The abuse of power must be controlled by an International Court of Economic Justice. To give poor countries a chance to develop, unbearable debt burdens must be written off. International cash flows must be controlled and imbalances in trade must be overcome. Aid should empower local potentials rather than enhance centre profits or repair the damage caused by international inequity. Food aid should make way for agricultural development. Free trade is not necessarily in the interest of weak trading partners, nor of centre agriculture, and should make way for a regulated balance of interests between all partners concerned.

Notes

[1] Post-war Germany had an explicit, Japan an implicit social contract. The Reconstruction and Development Programme in post-apartheid South Africa enjoyed widespread support, but it was virtually replaced by the more market-friendly GEAR programme.
[2] Cf Stackhouse 1987:52ff.
[3] Moll, Terrance 1991: Microeconomic redistributive strategies in developing countries, in Moll 1991:1ff.
[4] For a more detailed discussion see the companion volume, Nürnberger 1998 chapters 4, 5 and 8.
[5] I was told by an economist at the University of Nairobi that, in spite of ostensible official support for the informal sector, hawkers and artisans are regularly driven off the streets for phony reasons, such as hygiene or traffic flow.
[6] For a critique of the corporation see, apart from Korten 1995, Stackhouse 1987:128.
[7] Stackhouse is a theological ethicist who emphasises the positive aspect of the corporation as basic unit of cooperative production (1987:118ff).
[8] For the current exchange rate system see Goldstein 1995.
[9] For a detailed discussion refer to Nürnberger 1998.
[10] See Nürnberger 1990 for detail.
[11] Therborn 1986:23, cf David Lewis in Gelb 1991:252ff.
[12] This was even the case in the former socialist countries. Marxist theory postulated the solidarity of the international proletariat but this did not work out in practice. Stalin

concentrated the attention of the international communist movement on the Soviet Union and ultimately the "Comintern" was dissolved. See Nürnberger 1998, chapter 4, for detail.
[13] Küng 1997:160.
[14] To augment the following, see also Küng 1998:215ff.
[15] Küng 1997:167ff.
[16] For the development of the nation state and its transformation into a "megastate" see Drucker 1993:103ff. Of course, the megastate is not international. See pp 128ff for the latter.
[17] Drucker 1993:132ff.
[18] Drucker 1993:129ff; Korten 1990:156ff.
[19] Drucker 1993:138.
[20] See Korten 1990:135ff for detail.
[21] For instance the Red Cross, the international academic community with its multiple contacts and research facilities, the Vatican, the World Council of Churches, the World Evangelical Fellowship, the Lutheran World Federation, the World Alliance of Reformed Churches, the Anglican Communion, and so on.
[22] Von Weizsäcker 1990:208.
[23] According to von Weizsäcker it was "Reaganomics" which deflected the attention from the "limits to growth" issues raised by the Club of Rome studies to the cold war. Vast budget deficits in the First World caused interest hikes which led to the debt crisis in the Third World and to overexploitation of the environment to pay off debts. A decade was lost for ecology - and obviously for development as well (Von Weizsäcker 1990:57).
[24] For the following see Child in Paul, Miller & Paul 1992:114ff.
[25] Child *op cit* 135.
[26] Child *op cit* 139.
[27] Elliott 1994:55f, Simpson 1990:153.
[28] Cf Sider (1977:125ff).
[29] Cf Korten 1990:137ff.
[30] A very concrete sacrifice, which Duchrow suggests should be made, is the redistribution of excess wealth (Duchrow 1994:293)$.*
[31] In the 1970s, for instance, Brazilian agriculture was geared to soya beans for export to the US and Europe, where it was used as fodder, resulting in local food shortages.
[32] An example is the work of Dieter Senghaas, Weltwirtschaftsordnung und Entwicklungspolitik: Plädoyer für Dissoziation. Frankfurt/M: Suhrkamp, 1977.
[33] See Nürnberger 1998, chapter 6, for more detail.
[34] For a critical discussion of the theory of comparative advantage see Daly & Cobb 1989:209ff; Thirlwall 1994:366f, Johns 1985:149ff, Anzuck 1982, Krugman 1991.
[35] Let us demonstrate this with a crude example. Say you import a plastic plate for R 1.00, which takes the place of a wooddan plate which would cost R 3.00 to produce locally. There seems to be a saving of R 2.00 for the nation. In fact the nation has to raise R 1.00 in foreign currency by exporting something else, which could certainly be of value to the nation, and loses R 3.00 worth of local productive capacity. So the price of the plastic plate for the nation is R 4.00, not R 1.00.
[36] 1993:90. Krugman says it has not happened exactly that way. He argues that tariff protection worked in Canada only because Canada's population grew and the economy became self-sustaining. But why should it not have worked in India then? Krugman does not take real contrasts (North-South, Lesotho vs South Africa) into consideration.
[37] Krugman 1993:77ff.
[38] Krugman 1993:86.

Bibliography

Aaker, D A & Day, G S 1974. *Consumerism: Search for consumer interest.* New York: Free Press et al.
Abramovitz, Moses 1993. The search for the sources of growth: Areas of ignorance, old and new. *The Journal of Economic History* 53/1993 217-243.
Adams, W M 1990. *Green development: Environment and sustainability in* the Third World. London: Routledge.
Albrecht, P & Koshy, N (eds) 1983. *Before it's too late: The challenge of nuclear disarmament.* Geneva: WCC.
Albright D E (ed)s 1980. *Africa and international communism.* London: Macmillan.
Allen, T & Thomas, A (eds) 1992. *Poverty and development in the 1990s.* Oxford: Oxford Univ Press / Milton Keynes: Open University.
Alynsky, Saul D 1972. *Rules for radicals: A pragmatic primer for realistic radicals.* New York: Vintage Books.
Amin, Samir et al 1982. *Dynamics of global crisis.* London: Macmillan.
Amuzegar J 1981. *Comparative economics: National priorities, policies, and performance.* Cambridge Mass: Winthrop.
Anderson, K & Blackhurst, R 1992. *The greening of world trade issues.* New York: Harvester Wheatsheaf.
Andreski, Stanislav 1968. *The African predicament: A study in the pathology of modernisation.* London: Joseph.
Antoncich R 1987. *Christians in the face of injustice: A Latin American reading of Catholic social teaching.* Maryknoll, NY: Orbis.
Anyang Nyongo, Peter (ed) 1987. *Popular struggles for democracy in Africa: Studies in African political economy.* London: Zed Books.
Anzuck, B 1982. *Raumstrukturelle Fatoren der internationalen Arbeitsteilung.* Frankfurt/M: Peter Lang.
Apostolakis Bobby E 1992. Warfare-welfare expenditure substitutions in Latin America 1953-87. *Journal of Peace Research* 29/1992 85-98.
Argyle, Michael 1987. *The psychology of happiness.* London/New York: Methuen.
Arkhurst F A (ed) 1970. *Africa in the Seventies and Eighties: Issues in development.* New York et al: Praeger.
Asian NGO Coalition, IRED Asia, PCDForum 1993. *Economy, ecology and spirituality: Toward a theory and practice of sustainability.* Publication of the People Centred Development Forum (14 E 17th Street, Suite 5, NY 10003).
Athanasius, Padre OSB 1988. *Justice for the peasants.* Ndanda, Peramiho, Tanzania: Benedictine Publications.
Atkinson A B 1983. *The economics of inequality.* London: Oxford.
Bähr, J, Jentsch, C & Kuls, W 1992. *Bevölkerungsgeographie.* Berlin / New York: de Gruyter.
Balcomb, Anthony 1993. *Third Way theology: Reconciliation, revolution and reform in the South African church during the 1980s.* Pietermaritzburg: Cluster Publications.
Barde, J P & Pearce, D 1991. *Valuing the environment.* London: Earthscan.
Barney, G O (ed) 1980. *The global 2000 report to the president.* Washington: US Foreign Office.
Barratt Brown, M 1974. *The economics of imperialism.* Harmondsworth: Penguin.
Basalla, George 1988. *The evolution of technology.* Cambridge: Cambridge Univ Press.
Batista, Jorge C 1992. *Debt and adjustment policies in Brazil.* Boulder: Westview.
Bauer, P T 1981. *Equality, the Third World and economic delusion.* London: Weidenfels & Nicolson.
Bauer, P T & Yamey B 1984. *Reality and rhetoric: Studies in the economics of development.* London: Weidenfels & Nicolson.
Bedjaoui, Mohammed 1979. *Towards a new international economic order.* New York / London: Holmes & Meier for UNESCO
Bello, Walden 1989. *Brave new Third World? Strategies for survival in the global economy.* San Francisco: Inst for Food Development.

Benne R 1981. *The ethics of democratic capitalism: a moral reasssessment.* Philadelphia: Fortress.
Berg, R J, Whitaker, J S eds. 1986. *Strategies for African development,* sponsored by the Council on Foreign Relations and the Overseas Development Council. Berkeley: University of California Press.
Berger P L 1974. *Pyramids of sacrifice: Political ethics and social change.* New York: Basic Books.
Berger P L 1986. *The capitalist revolution: Fifty propositions about prosperity, equality, and liberty.* New York: Basic Books.
Berger, P L 1973. *The social reality of religion (= The sacred canopy).* Harmondsworth: Penguin.
Berger, P L, Berger, B & Kellner, H 1974. *The homeless mind: Modernization and consciousness.* Harmondsworth: Penguin.
Betten, Lammy 1993. *International labour law: Selected issues.* Deventer/Boston: Kluwer.
Birch, Charles, William R Eakin & Jay McDaniel (eds) 1990. *Liberating life: Contemporary approaches to ecological theology.* Maryknoll: Orbis.
Birdsall, Nancy 1980. *Population and poverty in the developing world.* Washington, DC: World Bank.
Block, Peter 1993. *Stewardship: Choosing service over self-interest.* San Francisco: Berrett-Koehler.
Block, Peter 1993. *Stewardship: Choosing service over self-interest.* San Francisco: Berrett-Koehler.
Boff, Leonardo 1995 (1993). *Ecology and liberation: A new paradigm.* Maryknoll, NY: Orbis.
Boff, Leonardo 1987 (1977). *Ecclesiogenesis: The Base Communities reinvent the Church,* London: Collins.
Bojö, J, Mäler, K-G & Unemo, L 1992 (1990). *Environment and development: An economic approach.* Dordrecht / Boston / London: Kluwer.
Bondestam L & Bergstrom S 1980. *Poverty and population control.* New York: Academic Press.
Bosch, David 1992. *Transforming mission.* Maryknoll, NY: Orbis.
Botkin, J W 1979. *No limits to learning* (Report to the Club of Rome). New York: Pergamon Press.
Boulding, Kenneth E 1970. *Economics as a science.* New York: McGraw-Hill.
Brackley, Peter (ed) 1990. *World guide to environmental issues and organizations.* Harlow Essex: Longman.
Brandt W *et al* 1980. *North-South: a program for survival.* Cambridge Mass: MIT.
Brandt W *et al* 1983. *Common crisis: North-South co-operation for World Revovery.* London: Pan Books
Brennan, Richard 1991 (1990). *Levitating trains and Kamikaze genes.* New York: Harper Perennial.
Brian Morris 1994. *Anthropology of the Self: The individual in cultural perspective.* London: Pluto Press
Broadley, E & Cunningham, R 1992. *Core themes in geography: Human.* Harlow Essex: Oliver and Boyd.
Brody H 1981. *Ethical decisions in medicine.* Boston: Little, Brown and Co., second edition.
Browett, J G & Fair, T J D 1974. South Africa, 1870-1970: A view of the spatial system. *SA Geographical Journal,* vol 58/1976, 118-129.
Brown L R *et al* (ed) 1987. *The state of the world 1987: A Worldwatch Institute report on progress for a sustainable society.* New York: W W Norton.
Brown, L & Kane, H 1995. *Full house: Reassessing the Earth's population carrying capacity.* London: Earthscan.
Brown, L R 1981. *Building a sustainable society.* New York / London: Norton (for World Watch Institute).
Brown, L R *et al* (eds) 1993 (annually). *The state of the world: A World Watch Institute Report on progress toward a sustainable society.* New York/London: Norton.
Brown-Weiss, Edith 1989. *In fairness to future generations: International law, common patrimony and intergenerational equity.* Dobbs Ferry NY: Transnational Publishers.
Brown-Weiss, Edith (ed) 1992. *Environmental change and international law: New challenges and dimensions.* Tokyo: United Nations University Press.

Bulletin of Concerned Asian Scholars 1983. *China from Mao to Deng : The politics of economics of socialist development.* London: Zed Press.
Bundy, C 1988. *The rise and fall of the South African peasantry.* Cape Town: David Philips.
Burton D F et al (eds) 1986. *The jobs challenge: Pressures and possibilities.* Cambridge Mass: Ballinger.
Cairncross, Frances 1991. *Costing the earth: What governments must do: What consumers need to know: How business can profit.* London: The Economist Publications.
Caldwell, B 1982. *Beyond positivism: Economic methodology in the 20th century.* London: Allen & Unwin.
Campbell, Colin 1987. *The Romantic ethic and the spirit of modern consumerism.* Oxford: Basil Blackwell.
Campolo, Tony 1992. *How to rescue the earth without worshipping nature: A Christian call to save creation.* Nashville TEN: Thomas Nelson.
Caplow, Theodore 1971. *Elementary sociology.* Englewood Cliffs, NJ: Prentice-Hall.
Caporaso, James A (ed) 1987. *A changing international division of labour.* Boulder: Lynne Rienner.
Capra, Fritjof 1983 (1982). *The turning point: Science, society and the rising culture.* London: Fontana Paperbacks.
Capra, Fritjof 1992. *Belonging to the universe: New thinking about God and nature.* Harmondsworth: Penguin.
Carbaugh, Robert J 1980. *International economics.* Cambridge, Mass: Winthrop.
Carlton, D & Schaerf, C (eds) 1989. *Perspectives on the arms race.* London: Macmillan.
Carson, Rachel 1962. *Silent spring.* Boston: Houghton & Mifflin.
Carson. Walter H (ed) 1990. *The global economy handbook.* Boston: Beacon Press.
Chambers, R 1983. *Rural development: Putting the last first.* London: Longman.
Checkland, P B 1981. *Systems thinking, systems practice.* Chichester: Wiley.
Chilcote R H, & Johnson, D L (eds) 1983. *Theories of development: Mode of production or dependency?* London et al: Sage publications.
Chisholm, Michael 1979. *Rural settlement and land use: An essay in location.* London: Hutchinson, 3rd ed.
Chodorow, Nancy 1978. *The reproduction of mothering.* Berkeley: Univ. of Calif. Press.
Cobb, John B 1992. *Sustainability: Economics, ecology and justice.* Maryknoll: Orbis.
Cochrane, James R. 1994. *Conversation or collaboration? Base Christian Communities and the dialogue of faith.* Paper presented at the Conference of the Centre for Contextual Hermeneutics, Stellenbosch, June 1994.
Cohen John M 1987. *Integrated rural development: The Ethiopian experience and the debate.* Uppsala: Scandinavian Institute of African Studies.
Coker, A & Richards, C 1992 (eds). *Valuing the environment: Economic approaches to environmental evaluation.* London et al: Belhaven.
Cole, Ken (ed) 1994. *Sustainable development in a democratic South Africa.* London: Earthscan.
Comaroff, J 1985 "Christianity and colonialism in South Africa" *American Ethnologist* 1985.
Comaroff, Jean & John Comaroff 1991. *Of revelation and revolution: Christianity, colonialism and consciousness in South Africa.* Volume I. Chicago / London: The Univ of Chicago Press.
Comaroff, Jean 1985. *Body of power, spirit of resistance.* Chicago, London: University of Chicago Press.
Commission of the European Communities 1992. *Towards sustainability: A European Community programme of policy and action in relation to the environment and sustainable development.* Brussels: EC Commission.
Congleton, Roger D (ed) 1996. *The political economy of environmental protection: Analysis and evidence.* Ann Arbor: Univ of Michigan Press.
Connerton, Paul (ed) 1976. *Critical Sociology: Selected readings.* Harmondsworth: Penguin.
Conroy, C & Litvinoff, M 1988. *The greening of aid: Sustainable livelihoods in practice.* London: Earthscan.
Cook, Guillermo 1985. *The expectation of the poor: Latin American Basic Ecclesial Communities in Protestant perspective,* Maryknoll: Orbis.
Coote, Belinda 1992. *The trade trap: Poverty and the global commodity markets.* Oxford: Oxfam.

Correa, Sonia 1994. *Population and reproductive rights: Feminist perspectives from the South*. London: Zed Books.
Cosmao V 1984. *Changing the world: An agenda for the churches*. Maryknoll: Orbis.
Cottrell, Alan 1978. *Environmental economics*. London: Edward Arnold.
Cropper, M L & Oates, W E 1992. Environmental economics: A survey. *Journal of Economic Literature,* vol 30/1992 pp 675-740.
Cropper, M L & Oates, W E 1992. *The economics of environment*. Aldershot, Hants.: Edward Elgar.
Crump, Andy (ed) 1991. *Dictionary of environment and development: People, places, ideas and organizations*. London: Earthscan.
Daly, H E 1981. *Energy, economics, and the environment: Conflicting views of an essential interrelationship*. Boulder Col: Westview Press.
Daly, H E & Cobb, J B 1989. *For the common good: Redirecting the economy toward community, the environment and a sustainable future*. Boston MA: Beacon Press.
Daly, H E & Townsend, K N (eds) 1993. *Valuing the earth*. Cambridge Mass: MIT Press.
Daly, Herman 1992. Is the entropy law relevant to the economics of natural resource scarcity? - Yes, of course it is! Raisin: *Journal of Environmental Economics and Management* vol 23/1992: 91-95.
Daly, Herman E 1980 (1973). *Essays toward a steady-state economy*. San Francisco: Freeman.
Daly, Herman E 1991. *Steady State Economics*. Second ed. with new essays. Washington DC: Island Pr.
Daly, Herman E (ed) 1980 (1973). *Economics, ecology, ethics: Essays toward a steady-state economy*. San Francisco: Freeman.
Davidson, Carl 1990. *Recent developments in the theory of involuntary unemployment*. Kalamazoo Mich: Upjohn Institute for Employment Research.
Davidson, Greg & Davidson, Paul 1988. *Economics for a civilized society*. New York & London: Norton.
Davies, P, 1983. *God and the new physics*. Suffolk: Richard Clay.
De Gaay Fortman, B & Klein Goldewijk (no date). *God and the goods: Global economy in a civilizational perspective*. Geneva: WCC.
De la Court, Thijs 1990. *Beyond Brundtland: Green development in the 1990s*. London: Zed Books.
De la Court, Thijs 1992. *Different worlds: Development cooperation beyond the nineties*. Utrecht: Netherlands International Books.
Deane, P 1978. *The evolution of economic ideas*. Cambridge: Cambridge Univ Press.
Deane-Drummond, Celia 1996. *A handbook in theology and ecology*. London: SCM.
Diefenbacher, Hans & Ratsch, Ulrich 1992. *Verelendung durch Naturzerstörung: Die politischen Grenzen der Wissenschaft*. Frankfurt: Fischer.
Dixon, C 1991. *Rural development in the Third World*. London: Routledge.
Dobkowski, Michael N & Isidor Walliman (ed) 1997. *The coming age of scarcity: Preventing mass death and genocide in the 21st Century*. Syracuse NY: Syracuse Univ Press.
Drakakis-Smith, David W 1990. *The Third World City*. London: Routledge.
Drakopoulus, S A 1994. Economics and the new physics: Some methodological implications. *South African Journal of Economics,* vol 62/1994, 333-353.
Drengson, A & Inoue, Y (eds) 1995. *The deep ecology movement: An anthology*. Berkeley Cal: North Atlantic.
Drucker, P F 1993. *Post-capitalist society*. London et al: Butterworth- Heinemann.
Du Plessis, J E & Smit, J N J 1992. *Methods of wealth and income distribution: A classification*. Occasional Paper No 4. Stellenbosch: Stellenbosch Economic Project.
Du Toit, J and Falkena, H B 1994. *The structure of the South African economy*. Johannesburg: Southern Books.
Duchrow Ulrich 1992. *Europe in the World System 1492-1992*. Geneva: WCC.
Duchrow Ulrich 1992. *Europe in the World System 1492-1992*. Geneva: WCC.
Duchrow, Ulrich 1987. *Global economy: A confessional issue for the churches?* Geneva: World Council of Churches.
Duchrow, Ulrich 1987. *Global economy: A confessional issue for the churches?* Geneva: World Council of Churches.
Duchrow, Ulrich 1995 (1994). *Alternatives to the global economy: Drawn from biblical history for political action*. Utrecht: International Books.

Duchrow, Ulrich 1995. *Alternatives to the global economy: Drawn from biblical history for political action.* Utrecht: International Books.
Dumont, R 1966. *False starts in Africa.* London: Deutsch.
Dupaquier J 1983. *Malthus past and present.* New York: Academic Press.
Durning, Alan B 1989. *Poverty and the environment: Reversing the downward spiral.* Washington: Worldwatch Institute, Paper 92.
Durning, Alan B 1989. *Poverty and the environment: Reversing the downward spiral.* Washington: Worldwatch Institute, Paper 92.
Durning, Alan T 1992. *How much is enough? The consumer society and the future of the earth.* New York: W W Norton.
Dussel, Enrique 1988 (1986). *Ethics and community.* Maryknoll: Orbis.
Easley, Brian 1973. *Liberation and the aims of science: An essay on obstacles to the building of a beauiful world.* Edinburgh: Scottish Academic Press.
Easley, Brian 1983. *Fathering the unthinkable: Masculinity, scientists and the nuclear arms race.* London: Pluto Press.
Ehrlich P R & Ehrlich A H 1990. *The population explosion.* New York: Simon & Schuster.
Ehrlich, P R, Ehrlich, A H & Daily, G C 1993. *Food security, population and environment. Population and Development Review,* 19/1993 1-32.
Ehrlich, P R & Ehrlich, A H 1996. *Betrayal of science and reason: How anti-environmental rhetoric threatens our future.* Washington DC: Island Press / Shearwater Books.
Ekins, Paul 1992. *A new world order: Grassroots movements for global change.* London & New York: Routledge.
Ekins, Paul (ed) 1986. *The living economy: A new economics in the making.* London: Routledge & Kegan Paul.
Ekpo M U (ed) 1979. *Bureaucratic corruption in sub-saharan Africa: Toward a search for causes and consequences.* Washington: University Press of America.
Elkington, J & Burke, T 1988. *The green capitalists: Industry's search for environmental excellence.* London: Gollancz.
Elkington, John 1990. *The green consumer: A guide for the environmentally aware.* New York: Viking Penguin.
Elliott Charles 1987. *Comfortable compassion? Poverty, power and the church.* London et al: Hodder and Stroughton.
Elliott, Jennifer A 1994. *An introduction to sustainable development: the developing world.* London / New York: Routledge.
Ellul J 1984. *Money and power.* Downers Grove Ill: Inter-varsity Press.
Encyclopaedia Britannica 1986. Economic Theory. Vol 17/1986:942-978.
Encyclopaedia Britannica 1986. Ecosystems. Vol 17/1986: 979-1036.
Etzioni A 1988. *The moral dimension: Towards a new economics.* New York: Free Press.
Faber, Gerrit (ed) 1990. *Trade policy and development: The role of Europe in North-South trade: A multidisciplinary approach.* The Hague: Universitaire Pers Rotterdam.
Fallman, David (ed) 1994. *Ecotheology: Insights from South and North.* Geneva: WCC.
Fields, Gary S 1980. *Poverty, inequality, and development.* Cambridge: Cambridge Univ Press.
Fitzgerald, Patrick, Anne McLennan & Barry Munslow (eds) 1995. *Managing sustainable development in South Africa.* Cape Town: Oxford Univ Press.
Fontaine, Jean-Marc (ed) 1992. *Foreign trade reforms and development strategy.* London/NY: Routledge.
Forgacs, D ed. 1988. *A Gramsci Reader,* London: Lawrence and Wishart.
Fountain, E Daniel (ed) 1990. *Let's build our lives: Church and community leaders working together for development.* Brunswick: MAP International Publication.
Frank, Robert H 1988. *Passions within reason: The strategic role of the emotions.* New York / London: Norton; Markham: Penguin.
Freeman, C & Jahoda, M 1978. *World futures: The great debate.* London: Robertson.
Freire P 1972 (1970). *Pedagogy of the oppressed.* Harmondsworth: Penguin.
French, Hillary 1992. *After the Earth Summit: The future of environmental governance.* Washington: Worldwatch.
Friedman M and Friedman R 1980. *Free to choose: A personal statement.* New York: Avon Books.
Friedman, John 1992. *Empowerment: The politics of alternative development.* Blackwell: Oxford.

Fröbel, Folker et al 1981 (1977). *The new international division of labour.* Cambridge: Cambridge Univ Press.
Froyen, Richard T 1996. *Macroeconomics: Theories and Policies.* Upper Saddle River NJ: Prentice Hall, fifth edition.
Fukuyama, Francis 1993 (1992/1990). *The end of history and the last man.* New York: Free Press / Avon (pb).
Galbraith, John Kenneth 1958. *The affluent society.* Boston: Houghton Mifflin.
Galtung, Johan 1971. A structural theory of imperialism. *Journal of Peace Research,* vol 8/1971, 81-117.
Gelb, Stephen (ed) 1991. *South Africa's economic crisis.* Cape Town: Philip / London: Zed Books.
George, A 1988. *Wealth, poverty and starvation.* New York: St. Martin.
George, S 1976. *How the other half dies: The real reasons for world hunger.* Harmondsworth: Penguin.
George, S 1988. *A fate worse than debt.* New York: Grove.
George, Susan 1992. *The debt boomerang.* Boulder Co: Westview.
Georgescu-Roegen, Nicholas 1971. *The entropy law and the economic process.* Cambridge Mass: Harvard Univ Press.
Georgescu-Roegen, Nicholas 1973. The entropy law and the economic problem. In: Daly, Herman (ed) 1973. *Toward a steady-state economy.* San Francisco: Freeman, 37-49.
Georgescu-Roegen, Nicholas 1976. *Energy and economic myths.* New York: Pergamon.
Georgesku-Roegen, Nicholas 1967. *Analytical economics - issues and problems.* Cambridge Mass: Harvard Univ Press.
Gerrard, Bill 1989. *Theory of the capitalist economy: Towards a post-classical synthesis.* Oxford et al: Basil Blackwell.
Gever, J, Kaufman, Skole, D & Vorosmarty, C 1987. *Beyond oil.* Cambridge Mass: Ballinger.
Gilder G 1981. *Wealth and poverty.* New York: Basic Books.
Gilroy, Bernard M 1993. *Networking in multinational enterprises: The importance of strategic alliances.* Columbia: Univ of South Carolina Press.
Goddard J B & Thwaites A T 1986. New technology and regional development policy. In: Nijkamp P (ed): *Technological change, employment and spatial dynamics.* Berlin: Springer.
Godfrey, Martin 1986. *Global unemployment: The new challenge to economic theory.* Brighton: Wheatsheaf.
Goldin, Ian & Winters, Alan (eds) 1995. *The economics of sustainable development.* Cambridge: Cambridge Univ Press.
Goldring, J, Maher, L & McKeough, J 1993 (4). *Consumer protection law.* Annandale NSW: Federation Press.
Goldsmith, Edward et al 1992. *The future of progress: Reflections on environment and development.* London: Internat Society for Ecology and Culture.
Goldstein, Morris 1995. *The exchange rate system and the IMF: A modest agenda.* Washington: Institute for international economics.
Goodland, R & Daly, H 1992. *Ten reasons why Northern income growth is not the solution to Southern poverty.* Washington: World Bank.
Goodland, Robert et al 1991. *Environmentally sustainable economic development: Building on Brundtland.* Paper No 46. Washington: World Bank.
Gorbachev M 1987. *Perestroika - New thinking for our country and the world.* London: Collins.
Gore, Al 1992. *Earth in the balance: Ecology and the human spirit.* Boston: Houghton Mifflin.
Gorringe, Timothy J 1995. *Capital and the Kingdom: Theological Ethics and economic order.* Maryknoll NY: Orbis / London: SPCK.
Goudzwaard, Bob & de Lange, Harry 1994. *Beyond poverty and affluence: Toward an economy of care.* Geneva: WCC / Grand Rapids: Eerdmans.
Gould, M & Goldman, B A 1990. *Deadly deceit: Low-level radiation, high-level cover-up.* New York: Four Walls Eight Windows.
Gramsci, A 1988. Some theoretical and practical aspects of 'Economism'. In Forgacs D ed. *A Gramsci Reader,* London: Lawrence and Wishart.

Granberg-Michaelson, Wesley (ed) 1987. *Tending the garden: Essays on the gospel and the earth.* Grand Rapids MI: Eerdmans.
Granberg-Michaelson, Wesley (ed) 1987. *Tending the garden: Essays on the gospel and the earth.* Grand Rapids MI: Eerdmans.
Gries, R, Ilgen, V & Schindelbeck, D (eds) 1995. *Ins Gehirn der Masse kriechen: Werbung und Mentalitätsgeschichte.* Darmstadt: Wissenschaftliche Buchgesellschaft.
Griffith-Jones, Stephany 1988. *Managing world debt.* Brighton: Wheatsheaf.
Grimshaw, J 1986. *Feminist philosophers.* London: Harvester Wheatsheaf.
Gross, Bertram 1985. *Friendly fascism: The new face of power in America.* Montreal: Black Rose.
Gupta, Anirudha 1988. *Politics in Africa: personalities, issues, and ideologies.* New Delhi: Vikas Publishing House.
Haber, Stephen H 1989. *Industry and underdevelopment: The industrialisation of Mexico.* Stanford: Stanford Univ Press.
Hagen, E E 1962. *On the theory of social change: How economic growth began.* Homewood Ill: Dorsey.
Hagen, E E 1968. *The economics of development.* Homewood Ill: Irwin.
Hall, D & Ravindranath, N H 1995. *Biomass, energy and the environment.* Oxford Univ Press.
Hallman, David G (ed) 1994. *Ecotheology: Voices from South and North.* Geneva: WCC / Maryknoll NY: Orbis.
Harcourt, W (ed) 1994. *Feminist perspectives on sustainable development.* London: Zed Books.
Harrison, Paul 1992. *The third revolution: Environment, population and a sustainable world.* London: Penguin et al.
Harvey, David 1973. *Social justice and the city.* Baltimore: John Hopkins Univ Press.
Hauchler, Ingomar (ed) 1993. *Globale Trends.* Frankfurt/M: Fischer.
Haught, John F 1993. *The promise of nature: Ecology and cosmic purpose.* Mahwah NJ: Paulist Press.
Hawking, Stephen W 1988. *A brief history of time: From the big bang to the black holes.* London et al: Bantam.
Hayek, F A von 1976. *Law, legislation and liberty, vol 2: The mirage of social justice.* London: Routledge & Kegan Paul.
Hayek, F A von 1983. *Knowledge, evolution and society.* London: Adam Smith Inst.
Heap, Hargreaves 1989. *Rationality in Economics.* Oxford / New York: Basil Blackwell.
Heilbroner R L & Thurow L C 1975. *The economic problem.* London et al: Prentice Hall, fourth or later edition
Heilbroner, Robert L 1980. *An inquiry into the human prospect: Updated and reconsidered for the 1980's.* New York: Norton.
Heilbroner, Robert L 1985. *The nature and logic of capitalism.* New York / London: Norton.
Heilbroner, Robert L 1988. *Behind the veil of economics: Essays in the worldly philosophy.* New York / London: Norton.
Heller, Walter H 1975. What's right with economics. *American Economic Review* 65/1975 1-26.
Henderson, Hazel 1978. *Creating alternative futures.* New York: Putnam.
Herbst, Jeffrey 1990. *State and politics in Zimbabwe.* Harare: Univ of Zimbabwe Publications.
Herrera, A O et al 1976. *Catastrophe or new society? A Latin American World Model.* Ottawa: Intern. Devel. Research Centre for the Fundacion Bariloche.
Hewitt T, Johnson H, Wield D eds 1992. *Industrialization and development,* Oxford: Oxford University Press.
Hillinger, Claude (ed) 1992. *Cyclical growth in market and planned economies.* Oxford: Clarendon Press.
Hinkelammert F J 1986. *The ideological weapons of death: A theological critique of capitalism.* Maryknoll: Orbis.
Hirsch, Fred 1977. *The social limits to growth.* London: Routledge & Kegan Paul.
Hirschman, Albert 1977. *The passions and the interests.* Princeton NJ: Princeton Univ Press.
Holland, Stuart 1987. *The global economy: From meso to macroeconomics.* New York: St Martin's Press.

Hoogendijk, Willem 1993 (1991). *The economic revolution: Towards a sustainable future by freeing the economy from money-making*. Utrecht: International Books.
Hoover, Kevin D 1988. *The new classical macroeconomics: A sceptical inquiry*. New York et al: Basil Blackwell.
Horton, P B & Hunt, C L 1976. *Sociology*. New York et al: McGraw-Hill, 4th ed.
Horton, Robin 1995. *Patterns of thought in Africa and the West: Essays on magic, religion and science*. Cambridge Univ Press.
Hunt E K & Sherman H J 1981. *Economics: An introduction to traditional and radical views*. London et al: Harper and Row, fourth ed.
Huntley, B, Siegfried, R & Sunter, C 1989. *South African environments into the 21st Centuryl*. Cape Town: Human & Rousseou / Tafelberg.
Huppes, Gjalt et al 1992. *New market oriented instruments for European environmental policies*. London: Graham & Trotman.
Hurst, Charles E 1995. *Social inequality: Forms, causes, and consequences*. Boston et al: Allyn & Bacon, second ed.
Illich, Ivan 1992. Needs. In: Wolfgang Sachs (ed): *The Development Dictionary*. London: Zed Books, pp 88ff.
Issar, Devendra 1991. *Earthkeeping in the nineties: Stewardship and creation*. Grand Rapids: Eerdmans.
Jackson, B 1990. *Poverty and the planet: A question of survival*. London: Penguin.
Jacobson, Jodi L 1988. *Environmental refugees: A yardstick of habitability*. Washington: Worldwatch Inst.
Jensen, Kenneth M (ed) 1990. *A look at `The end of history'*. Washington: US Institute of Peace.
Johns, R A 1985. *International trade theories and the evolving international economy*. London: Frances.
Jonas, Hans 1985. *The imperative of responsibility: In search of an ethics for the technological age*. Chicago: Univ of Chicago Press.
Jones, Arthur 1993. *Capitalism and Christians: Tough gospel challenges in a troubled world economy*. Mahwah NJ: Paulist Press.
Jones, B 1982. *Sleepers, wake! Technology and the future of work*. Brighton: Wheatsheaf.
Joranson, P N & Butigan, K (eds) 1984. *Cry of the environment: Rebuilding the Christian creation tradition*. Santa Fe, NM: Bear.
Jordaan, J 1991. *Population growth - our time bomb*. Pretoria: Van Schaik.
Kahn, H & Wiener, A J 1967. *The year 2000*. London: Macmillan.
Kahn, H et al 1976. *The next 200 years - a scenario for America and the World*. New York: Morrow.
Katsoulacos Y S 1986. *The employment effect of technical change : A theoretical study of new technology and the labour market*. Wheatsheaf Books, Univ of Nebraska Press.
Keesing, Donald B 1988. *The four successful exceptions: Official export promotion and support for export marketing in Korea, Hong Kong, Singapore, and Taiwan*. Washingtom: World Bank, Policy Trade Division.
Kelvin P & Jarret J E (eds) 1985. *Unemployment: Its psychological effects*. London: Cambridge Univ Press.
Kennedy, Paul 1993. *Preparation for the 21 Cent*. London: Harper Collins.
Keohane, Robert O 1984. *After hegemony: Cooperation and discord in the world political economy*. Princeton NJ: Princeton Univ Press.
Kessler, Hans (ed) 1996. *Ökologisches Weltethos im Dialog der Kulturen und Religionen*. Darmstadt: Wissenschaftliche Buchgesellschaft.
Kieffer, George H 1979. *Bioethics: A textbook of issues*. Reading Mass et al: Addison-Wesley.
Kindleberger, C P 1958. *Economic development*. New York: McGraw-Hill.
King, Alexander (ed for Club of Rome) 1991. *The first global revolution*. New York: Pantheon.
Kinlaw, Denis C 1993. *Competitive and green: Sustainable performance in the environmental age*. Amsterdam / Johannesburg / London: Pfeiffer.
Kiriswa, Benjamin 1982. "Small Christian Communities in a Kenyan parish" AFER Vol. 24, April 1982:90-93.
Klay, R K 1986. *Counting the cost: The economics of Christian stewardship*. Grand Rapids, Mich: Eerdmans.

Klitgaard, Robert 1988. *Controlling corruption.* Berkeley et al: University of California Press.
Klitgaard, Robert 1991. *Tropical gangsters.* London/New York: Tauris.
Knight, David 1986. *The age of science: The scientific world-view in the nineteenth century.* Oxford/New York: Basil Blackwell.
Konings, Nicholas 1982. *Christian Communities,* Nangina Parish, 1974-1979. *AFER,* vol. 23, August 1981:247-254.
Korten, David C 1990. *Getting to the 21st Century: Voluntary action and the global agenda.* West Hartford, Conn: Kumarian Press.
Korten, David C 1995. *When corporations rule the world.* West Hartford: Kumarian.
Kotecha, K & Adams, R 1981. *African politics: The corruption of power.* Wasgington DC: University Press of America.
Krelle W & Shorrocks A F (eds) 1978. *Personal income distribution.* Amsterdam - New York - Oxford: North-Holland Publ. Co.
Krishnan, R et al (eds) 1995. *A survey of ecological economics.* Washington: Island Press.
Krugman, Paul 1991. *Geography and trade.* Leuven: Leuven Univ Press / Cambridge Mass: MIT Press.
Krugman, Paul 1994. *Peddling prosperity: Economic sense and nonsense in the age of diminished expectations.* New York / London: Norton.
Kubálková, V & Cruickshank, A A 1981. *International inequality.* London: Croom Helm.
Kuhn, T S 1970. *The structure of scientific revolutions.* Chicago: Univ of Chicago Press, 2 ed.
Kulessa, Manfred 1990: *Europe in the nineties and the Third World.* Berlin: ICT.
Küng, Hans 1991. *Global responsibility.* London & New York: Crossroads.
Küng, Hans & Kuschel, Karl-Josef 1994. *A global ethic: The Declaration of the Parliament of the World's Religions.* London: SCM.
Küng, Hans 1997. *A global ethic for global politics and economics.* London: SCM.
Lampman R J 1971. *Ends and means of reducing income poverty.* Chicago: Markham.
Larrain, Jorge 1979. *The concept of ideology.* London et al: Hutchinson.
Laszlo Ervin (ed) 1977. *Goals for Mankind : A report to the Club of Rome on the new horizons of global community..* London: Hutchinson.
Laszlo, E 1972. *Introduction to systems philosophy.* New York: Harper Torchbooks et al.
Lauterbach, A 1959. *Man, motives and money: Psychological frontiers of economics.* Westport Conn: Greenwood.
Lauterbach, A 1974. *Psychological challenges to modernization.* New ork et al: Elsevier Scientific.
Lazonick, William 1991. *Business organization and the myth of the market economy.* London: Cambridge Univ Press.
Leakey, R & Lewin, R 1996. *The sixth extinction: Biodiversity and its survival.* London: Phoenix.
Leatt J, Kneifel T & Nürnberger K (eds) 1986. *Contending ideologies in South Africa.* Cape Town: David Philip.
Lee, Jung Young 1995. *Marginality: The key to multicultural theology.* Minneapolis: Fortress.
Leggett, Jeremy (ed) 1990. *Global warming: The Greenpeace Report.* Oxford: Oxford Univ Press.
Lehman, Howard P 1993. *Indebted development: Strategic bargaining and economic adjustment in the Third World.* London: Macmillan.
Leibenstein, Harvey 1980 (1976). *Beyond economic man: A new foundation for microeconomics.* Cambridge MA / London: Harvard Univ Press.
Leiser, Burton M 1986 (1973). *Liberty, justice, and morals:* Contemporary value conflicts. New York/London: Macmillan.
Leiss, William 1988. *The limits to satisfaction: An essay on the problem of needs and commodities.* Toronto: Univ of Toronto Press.
Leonard, H J 1989. *Environment and the poor: Development strategies for a common agenda.* Oxford: Transaction Books.
Leontief, W et al 1976. *The future of the world economy.* New York: United Nations.
Levenson, Jon D 1988. *Creation and the persistence of evil: The Jewish drama of divine omnipotence.* San Francisco: Harper & Row.

Lewis, Martin W 1992. *Green delusions: An environmentalist's critique of radical environmentalism*. Durham NC: Duke Univ Press.
Lichtheim G 1975. *A short history of socialism*. London: Fontana.
Lindblom C E 1977. *Politics and markets: The world's political-economic systems*. New York: Basic Books.
Lindenberg, M & Ramitez, N 1989. *Managing adjustment in developing countries*. ICS Press.
Linthicum, R C 1991. *City of God, city of Satan*. Grand Rapids: Zondervan.
Linthicum, R C 1991. *Empowering the poor: Community organizing among the city's 'rag, tag and bobtail'*. Monrovia, Ca: MARC.
Liston, D & Reeves, N 1988. *The invisible economy: A profile of Britain's invisible exports*. London: Pitman.
Litvinoff, Miles 1990. *The Earthscan action handbook for people & the planet*. London: Earthscan.
Livingstone Ian (ed) 1981. *Development Economics and policy readings*. London: London / Boston: Allen & Unwin.
Linzey, Andrew 1987. *Christianity and the rights of animals*. London: SPCK.
London, Bruce & Williams, Bruce 1990. National politics, international dependency, and basic needs provision: a cross-national analysis. *Social Forces*, 69/1990, pp 565-584.
Lovett, William A 1987. *World Trade Rivalry: Trade equity and competing industrial policies*. Lexington Mass: Lexington Books.
Lutz M A & Lux K 1979. *The challenge of humanist economics*. London et al: Benjamin/Cummings.
Lux, Kenneth 1990. *Adam Smith's mistake: How a moral philosopher invented economics and ended morality*. Boston: Shambala.
Lydall H 1984. *Yugoslav socialism: Theory and practice*. Oxford: Clarendon Press.
MacNeill, J et al 1991. *Beyond interdependence: The messing of the world's economy and the earth's ecology*. Oxford & New York: Oxford Univ Press.
MacNeill, J, et al 1989. *CIDA and sustainable development*. Halifax (Canada): Institute for Research on Public Policy.
Maduro, Otto 1982. *Religion and social conflicts*. Maryknoll: Orbis Books.
Malthus, T R 1970 (1798). *An essay in the principle of population*. Harmondsworth: Penguin.
Mankiw, N G & Romer, D (eds) 1991. *New Keynesian Economics:* Vol I *Imperfect competition and sticky prices*. Vol II *Coordination, failures and real rigidities*. Cambridge Mass / London: MIT Press.
Mannheim, Karl 1972. *Ideology and utopia*. New York: Routledge and Kegan Paul.
Mannoni, D O 1956. *Prospero and Caliban: The psychology of colonization*. London: Methuen.
Marcuse, Herbert 1968. *One dimensional man*. Gray's Inn Rd: Sphere Books.
Marwell, G & Oliver P 1993. *The critical mass in collective action: A micro-social theory*. Cambridge Univ Press.
Maslow A 1970. *Motivation and personality*. New York: Harper and Row.
Max-Neeff, M 1991. *Human scale development: Conceptions, applications and further developments*. London: The Apex Press.
Mazrui, Ali 1980. *The African condition: a political diagnosis*. London et al.: Cambridge University Press.
McClelland, D C 1961. *The achieving society*. Princeton: Van Nostrand.
McGovern A F 1981. *Marxism: an American Christian perspective*. Maryknoll NY: Orbis.
McGrath, M D 1983. *The distribution of personal income in South Africa in selected years over the period from 1945 to 1980*. Durban: Univ of Natal PhD thesis.
McFague, Sallie 1993. *The body of God: An ecological theology*. London: SCM.
McKenzie, R B & Tullock, G 1985 (1978). *Modern political economy: An introduction to economics* (Internat. Student Edition). Auckland et al: McGraw-Hill Intern Book Co.
Meadows, D et al (eds) 1972. *The limits to growth*. New York: Universe Books.
Meadows, D H, Meadows & D L, Randers, J 1992. *Beyond the limits: Confronting global collapse, envisioning a sustainable future*. Post Mills VT: Chelsea Green.
Meadows, D L ed. 1977 *Alternatives to growth: A search for sustainable futures*, Cambridge Mass: Balinger Publishing.

Meadows, Donella H 1992. *Beyond the limits: Global collapse or a sustainable future.* London: Earthscan.
Meeks, M Douglas 1989. *God the Economist: The doctrine of God and Political Economy.* Minneapolis: Fortress.
Mende, T 1973. *From aid to re-colonization: Lessons of a failure.* London: Harrap.
Merton, R K & Nisbet, R (eds) 1971. *Contemporary social problems.* New York et al: Harcourt Brace Jovanovich.
Mesarovic, M D & Pestel, E 1975 (1974). *Mankind at the turning point: The Second Report to Club of Rome.* London: Hutchinson.
Meyer, Art 1991. *Earth keepers.* Kitchener Ont: Herald.
Mieth, D & Pohier, J (eds) 1980. *Christian Ethics and Economics: The North-South Conflict.* (Concilium). Edinburgh: T & T Clark / New York: Seabury.
Mikesell, R F & William L F 1992. *International Banks and the environment: From growth to sustainability: An unfinished agenda.* San Francisco: Sierra Club Books.
Milbank, John 1993 (1990). *Theology and social theory: Beyond secular reason.* Oxford: Blackwell.
Mirowski, P 1988. *Against mechanism: Protecting economics from science.* Totowa NJ: Rowman and Littlefields.
Mirowski, P 1989. *More heat than light.* Cambridge: Cambridge Univ Press.
Mishan, E J 1977. *The economic growth debate.* London: Allen & Unwin.
Mishan, E J 1993 (1967). *The costs of economic growth* (revised edition). London: Weidenfeld and Nicolson.
Moll, Peter, Nattrass, Nicoli & Loots, Lieb 1991. *Redistribution: How can it work in South Africa?* Cape Town: David Philip, pp 1ff.
Moltmann, Juergen 1984. *On Human Dignity: Political Theology and Ethics.* London: SCM.
Moltmann, Juergen 1985. *God in Creation: An Ecological Doctrine of Creation.* London: SCM.
Momsen J H 1991. *Women and development in the Third World.* London: Routledge.
Monsma S V et al 1986. *Responsible technology: A Christian perspective.* Grand Rapids Mich: Eerdmans.
Morishima, Michio 1982. *Why has Japan succeeded? Western technology and the Japanese ethos.* Cambridge et al: Cambridge Univ Press.
Morris, Brian 1994. *Anthropology of the Self: The individual in cultural Perspective.* London: Pluto Press.
Mosley, P, Harrigan, J. and Toye, J 1991. *Aid and power: The World Bank and policy-based lending in the 1980s.* London: Routledge.
Mowoe I J 1980. *The performance of soldiers as governors: African politics and the African military.* Washington: Univ Press of America.
Mulholland C (ed) 1988. *Ecumenical reflections on political economy.* Geneva: WCC.
Müller-Fahrenholz 1995. *God's Spirit: Transforming a world in crisis.* Geneva: World Council of Churches.
Murphy, Nancey & Ellis, George F R 1996. *On the moral nature of the universe: Theology, cosmology, and ethics.* Minneapolis: Fortress.
Myers, Norman 1984. *The primary source: Tropical forests and our future.* New York: W W Norton.
Myrdal, G 1953. *The political element in the development of economic theory.* London: Routledge and Kegan Paul.
Nanjundan S 1993. The re-emphasis on small enterprises: a review article. *Industry and Development* vol 33/1993 109ff.
Nkrumah, Kwame 1965. *Neo-colonialism: The last stage of imperialism.* London: Panaf.
Newbigin, Leslie 1994. *A word in season: perspectives on Christian world mission.* Grand Rapids: Eerdmans.
Nicholls, J Bruce 1986. *The church: God's agent of change for change.* Exeter: Paternoster Press.
Northcott, Michael S 1996. *'The environment and Christian ethics.* Cambridge: Cambridge Univ Press.
Novak M (ed) 1979. *The denigration of capitalism.* American Enterprise Institute.
Novak M 1982: *The spirit of democratic capitalism.* New York: Simon & Schuster.
Nurkse, Ragnar 1955. *Problems of capital formation in underdeveloped countries.* Oxford: Blackwell.

Nürnberger, K 1975. The Sotho notion of the Supreme Being and the impact of the Christian proclamation. *Journal for Religion in Africa* VII/1975 174-200.
Nürnberger, K 1982. *Die Relevanz des Wortes im Entwicklungskonflikt.* Frankfurt/Bern: Peter Lang.
Nürnberger, K 1983. Socio-political ideologies and church unity. *Missionalia* 10/1983 42-53 and *JTSA* 44/1983, 47-57.
Nürnberger, K 1987. *Ethik des Nord-Süd-Konflikts.* Gütersloh: Mohn, 1987.
Nürnberger, K 1987a. Ecology and Christian ethics in a semi-industrialised and polarised society. In: W S Vorster (ed): *Are we killing God's earth? Ecology and theology.* Pretoria: Unisa, 45-67.
Nürnberger, K 1988. *Power and beliefs in South Africa: Economic potency structures in SA and their interaction with patterns of conviction in the light of a Christian ethic.* Pretoria: Unisa.
Nürnberger, K, Tooke, J & Domeris, W (eds) 1989. *Conflict and the quest for justice.* Pietermaritzburg: Encounter.
Nürnberger, K 1990. *The scourge of unemployment in SA.* Pietermaritzburg: Encounter Publications.
Nürnberger, K 1991a. Democracy in Africa - the raped tradition. In: Nürnberger, K (ed) 1991. *A democratic vision for South Africa.* Pietermaritzburg: Encounter Publications pp 304-318.
Nürnberger, K 1991b. Subscribing to confessional documents today. *Journal of Theology for Southern Africa* 75, June 1991, pp 37-47.
Nürnberger, K 1991c. The seed that cracks the rock: The biblical thrust towards democratic assumptions. In: Nürnberger, K (ed) 1991. *A democratic vision for South Africa.* Pietermaritzburg: Encounter Publications, pp 75-84.
Nürnberger, K (ed) 1991. *A democratic vision for South Africa: Political realism and Christian responsibility.* Pietermaritzburg: Encounter Publications.
Nürnberger, K 1994. *An economic vision for South Africa. The task of the church in the post-apartheid economy.* Pietermaritzburg: Encounter Publ.
Nürnberger, K 1995. *Making ends meet: Personal money management in a Christian perspective.* Pietermaritzburg: Encounter Publications.
Nürnberger, K 1998. *Beyond Marx and market: Outcomes of a century of economic experimentation.* Pietermaritzburg: Cluster Publications / London: Zed Books.
Okeyo, Vitalis 1983. "Small Christian Communities in Kish - A new way of being Church" *AFER* Vol. 25 August 1983:226-229.
O'Brian, Niall 1987. *Revolution from the heart.* New York: Oxford Univ Press.
O'Dowd, Michael 1991. *South Africa - The growth imperative.* Johannesburg: Jonathan Ball.
Occhiolini, Michael 1990. *Debt-for-nature swaps.* Washington DC: Debt and Internation Finance Division.
Ogle, George E 1990. *South Korea: Dissent within the economic miracle.* London: Zed Books.
Omari, C K (ed) 1989. *Persistent principles amidst crisis.* Nairobi: Uzima Press.
Orr, D & Soroos, M (ed) 1979. *The global predicament.* Chapel Hill: Univ of North Carolina Press.
Pacey, Arnold 1992. *The maze of ingenuity: Ideas and idealism in the development of technology.* Cambridge Mass: MIT Press.
Packard, Vance 1985. *The hidden persuaders.* Harmondsworth: Penguin.
Paul, E F, Miller, F D & Paul, J (eds) 1992. *Economic rights.* New York et al: Cambridge Univ Press.
Peacocke, Arthur 1994. *Theology and the scientific age: Being and becoming - natural and divine.* Oxford: Blackwell.
Pearce, D, Markandya, A & Barbier, E B 1989. *Blueprint for a green economy.* London: Earthscan.
Peck, M S 1993. *A world waiting to be born: Civility rediscovered.* Bantam Books, NY.
Philpott, Graham 1993. *Jesus is tricky and God is undemocratic: The kin-dom of God in Amawoti.* Pietermaritzburg: Cluster Publications.
Phiri, I A, Ross, R K and Cox L J (eds) 1996. *The role of Christianity in development, peace and reconstruction: Southern perspectives.* Nairobi: Kolbe Press.
Power, Thomas M 1988. *The economic pursuit of quality.* Armonk NJ: Sharp.

Prebisch, R 1950. *The economic development of Latin America and its principal problems.* New York: ECLA, UN Dept of Econ Affairs.
Preston R H 1983. *Church and society in the late Twenthieth Century: The economic and political task.* London: SCM.
Preston R H 1991. *Religion and the ambiguities of capitalism.* London: SCM.
Prinx, Gwyn (ed) 1984. *The choice: Nuclear weapons versus security.* London: Chatto & Windus / Hogarth.
Purcell, Randall B (ed) 1989. *The newly industrializing countries in the world economy: Challenges for US policy.* Boulder/London: Lynne Rienner.
Ranis, G 1990. Asian and Latin American experience: Lessons for Africa. *Journal of International Development,* 2/1990 (April).
Rasmussen L L 1981. *Economic anxiety and Christian faith.* Minneapolis: Augsburg.
Ravallion, Martin 1991. *Quantifying the magnitude and severity of absolute poverty in the developing world.* Washington DC: Office of the Vice President.
Rawls, John 1971. *A theory of justice.* Cambridge Mass.: Belknapp (Harvard Univ. Press).
Redclift M R 1984. *Development and the environmental crisis: Red or green alternatives.* London: Methuen.
Reeck D 1982. *Ethics for the professions: A Christian perspective.* Minneapolis: Augsburg.
Reed, David (ed) 1992. *Structural Adjustment and the environment.* Boulder / Oxford: Westview.
Renner, Michael 1991. *Jobs in a sustainable economy.* Washington: Worldwatch.
Repetto, Robert et al 1992. *Green fees: How a tax shift can work for the environment and the economy.* Washington: World Resources Inst.
Reynolds, Bruce L (ed) 1988. *Chinese economic policy: Economic reform at midstream.* New York: Paragon House.
Rhodes, R I 1971. *Imperialism and underdevelopment: A reader.* New York / London: Monthly Review Press.
Rifkin, Jeremy 1990. *Entropy: Into the greenhouse world.* New York: Bantam Books.
Robertson, James 1990. *Future wealth - A new economics for the 21st century.* New York: Bootstraps Press.
Rodney, W 1974. *How Europe underdeveloped Africa.* London: Bogle-L'Ouverture / Dar es Salam: Tanzania Publ House.
Rohrlich, Fritz 1987. *From paradox to reality: Our basic concepts of the physical world.* Cambridge: Cambridge Univ Press.
Rostow, W W 1990 (1960). *The stages of economic growth: A non-communist manifesto.* Cambridge: Cambridge Univ Press, 3rd edition.
Roszak, Theodore 1992. *The voice of the earth.* New York: Simon & Schuster.
Rothschild, K W (ed) 1971. *Power in economics.* Harmondsworth: Penguin.
Ruether, Rosemary Radford 1993. *Gaia and God.* London: SCM.
Ruether, Rosemary Radford (ed) 1996. *Women healing earth: Third-World women on ecology, feminism, and religion.* Maryknoll: Orbis.
Ruffin, R J & Gregory, P R 1983. *Principles of economics.* Glenview Ill: Scott, Foresman and Co.
Russel, James C 1994. *The Germanization of early medieval Christianity: A sociohistorical approach to religions transformation.* New York / Oxford: Oxford Univ Press.
Sachs, Wolfgang (ed) 1992. *The development dictionary: A guide to knowledge as power.* London: Zed Books, 2nd ed.
Sachs, Wolfgang (ed) 1993. *Global ecology: A new arena of political conflict.* London: Zed Books.
Sachsse, Hans 1984. *Ökologische Philosophie: Natur - Technik - Gesellschaft.* Darmstadt: Wissenschaftliche Buchgesellschaft.
Samuel, V & Sugden, C 1987. *The church in response to human need.* Grand Rapids: Eerdmans.
Samuel, V & Sugden, C (eds) 1987. *The church in response to human need.* Grand Rapids: Eerdmans.
Samuels, Warren J (ed) 1990. *Economics as discourse: An analysis of the language of economists.* London / Dordrecht: Kluwer Academic.
Samuelson, Paul A & Nordhaus, William D 1989. *Economics.* New York et al: McGraw-Hill, 13th edition.

Sanderson, Fred H (ed) 1990. *Agricultural protectionism in the industrialised world*. Washington: Resources for the Future.
Sandison, A 1967. *The wheel of empire: A study of the imperial idea in some late 19th and 20th century fiction*. London: Macmillan.
Sassen, Saskia 1988. *The mobility of labor and capital: A study in international investment and labor flow*. Cambridge: Cambridge Univ Press.
Schäfer, G (ed) 1993. *Basic Human Needs: An interdisciplinary and international view*. NY: Peter Lang.
Schmidheiny, Stephan et al 1992. *Changing course: A global business perspective on development and the environment*. Cambridge MS: MIT Press.
Schneider, Stephen H 1989. *Global warming: Are we entering the greenhouse century?* San Francisco: Sierra Club Books.
Scholtes, Philippe R 1993. Formulating industrial strategies and policies in the context of restructuring economies: some preliminary thoughts. *Industry and Development* 33/1993 43-51.
Schramm, Michael 1994. *Der Geldwert der Schöpfung: Theologie - Ökologie - Ökonomie*. Paderborn et al: Ferdinand Schöningh.
Schreiber, Helmut 1989. *Debt-for-nature-swap: An instrument against debt and environmental destruction?* Bonn: Institut für Europäische Umweltpolitik.
Schrire, Robert (ed) 1992. *Wealth or poverty? Critical choices for South Africa*. Cape Town: Oxford Univ Press.
Schudson, Michael 1984. *Advertising: The uneasy persuasion*. New York: Basic Books.
Schumacher E F 1974 (1973). *Small is beautiful: A study of economics as if people mattered*. London: Sphere Books.
Schumpeter, Joseph A 1943. *Capitalism, socialism and democracy*. London: Allen & Unwin.
Schwartzman, David 1989. *Economic policy: An agenda for the nineties*. New York: Praeger.
Scott Maynes, E 1976. *Decision-making for consumers: An introduction to Consumer Economics*. NY /London: Macmillan.
Scott Peck, M 1978. *The road less travelled*. London: Arrow.
Scott, James C 1990. *Domination and the arts of resistance: Hidden transcripts*. New Haven / London: Yale Univ Press.
Seeger, Joni 1995. *The new state of the earth atlas*. 2nd ed. New York et al: Simon Schuster.
Seifert, E K & Marinez-Alier (eds) 1992. *Entropy and bioeconomics*. Milan: Nagard.
Seligson, Mitchell A 1984. *The gap between rich and poor: Contending perspectives on the political economy of development*. Boulder / London: Westview.
Sen M 1987. *On ethics and economics*. London: Blackwell.
Senge, Peter 1990. *The fifth discipline: The art and practice of the learning organisation*. New York et al: Doubleday.
Shiva, Vandana 1989. *Staying alive: Women, ecology and development*. London: Zed Books.
Shrader-Frechette, K S 1985. *Science policy, ethics, and economic methodology*. Dordrecht / Boston / Lancaster: D Reidel.
Shriver, Donald W 1995. *An ethic for enemies: Forgiveness in politics*. New York et al: Oxford Univ Press.
Sider Ron J (ed) 1981. *Evangelicals and development: Towards a theology of social change*. Exeter: Patenoster.
Sider, Ron J 1978. *Rich Christians in an age of hunger*. London: Hodder & Stoughton.
Sider, Ron J 1993. *One-sided Christianity? Uniting the church to head a lost and broken world*. Grand Rapids: Zondervan.
Sider, Ron J 1993. *Evangelism & social action in a lost and broken world*. London: Hodder & Stoughton.
Sider, Ron J 1993. *Evangelism & social action in a lost and broken world*. London: Hodder & Stoughton.
Sider, Ronald J. 1977. *Rich Christians in an age of hunger*. London: Hodder and Stoughton.
Simpkins, C & Lipton, M (eds) 1993. *The state and market in post apartheid South Africa*. Johannesburg: Witwatersrand Univ Press.
Simpson, Struan 1990. *The Times Guide to the environment: A comprehensive handbook to green issues*. London: Times Books.

Sine, Tom (ed) 1983. *The church responds to human need.* Monrovia: Marc.
Skitovsky, T 1976. *The joyless economy: An inquiry into human satisfaction and consumer dissatisfaction.* New York: Oxford Univ Press.
Skousen, Mark 1990. *The structure of production.* New York / London: New York Univ Press.
Sleeman, John E. 1976. *Economic Crisis: A Christian Perspective.* London: SCM.
Smart, Bruce 1992. *Beyond compliance: A new industry view of the environment.* Washington: World Resources Inst.
Smith, D M 1977. *Human geography: A welfare approach.* London: Edward Arnold.
Smith, D M 1979. *Where the grass is greener. Geographical perspectives on inequality.* London: Croom Helm.
Smith, D M 1994. *Geography and social justice.* Oxford / Cambridge: Basil Blackwell.
Sobrino, Jon 1989. *Spirituality of liberation: Toward political holiness.* Maryknoll, NY. Orbis.
Sonko, Karamo N M 1994. *Debt, development and equity in Africa.* New York et al. Univ Press of America.
Speth, James G 1990. *The greening of technology.* Washington: World Resources Inst.
Spurway, Neil (ed) 1994. *Humanity, environment and God.* Oxford: Blackwell.
Stackhouse M L 1987. *Public theology and political economy; Christian stewardship in modern society.* Grand Rapids: Eerdmans.
Starke, L 1990. *Signs of hope: Working towards our common future.* Oxford: Oxf Univ Press.
Stokes, E 1960. *The political ideas of English imperialism.* London: Oxford Univ Press.
Storkey, Alan 1986. *Transforming economics: A Christian way to employment.* London: SPCK.
Sunter, Clem 1987. *The world and South Africa in the 1990s.* Cape Town: Human & Rousseau / Tafelberg.
Swimme, Brian 1996. *The hidden heart of the cosmos: Humanity and the new story.* Maryknoll: Orbis.
Tan, M T & Soeradji, B 1986. *Ethnicity and fertility in Indonesia.* Singapore: Institute of Southeast Asian Studies.
Tawney, R H 1948. *The acquisitive society.* New York: Harcourt, Brace.
Tawney, R H 1969 (1926). *Religion and the rise of capitalism.* Harmondsworth: Penguin.
Taylor, John V 1963. *The primal vision.* London: SCM.
The Council of The Club of Rome 1991. *The first global revolution.* New York: Simon & Schuster.
The Worldwatch Institute 1990. *State of the world 1990: A Worldwatch Institute report on progress towards a sustainable society.* New York: Norton.
Therborn, G 1986. *Why some peoples are more unemployed than others.* London: Verso.
Thirlwall, A P 1994. *Growth and development: with special reference to developing economies.* London et al: Macmillan Press, fifth edition.
Thomas J J 1992. Informal economic activity. Ann Arbor: Univ of Michigan Press.
Thurow, Lester 1983. *Dangerous currents.* New York: Random House.
Thurow, Lester 1985. *The zero-sum solution: Building a world-class American economy.* New York: Simon & Schuster.
Timberlake, L & Holmberg, J 1991. *Defending the future: A guide to sustainable development.* London: Earthscan.
Timberlake, Lloyd 1984. *Natural disasters - acts of God or acts of man?* LondonL Earthscan.
Tinbergen, J (ed) 1976. *Reshaping the international order: A report to the Club of Rome.* New York: Dutton.
Tinbergen, Jan 1987. *Warfare and welfare: integrating security policy into socio-economic policy.* Brighton: Wheatsheaf Books.
Tisdell, Clement A 1991. *Economics of environmental conservation: Economics for environmental and ecological management.* Amsterdam et al: Elsevier.
Todaro, Michael P 1989. *Economic development in the Third World.* New York: Longman.
Toffler, Alvin 1983. *Future shock.* London: Pan Books.
Townsend, K 1992. Is the entropy law relevant to the economics of natural resource scarcity? Comment. Raisin: *Journal of Environmental Economics and Management* vol 23/1992: 96-100.

Trzyna, T C & Childers, R 1992. *World directory of environmental organizations.* Sacramento: Inst of Public Affairs.
Tudor, D J & Tudor I J 1995. *Systems analysis and design: A comparison of structured methods.* Oxford: NCC Balckwell.
Turner, R K, Pearce, D & Bateman, I 1994. *Environmental Economics: An elementary introduction.* New York et al: Harvester-Wheatsheaf.
Tussie, Diana & Glover, David (eds) 1993. *The developing countries in world trade.* Boulder: Lynne Rienner.
UN Environmental Programme 1985. *Environmental refugees.* New York: United Nations.
UN Development Programme 1992. *Human Development Report 1991-1992.* New York: Oxford Univ Press.
UNFPA 1992. *World Population Report 1992.* New York: United Nations.
United Nations 1993. *Report of the United Nations Conference on Environment and Development: Rio de Janeiro, 3-14 June 1992: Vol I: Resolutions adopted by the Conference.* New York: United Nations.
United Nations 1993. *The global partnership for environment and development: A guide to Agenda 21: Post Rio edition.* New York: United Nations.
United Nations Development Programme (UNDP) 1993. *Human Development Report 1993.* New York / Oxford: Oxford University Press.
Uphoff, N 1986. *Local institutional development.* West Hartford, Conn: Kumarian Press.
Uphoff, N 1992. *Possibilities for participatory development and post Newtonian social science.* Cornell Univ Press, Ithaca.
Van Bergen, Annegreet *et al* 1993. De vierdaagse werkweek. *Elsevier* vol 49/1993:97).
Van der Riet, P 1980. *Co-operative water resources development in Southern Africa. Hydrological Reserach Unit Report 5/80.* Johannesburg: University of the Witwatersrand.
Van Marrewijk Charles & Verbeek Jos 1993. On opulence driven poverty traps. *Journal of Population Economics* vol 6/1993 67-81.
Vestal, James E 1993. *Planning for change: Industrial policy and Japanese economic development, 1945-1990.* Oxford: Clarendon.
Villa-Vicencio, C 1992. *A theology of reconstruction: Nation building and human rights.* Cambridge: Cambridge Univ Press.
Von Schelling, Vivian 1992. "Culture and industrialization in Brazil" in T Hewitt, H Johnson, D Wield (eds) *Industrialization and development,* Oxford: Oxford University Press 248-276.
Von Weizsäcker's Carl F 1973. *Die Tragweite der Wissenschaft.* Stuttgart: Hirzel, fourth ed.
Von Weizsäcker, Ernst U 1992. *Earth politics: After the earth summit.* New York: Apex Pr.
Von Weizsäcker, Ernst U 1994. *Earth politics.* London & Atlantic Highlands NJ: Zed Books.
Von Weizsäcker, Ernst U & Jesinghaus, Jochen 1992. *Ecological tax reform: A policy proposal for sustainable development.* Antlantic Highlands NJ: Humanities International / London: Zed Books.
Wallerstein, Immanuel 1974. *The modern world system: Capitalist agriculture and the origins of the European world economy in the 16th Century.* New York: Academic Press.
Walsh, Kenneth 1987. *Long-term unemployment: An international perspective.* Basingstoke: Macmillan.
Walton C C 1969. *Ethos and the executive: Values in managerial decision making.* Englewood Cliffs NJ: Prentice-Hall.
Watson, S & Gibson, K (eds) 1995. *Postmodern cities and spaces.* Oxford et al: Blackwell.
Watts A G 1983. *Education, unemployment and the future of work.* Stratford: Open University Press.
Weber, Max 1976. *The Protestant Ethic.* London: Allen & Unwin.
Weigel, George & Royal, Robert 1994. *Building the free society: Democracy, capitalism and Catholic social teaching.* Grand Rapids: Eerdmans.
Weisskopf, Walter 1971. *Alienation and economics.* New York: Dutton.
Wells, Harold 1995. *A future for socialism? Political theology and the 'triumph of capitalism'.* Valley Forge PA: Prinity Press International.
Wilkinson, Loren 1991. *Earthkeeping in the nineties: Stewardship and the renewal of creation.* Grand Rapids: Eerdmans.

Wilkinson, Loren (ed) 1980. *Earthkeeping: Christian stewardship of natural resources.* Grand Rapids MI: Eerdmans.
Wilkinson, Loren (ed) 1980. *Earthkeeping: Christian stewardship of natural resources.* Grand Rapids MI: Eerdmans.
Williams, Robert 1987. *Political corruption in Africa.* Vermont, England: Gower.
Williamson, Jeffrey G 1985. *Did British capitalism breed inequality?* Boston: Allen & Unwin.
Williamson, Jeffrey G & Lindert, Peter H 1980. *American inequality: A macroeconomic history.* New York: Academic Press.
Wilson, F & Ramphele, M 1989. *Uprooting poverty: The South African challenge.* Cape Town/Johannesburg: David Philip.
Winter, Georg 1989. *Business and the environment.* New York: McGraw-Hill.
Wisner, Ben 1988. *Power and need in Africa: basic human needs and development policies.* London: Earthscan.
Wogaman J P 1986. *Economics and ethics: A Christian inquiry.* Philadelphia: Fortress.
Wolff, R D 1974. *The economics of colonialism: Britain and Kenya, 1870-1930.* New Haven / London: Oxford Univ Press.
World Bank 1989. *The long-term perspective for Sub-Saharan Africa: A strategy for recovery and growth.* Washington DC: World Bank
World Bank 1990. *World Development Report 1990: Poverty.* Washington: World Bank.
World Bank 1991. *World Development Report 1991: The challenge of development.* Washington: World Bank.
World Bank 1992. *World Development Report 1992: Development and the environment.* Washington: World Bank.
World Bank 1993. *World Development Report 1993: Investing in health.* Oxford / New York et al: Oxford Univ Press (for World Bank).
World Bank 1997. *World Development Report 1997: The state in a changing world.* Oxford et al: Oxford Univ Press.
World Commission on Environment and Development 1990 (1987): *Our common future: The Brundtland Report.* Oxford, New York: Oxford Univ Press.
World Council of Churches 1993. *Christian faith and the world economy today.* Geneva: WCC.
World Wildlife Fund 1987. *Debt-for-nature-swaps.* Washington DC: Background Information.
Wright, Nancy G & Donald G Kill (ed) 1993. *Ecological healing: A Christian vision.* Maryknoll NY: Orbis.
Yamamori, T, Myers, L B, Bediako K & Reed, L (eds) 1996. *Serving with the poor in Africa,* Monrovia: MARC.
Yergin, Daniel 1991. *The prize - the epic quest for oil, money and power.* London: Simon & Schuster.
Young, Jeffrey T 1991. Is the entropy law relevant to the economics of resource scarcity? *Journal of Environmental Economics and Management* vol 21/1994:169-179.
Young, Jeffrey T 1994. Entropy and natural resource scarcity: A reply to my critics. *Journal of Environmental Economics and Management* vol 26/1994.
Zuvekas C 1979. *Economic development: An introduction.* London: Macmillan.

Journals and Indices

American Economic Association (annual): *Index of Economic Articles.* Nashville: Tenn: Americ Econ Assoc.
American Economic Review
American Journal of Economics and Sociology
British Review of Economic Issues
Canadian Journal of Development Studies
Challenge
Conflict Management and Peace Science
Defence Economics
Demography
Ecologist
Economic and Industrial Democracy

Economic Development and Cultural Change
Economic Geography
Economic Impact
Economics and Philosophy
Economics and Planning
Economics and Politics
Energy Economics
Environment and Development Economics (Cambr Univ Pr)
Environment and Planning
European Economic Review
Finance and Development
Intereconomics
International Bibliography of Economics (annual). London & New York: Routledge.
International development abstracts (annual): Norwich UK: Elsevier.
International Journal for Sustainable Development
International Journal of Social Economics
International Labour Review
International Review of Applied Economics
International Spectator
Journal of Comparative Economics
Journal of Conflict Resolution
Journal of Cultural Economics
Journal of Developing Areas
Journal of Development Economics
Journal of Development Studies
Journal of Economic Development
Journal of Economic Literature
Journal of Environmental Economics and Management
Journal of Interdisciplinary Economics
Journal of Law and Economics
Journal of Political Economy
Journal of Population Economics
Journal of Population Economics
Journal of the Society for International Development
Natural Resources Journal
Nature
Social Choice and Welfare
South African Journal of Economics
The Economic Journal
The New Internationalist
World Bank Economic Review
World Development
World Development Report

Index

Abortion 400 401
Absolutism 194
Abuse of Power 130
Academic Isolation 283
Academy 9 269
Acceleration 334 344
Acceptability 201 218 219 Social 286 Un/conditional 177
Acceptance 165 248 Conditional 173 Unconditional 174 218
Acculturation 208 211
Acid Rain 86
Acquiescence 207
Adaptation 209
Advertising / Advertisers 131 158 190 222 287 289 306 309 310 425-430 and Entertainment Industry 286
Affirmative Action 241 248f
Affluence / Affluent 48 49 54 56 60 61 70 73 97 110 119 189 213 277f 304 307-309 319 344 350 Gap 57 59 304 313
African Initiated Churches 217
Age to Come 171 172 175 177 220
Agenda Economic 383ff
Agents of Change 151 360
Agriculture 72 80 120 345 350 439 Imports 452 Land 78 101 107 110 119 129 178 Overproduction 452 Production / Products 33 46 77 79 120 124 127 403 406 451 Sector 132 316 Subsistence 45 101 298
Aids 56 105 107 108 445
Ambiguity 173f
American Dream 277
Ancestor 20 160 177 197 Veneration 198 200 220
Animal 26 27 33 Husbandry 298
Anomie 196 208 222
Anthropology 197 283f 285ff
Anti-Semitism 207
Apartheid 131 159 256 261
Apocalyptics 160 175 277
Appropriate Technology 114 451
Arms: Industry 29 90 316 Production 433 434 Race / Buildup 3 29 71 89 90 213 Reduction 433 434 Regulation 434 Third World Sales 433 Trade / Military Expenditure (Milex) 132 133 390
Army 107
Art 48
Ascending Party 247
Asian Disease 56
Asian Tigers 56 125 126 322-324 343 373 386 404 420 443
Assumptions 160 345
Atmosphere 338
Attitudes 48 Change in 347
Authentic Humanity 177 179 Life / New Life in Christ 161 164 165 166 175 220
Authenticity 172 173 178 221
Authoritarian Rule 179 232
Authority / Group or Individual 145 198 199 202 204 216 Legitimate 179
Automation 317 323 324 419 to Act 188 286
Autonomy 189 281
Avarice 289
Awareness, Public 4ff

Babylonian Exile 172
Balance of Payments 84 127 129
Balanced Trade 450 452
Bankruptcy Law 449
Base Desires 352
Basic Essentials 59
Battle between the Gods 147
Behaviour 145 173 285 288 Patterns 101 103
Beliefs (see also Conviction) 98 142 150 189
Belonging (see also Right of Existence) 101 105 145 146 148 199 217 Social 286
Benefits and Sacrifices 354
Benevolence of God 164 of State 437
Biblical Faith 14 143 157 163 164 167 168 171 174-178 214 218 219 269 Witness 163 168
Big Bang 335 336 342
Big Business 286 353

Biological Factors 105
Biology 190
Biomass 82 83
Biosphere 81 90 338 342 345 347
Birth Rate 210
Black Hole 335
Black Theology 218
Body of Christ / Community of Believers 165 166 168 177
Body vs. Soul 168
Borderline Actions 235 Situations 235
Brain Drain 118 314
Bride Price 410
British Empire 40
Brundtland Report 416
Buddhism 144 160
Budget Deficit 213 306 324 418
Bureaucracies 102 192 316 442
Business 101 Cycle 307 Takeover 130
Buyer 300

C-factors 98f
C/P-factors, 98f
Cancer 87 169
Capacity 7 23 300 307 319 Building 301-303 Lack of 318 Ownership 309 Surplus 318
Capital 6 22 24 25 28 45 63 83 89 90 98 99-114 120-128 312 314 347 421 Access to 309 Accumulation 430 Cost 320 Drain 324 Flight 120 125 Free Floating 420 441 Goods 301 Human 113 Investment 303 307 421 442 International 320 322 Movement of 317 Output Ratio 113
Capitalism 40 111 146 149 158 188 Liberalism 152 156 201-203 209 222 258 273 274 278 373
Carbon Dioxide 347
Caste System 56 101
Catalyst 243 346
Cause and Effect 158 159 191 215
Causes of discrepancies 98ff Non-economic 100ff Economic

109ff
Central Planning (Controlled Economy) 373
Centre: And Periphery 28ff 40ff 50-57 98ff 118 193 210 353 257f 297ff 308 385ff 396ff 416 National 438ff International 446ff Colonial 44 117 Dependent 45 Hierarchy of 43 44 Population 425
Century of Ecology 268
Change 15 143 199
Chaos 341 342
Character Deficiencies 351
Checks and Balances 203 439
Childhood 202 212
Children 106 107 177
Chlorofluorocarbons 85 86
Chosen People 172
Christ 75 160 164 165 172 177 179 194 217 220 Crucifixion 165 220 Death 168 217 Event 217 Life of 168 Risen Lord 165 Resurrection 217 Redeemer 240
Christianity / Christian Faith 14 15 144 148 149 151 152 156-167 170 171 178 193 194 202 216 217 220 222 253 261 264 Assumptions 143 240 Community 156 240 262 263 Creed 160 Message 261
Church 160 193 195 217 218 260 262 371 445 Agent of Social Change 371 Strengths 371 Weaknesses 372
Civil Society 445
Civil War 89
Civil Religion 150
Class: Dependency 209 Discrimination 217 Dominant 151 Structure 62 Struggle 258 System 202
Climate 112
Clothing 129
Cognitive Dissonance 254
Cold War 88 124 128 192 322
Collective: Consciousness 13 14 98 103 111 142 143 146 151 158 159 160 170 190 206 222 232 238 239 245 269 284 286 390 409 451 Interest 10 111 143 268 287 Personality 229 232 Prosperity 289 Responsibility 158 Survival 27 Wellbeing 179
Colonial Personality Type 14
Colonialism 102 116 117 143 195 196 205 207 339 Japanese 322
Commerce 188
Commercial Farm / Farmer 119 124 148 440
Commercial Undertakings 114
Commercialisation 282
Commodity 21
Common Good 430
Communal Cohesion 352 Interests 277 Responsibility 352 Sharing 372
Communication 29 44 79 110 114 120 150 Networks 101 121 Sector 350 Systems 275
Communism 256 Communists 260 Governments 311
Community 5 Development 5 Spirit 25 32
Comparative Advantage, Theory / Law of 116 125 126
Competence, Levels of 361
Competition 25 48 101 103 110 112 119 120 130 187 307 International 322
Competitiveness 29 103 123 125 196 201 324
Computer 79 103 112 Games 306
Concern for the Weak 176
Conflict: Armed 280 388 Prevention 408 Violent 408 Potential 3 13 34 71 73 80 88 89 91 War 5 29 55 89 90 106 133 274 307
Conquest 339 of Reality 189 Reaction to 206ff
Conscience 351
Conscientisation 238 244 426 433
Conservation (See Ecology)
Conservatism 213
Construction 351
Constructs 336 337 338 342 346
Consumer 32 111 114 285 300 430 Demand 301 302 324 Rights 426 427
Consumerism 286
Consumption 20 21 35 59 82 127 128 167 278 299 301 303 304 305 307 309 321 347 416 423 424 Conspicuous 201 Current 419 Deferred 419 Levels of 347 384
Contentment 304 305 307
Contraception 400 Policy 401 Vatican 400
Convictions 143 146-149 151 152 156 157 202 204 212 213 283 287 336 367 391
Cooperation 25 35 150 202
Corruption 46 86 90 102 121 122 131 158 208 302 312 318 322 407
Cosmic Order 174
Cosmology 280
Cost of Living 58
Cost-benefit Analysis 20 31 32 33 91 283 424
Countervailing Measures 324
Cowboy Economics 394
Creation 161 168 215
Creative Authority 163 168 from God 164 166
Credit 119 124 127 131 Access to 320 Cards 430 Facilities 441 445
Crime 108 119 158 194f 311
Critical Mass 159
Cross 166 221
Crown of Creation 170 171
Cults 147 222
Culture 5 32 48 50 103 104 106 131 133 145 148 195 198 287 336 351 Autonomy 213 Change 352 Clash 391 Conditioning 212
Context 144 Degeneration 304 352 Disempowerment 210 Evolution 200 Failure 352 Modern 6 101 Norms 192 Synthesis 235 Tradition 198 275 Transformation 212
Currency Speculation 125 126 222 441

Death 167 175 177 220
Debt Burden 24 25 84 114 116 121 126-129 312 Foreign 325 Relief 275 Servicing 129 388 Third World 447 448
Declining Marginal Productivity / Utility 452 428 306 423ff
Deconstruction / Dissipation of Energy 221 336 337 351
Deconstructionism 146 158
Deforestation 29 72 79 84 88 113 210 213 449
De-ideologisation 259ff 261ff
Demand 302 305 318 for Expertise 314 Demand Side 423
Democracy / Democratisation 29 90 102 128 179 187 199 244 352 388 419 430 439 of Economy 390
Democratic 243 Institutions 64 115 State 192 Values 132 147
Demography 50 56 Transition 56 74ff 107
Demonstration Effect 60 309f
Demythologisation 162 218 353
Dependency 25 29 63 65 104 108 122 126 143 149 179 206 211 213 227ff 237 Theory 14 63 116 Theorists 258 Syndrome 211 227ff 391
Dependent Personality 236 247
Depletion of Natural Resources 3 4 13 79ff 88 90 133 194 344
Deprivation, Relative 61
Depth Dimension 145 287
Descending Party 247-49
Desertification 88
Despair 48 159
Developing Nations 279 (see also Third World)
Development 8 9 12 89 100 103 128 133 289 Agricultural 323 Aid 205 Projects 198 451 Sustainable 386 Theory 14 100 116
Dignity 91 104 105 159 169 214 217 218 221 289 290
Discrepancy in Potential 28 103 203 in Prosperity 20 33
Discrimination 339

Disguised Unemployment 107 110 115
Disillusionment 311
Disinformation 131
Distribution 20 21 24 114 121 305 319 of Earnings 114 Equitable 384
Divide and Rule 101
Divine 58 197 198 Benevolence 171 Gift 216 Mastery 171 Omnipotence 280
Diviners 197
Dominance 143 227ff
Downsizing 418
Drugs 215 222 307 Abuse 131
Dualism 174

Earnings 302 305
Earth 334 338
Ecologists 113 171 351 352
Ecology 97 191 383 Awareness 158 Concerns 88 190 Destruction 20 25 84 90 108 133 192 210 Economics 191 Imbalance 97 Impact / Deterioration 4 30 70 71 81 83 84 91 277 309 Problem 13 56 84 Refugees 71 88 Responsibility 275 353 Sustainability 5 Theology 178
Economic / Geographic Centre 4 9 29 40 41 45 46 50 56 58 59 62-64 70 84 91 98 101 103 106 108-133 165 179 257 279 303 305 307 309-313 315 324 343-345
Economic / Geographic Periphery 4 7 9 29 40 41 46 50 52 54 56 58 62 63 64 70 72 84 91 98 101-133 165 179 279 303 305 309-315 324 343-345 Outer 325 Semi-periphery 325
Economic: Activity 50 91 Behaviour 274 Boom 56 301 308 323 Capacity 318 Competition 206 Concentration 117 130 Dependence 25 63 Development 99 113 133 297 343 Differentiation 45 52 54 Discrepancies 325 Disintegration / Decay 4 56 123 Efficiency

23 31 Freedom 102 418 Geography 191 272 Growth 6 13 56 71 88 89 98 108 113 117 133 274 279 301 305 315 317 318 319 322 340 345 416 Growth Rate 343 Imbalance 9 97 123 Inefficiency 23 45 Integration 43 45 57 444 Interests 86 131 Justice 275 Paradigm 6 280 Policies 279 383 Potency / Income 50 51 56 57 71 80 98 102 106 111 112 116 117 133 274 302 304 Potential 55 103 202 272 289 303 Processes 272 Psychology 273 Recession 302 Sector 85 Sufficiency 178 Theory 14 272 313
Economic Man 157 178 285 287
Economics 10 11 15 31 32 36 50 58 84 97 109 269 281 282 283 284 287 297 348, Classical 273 Mainline 279
Economy/ies 21 22 55 100 102 112 116 119 130 168 344 Balanced 321 Command 418 Emerging 441 Formal 441 Global 70 122 123 125 128 129 133 258 Modern 298 305 of Scale 123 of Scope 123 Steady-state 321 416
Economists 5 10 70 85 116 279
Ecosystem 282 Entropic Deconstruction of 344
Ecumenism 263 264
Education / Training 12 50 84 101 103 104 106 107 113 118 127 190 315 319 Access 309 Formal 445 Spending 133
Efficiency / Inefficiency 3ff 82 88 101f 110 114 117 120 125 191 347 350 354 Gains 349
Ego 351
Elderly 103 107
Elders 197
Elites 8 101 102 103 106 111 130 131 133 149 151 177 192 304 312 Colonial 150 Global 325 Technological 314
Emancipation 201
Emissions 85 Control 431

Emotion 57 Security 146
Empiricism 189 194 220 281
Employment / Unemployment 3 65 90 106 107ff 119 122 302 308 318 319 Generation 320 442 Targeted 320
Enculturation 175
Endangered Species 275
Energy 58 77 79-82 86 88 110 112 113 115 200 201 306 320 334 336 350 Alternative Forms 83 423 Balance 338 Concentration 336 337 339 Consumption 350 Dissipation 335 338 Distribution 334 335 High Entropy 336f Low Entropy 336 337 345 Sources 347 Sustainable 5 Throughput 351 Tidal 345 Transformation 334
English Language 104
Enlightenment 144 177 187 190 237 280 289
Entertainment 222 Industry 190
Entrepreneurial Initiative / Entrepreneurship 46 102 111 121 312 349
Entropy 14 280 282 333ff 338 341 342 344 346 349 352-354 420 Law 334 Cultural 352
Environment/al 133 258 325 Cost 31 125 Destruction 35 89 108 133 274 307 Economics 273 274 Laxity 323 Safeguards 323 Sustainability 416
Environmentalists 70 352
Epidemics 55 210
Equality 188 Dignity 7 Distribution 439 Policies 319, Principle of Equality of Opportunity 26 319 349
Equilibrium, Theory of 116
Equity 6 24 56 79 82 160 176ff 203 273 274 279 347 353 354 378 418
Erosion 210
Eschatology 160 161
Eternity 174 280
Ethics 9 11 15 20 99 190 283 284 287f 347 353 Biblical 382
Ethnic: Barriers 175 177 Conflict 408 Hostilities 88

Ethnicity 288
Ethnocentrism 392 446
European Union 446 454
European Community 361 447
Evil 161 174
Evolution 14 162 170 192 281 282 287 289 334 341ff 342 346 351-353 Cultural 343 352 Economic 343 of Matter 341 342 Organic 343 Social 102
Evolutionary Process 160
Exile 175 177
Existentialism 178
Expatriates 121
Experience 152 153 162
Experienced Reality 169
Expertise 48 63 102 111 112 113 115 121 127 314 Supply 314 Traditional 315
Exploitation 176 316f 339 340 colonial 188 People 48 117
Exploitation / Overexploitation of Natural Resources 3 7 13 28 29 32 35 72 79 113 133
Exponential Growth 56 Theory 342
Export 127-129 302 319 322ff Market 129 443 Goods 448 Earning 128 129
Extended Self 352
External Costs 417 Internalised 422
Extinction of Species 340
Extra Market Activity 130 131
Extraction 20 21

Factors of Production 109-111 113 114 116 119 120 303 312 314 315 320 349 419
Factors / Mechanisms, Structural 49 99 111 Volitional 98 99
Faith 143f 156ff 204 215 221 368 in Christ 218
Family 47 48 105 106 107 160 194 Extended 178 Planning 107 as Primary Group 365
Famine 6 33 55 76 77 80 106
Fascism 147 151 156 160 212
Fatalism 157ff 173 200 278 381 Economic 381 Ecological 381

Metaphysical 381 Moral 381 Religious 381
Fate 169 197 198
Feedback Loop 98 213 284
Fellowship: with God 164 166 168 171 177 of Christ 217
Female Emancipation 107
Fertilizer 77
Feudalism 28 46 56 101 130 192 193 207
Financial Institutions 46 110 114 119 121 128
Fish Resources 78 88
Flesh 175 177 220
Flexibility 29
Flexible Automation 123
Food 13 58 73 77 79 107 148 Aid 451 Production 77 89 108 129 Subsidies 452 Surplus 77 78 124 132 Gap 73
Foreign / Development Aid 64 89 90 124 127 129 132
Foreign Business 120 121
Foreign Investors 318
Forgiveness 165
Formal Sector 444
Fossil Fuels / Energy 3 4 77 79 80 81 83 85 86 320 344 345 422 Depletion 347 End of 345
Free Enterprise / Open Market 102 116 124 126 130 194 256 258 309 323 373 418 420 428 441 System 349 362
Freedom 171 187 213-217 379 of Choice 401
Fuel Efficiency 431
Fundamentalism 147 160
Funerals 198
Future 32 36 70 71 80 83 89 91 98 103 111 133 143 157-159 167 174 179 191 192 199 201 207 217 219 344 Consumption 419 Generations 352
Futurology 191

Galtung Model 62 63 64
Gangs 107
GATT 454
GEAR (South Africa) 457
Gene Manipulation 190
Generative Themes 235

Genetic Engineering 188 288
Genetic Pool 351
Geography 40ff 283
Germany 418 419
Gifts of the Spirit 177
Give and Take 353
Global Empires 188 Network 111 132 Catastrophe 221
Global Warming 71 88 275 338
Globalization 55 70 79 446
Gnosticism 161
Goals 98 111 126 142 397f
God 15 159 162 163 164 167 168 173 174 179 193 214 216 Benevolence 173 Comprehensive Vision 219 Concept of 162 177 Creative Authority 219 Hidden 188 Image 179 216 Intentions 164 167 Mastery 164 173 215 Mission 164 166 168 Redemptive Concern 219 261 Redemptive Love 262 Spirit 165
Goods and Services 23ff 46 110 114 121 130 299-302 306 312
Gospel 165 169
Government 86 102 125 Corruption 407 Deficit 127-129
Grace 178 262
Grass Roots 5 151
Gravity 335 Law of 336
Greed 15 177 321
Green Lobby 88
Green Revolution 77
Green Taxes 395
Greenhouse Effect 85 338
Gross Domestic Product (GDP) Gross National Product (GNP) 40 52 56 82 129 219 289 316
Group Relationships 48 50 62
Growth 82 84 117 Accelerating 3 4 Economic 378 in Quality 344 Infinite 350 351 Rate 55 56 84 vs. Equity 349

Habitat Destruction 3 351
Handouts 319
Happiness 307
Harmony of Interests 63-65
Health 59 127 307 Spending 133

Heat Radiation 338
Heaven 172 193 280
Hebrew Religion 288
Hebrews 174
Hedonism 178 203
Hegemony 151 152
Hellenism 160 161 174
Hereditary Factors 341
Hermeneutics 149
Hidden Transcripts 149
Hierarchy 43 187 192 199
Hinduism 144
Hire Purchase 430
Historical Consciousness 191
Historical Factors 49 Processes 191 Acceleration of 342
History 99 159 167 172 173 280 Flow of 192
HIV 445
Holy Spirit 168
Hope 5 83 152 159 162 167 170 175 188
Hopelessness / Fatalism 157ff 268 381f
Horizon Universal 288f Shifting 277
Horizontal Relationships 180 262 Mobility 101
Horizontalising Vertical Relationships 239ff 244ff
Human: Agent 143 Appetite 304 Autonomy 162 Dignity 177 Freedom 157 167 221 Initiative 351 Nature 170 175 283 285 351 Potential 189 321 Responsibility 157 Species 170 Under-standing 197
Human Rights 147 178 187
Humanity, Single 177
Hunger / Starvation 77 90 165
Hungry 169
Hunting and Gathering 297

Idealism 14 99 282 284 370
Identity 104 107 147 191 208 212 213
Ideology 10 14 148ff 153 158 188 192 254 255 259 282 284 Biases 99 Legitimation 143 253ff 257ff 269 289 350 369 Power Struggle 159

Idolatry 177 221 215
Illiteracy 104
Imagination 189
Imbalances 26ff
Immigration / Migration 89 106 118 119 322
Imperialism 27 116 149 179 195 339 353
Imports 324 325 Restrictions 450 Substitution 322 323 404
Inauthentic Life 166 220 (see also authentic life)
Incentive 46
Inclusivity 14 288
Income / Salary / Wage 24 50 52 57-59 61 62 64 88 115 122 124 127 298 308 312 313 Discrepancies 46 51 52 59 80 115 118 120 308 348 Illicit 102 Per Family 51
Indigenous Peoples 207 Industry 122 Leadership 244 246
Individual: Accountability 177 Experience 177 281 Freedom 14 48 Mastery 200 Mindset 146 Roles 48 142 145 148
Individualisation 288
Individualism 178 187 193 194 220 390 Modern 170
Industrial: Age 103 Civilisation 345 Democracy 440 Economy 160 315 Growth 33 70 72 79 83 84 87 91 288 443 Overdevelopment 398 Revolution 40 188 314 Sector 440 442
Industrialisation 4 72 79 82 83 85 192 213 Export-orientated 322 323 324 453
Industrialised Nations / First World 82 84 86 89 111 113 115 123 128 129 133 279 316 350 442 Societies 101
Inefficiency 302 308 319
Inequity 321
Infant Mortality 210
Inferiority 206 228ff
Inflation 127 129 307f 442
Informal Settlements 344
Informal Sector 316 439 441 Work Sector 118
Information 11 29 111 117

Explosion 288 352 Technology 188
Infrastructure 45 46 109 110 117 119 120 121 127 132 133 301 312 Development 451
Inhibitions 286
Initiative 22 101 104 111 112 118 187 378
Insider Trading 131
Instinct 143 289
Institutions 351 Saving 301 Supranational 446
Insurance, Life 440
Integration 209 456
Integrity 172 173
Intellectual Discovery 336
Inter-faith Discussion 160
Interdisciplinary 8 9 Efforts 276 333 Accountability 282
Interest Rates 98 127 128
Interests 148 149 253ff 448 Collective 346 Private 346
Intermediate Technology 275
International Aid 450f
International Court of Justice 434 447
International Law 447 449 Regulation 420
International Debt 259 347 Loans 448
International Trade 439 449 450
International Monetary Fund 127 128 322 447
Internet 104 288
Intolerance 176
Inventions 189
Investment 113 120 121 127 302 308 Capital 314
Irrational Behaviour 157
Islam 151 152 160 188
Isolationism 453 456
Israel 175 176 178 219

Jerusalem 172
Jesus (see Christ)
Jewish-Christian Heritage 277 Assumptions 156
Jews 150f 172 173 176 177
Job Creation 314
Justice 5 99 143 172 173 176 178 179 190 281 348

Intergenerational 345
Justification 165
Kingdom of God 160f 164 174
Knowledge 111 200 270 333

Labour / Workers 22 24 25 45 56 89 103 106 110 111 113 114 117 118 122 125 126 131 303 314-316 421 Centre 317 Cheap 315-317 323 Controls 132 Costs 56 113 123 320 324 Division 22 Labour Intensive Methods 443 Labour-intensive Production 114 122 Labour Saving Technology 314 443 Legislation 317 Market 443 Saturated Labour Market 317 Peripheral 317 Skilled / Semiskilled / Unskilled 113 115 118 121 122 Surplus / Redundant 316 Unskilled 317
Land 111 112 119 314 Access 309 Reform 322 323 439
Language 104
Large Scale Organisations 192
Last Judgement 173 177 220
Law 173 178 of Declining Marginal Productivity 77 Enforcement 100 108 131 of Jubilee 178 of Nature 168 214 215 of Thermodynamics 334 of Entropy - economic relevance of 337
Laws / Legislation 88 100 123 130
Leadership / Authority 36 90 100 101 104 120 122 145 208
Legitimacy 102
Legitimating Ideology 274
Lethality Index 89
Liberal 5 70 Capitalism 151 352 Democracy 201 Economy 286 Mind Set 5 130 School 99 126 130 Society 286 430
Liberalism 187 378 Liberals 258
Liberation Theology 178 218
Life: Chances 88 189 274 Expectancy 106 Forms of 339 Un/inauthentic 164 165 175 World 214
Lifestyle 275

Lineage 197
Living Wage 316
Living Standards 203
Loan 121 126 127 129
Lobbies 6 115
Local Technology 406
Local Production for Local Needs 325
Loneliness 218
Long Term Thinking 288
Love 282 Act of 169
Lower Classes 50 115
Luxury: Consumption 417 Demand 309 Goods 308 Needs 129

Machines 24 31
Macro-economics 5
Mafia 131
Malnutrition 105
Management 112 115 119 121
Management Science 179
Manipulation 426
Manufactured Goods 113 123 124 128 314 450
Manufacturing 316 345
Marginal Utility 424
Marginalisation 4 13f 24 65 103 133 165 189 208 272 289 297ff 316 344 419 442 444ff
Market 5 6 7 25 43 46 57 81 82 90 111 114 116 117 119 121 123 126 127 130 202 300 Asymmetrical Interaction 123 125-127 130 Demand 298 307 308 Dominance 131 Economy 301 Entry 131 Equilibrium 7 Forces 302 307 International 325 Mechanism 272 299 307 308 349 Penetration 126 Rigidities 320 Saturation 324 Shortage 131
Marketing 11 120 121 132 149 219 281 306 310 425 425ff
Markets, 371 Domestic 322
Marriage 148
Martyrdom 175
Marx, Karl 418
Marxism 5 56 70 99 116 149 151 152 159 160 179 192 193 209 232 242 258 277 309 322

441ff 454 Socialism 273
Mass Starvation 353
Mass Media 190 363 Communication 288
Material World vs. Spiritual World 168
Material 32 Craving 307 Factors 109 111 112 Throughput 350
Materialism 14 32 99 142 188 284 370 390
Mathematical Laws 336
Matter 145
Meaning 168 191 219 286 of Life 36 System of 367 377 393
Meaninglessness 218
Means of Production 309
Mechanisation 320 324 443
Media 445
Medicine 190 Traditional 211
Mental Constructs 158 160 221 222 Structures 98 133 142 145 151 166 213
Mental Factors of Production 109 111 118
Messianism 179 234
Metaphysics 187
Methodological Assumptions 6
Method, Model Construction 11
Methods of Production 320
Microprocessors 130
Middle Class 102 149 208
Migrant Labour 118 315
Migration 323 Illegal 307
Military 407 Expenditure 306 312 Military-industrial Complex 90 433
Milky Way 335
Mindset 9 14 15 98 103 142 145 204 213 276
Mining Sector 316
Mismanagement 129
Mission 167 168
Missionary 244 247 262 Movement 188
Mobility, Upward 317 Vertical 56 101
Model: of Causation 97 98 Collective Consciousness 148 Factors of Production 109 Marginalisation 313 Obstacles 391ff Potency 50ff Need 57ff Policy 385ff Reactions to Conquest 209 See Centre-Periphery
Model Construction 11 12
Moderation 352
Modern Scientific Thought 168
Modern Society 378
Modernist Revolution 144 World View 220
Modernity 14 143 151 160 186 196 200 201-204 206 212 213 215 222 271
Monetary Policy 272
Money 22 114 298 Flow 298ff
Monopolies 130 132 300 440
Monopsonies 130 300
Moral Chaos 218
Morality 3 11 131 147 170 190 196 218 281f
Mortality Rate 107
Mosaic Law 219
Motivations 286 368 382 Redemption of 369
Multi-dimensional Problems 333
Multinational Corporations 29 54 55 63 130 273 388 441 447 455
Mutations 341
Mutual Acceptance 240 241
Mutually Assured Destruction 133
Mysticism 281
Mythology 197 198

Nation State 194 362 446 Agent of Change 372
National Service Programs 445
Nationalism 147 187 192 194
NATO 434
Natural Sciences 170 333 334
Natural Law 200 280
Natural Disasters 27
Natural Environment / Nature 5 7 22 25 70 72 75 133 167 189 190 192 194 210 214 280 282 287 288 290 338 339 344 345 353 Capital 73 345 Debt Swap 449 Preservation of 390 Resources 194 201 349 Sinks 347 Need 7 8 20 23 24 32 57 58 59 77 80 81 84 114 116 128 131 144 148 159 164 168 171 174 176 218 221 279 285 286 290 299 300 302 304 305 307 312 313 318 319 321 and Capacity 14 297 298 Artificially Created 306 Basic 219 297 306 Curve 58 Fulfilment 320 Immanent 144 145 168 169 254 Pattern 311 Satisfaction and Balance 347 420 424 Saturated 309 Structure 310 Immanent and Transcendent 144 145 168 169 Unfulfilled 310
Negative-sum Game 337
Neo-liberalism 436
Nepotism 442
New International Economic Order 449
New Life in Christ 165 166 171
New Humanity (See Authentic Humanity) 177
New Heaven / New Earth 174
New Testament 148 161 172 193 220 289
Nirvana 147
Non-governmental Organizations 363 445 446
Non-military Expenditure 133
Non-renewable Resources 72 79 416
Norms 48 98 111 142 145 149 158 173 196 198 212 219 271 351 352 Social 287
Nuclear 222 Energy 80 82 83 86 Fission 83 Holocaust 188 192 213 Power 188 190 288 Radiation 71 Testing 90 91 275 Weapons 79 83 133 194

Objectivity 285
Oil 24 33 83 87 88 123 128 213 320 340 346
Old Testament / Hebrew Scriptures 161 288
Oligopolies 130
Open System 350 351
Open Society 56
Opportunity Cost 278
Oppression 48 116f 150 176 232 242
Organisational Breakdown 380

Expertise 46 112 119 131 444
Organised Crime 89 90
Orthodoxy 273
Overconsumption 219 420 427
Overexploitation of Natural Resources 203 338 340
Overdevelopment 318 319 323
Overgrazing 72 78
Overpopulation 347 (see also Population)
Ownership, Joint 440
Ozone Layer 71 72 85 86

P-Factors 98ff
Pace 342
Packaging 431-2
Paradigm 6ff 9 10 11 12 15 152 270 271 275 277 278 284
Paradigm Shift 268ff
Participation 24 91
Partnership 179
Passage Rites 197
Patriarchy 56 197 199 202
Patterns of Consumption 311
Patterns of Behaviour 200 207
Peace 32 73 90 172 179 190 Research 191 with God 163ff
Peasant 49 64 119
Pedagogy of the Oppressed 227 232-35 of the Oppressor 242 of the Catalyst 243
Penetration of Periphery 121ff
Pension Fund 320 440
Perception of Reality 146 187
Periphery 28ff 40ff 98ff 210 257f 297ff 385ff 396ff 454 Countries 308 443 448 Economy 385 456 Population 425 427 444 450 454
Personal Satisfaction 220 281
Philosophy 192ff 281ff 287 352
Photosynthesis 338 344 351
Physics 270 282 284 335
Pietism 150 161 177f 194 281
Planned Obsolescence 131 306 432
Pluralism 48 158
Political Agenda 348
Political Order 11 151
Political Stability / Instability 120 128

Political Theology 178
Political Will 351
Politics 50 55 101 151 Influence 131 Parties 64
Policy Model 385f
Poll Tax 205
Pollution 4 5 13 25 29 31 33 72 73 76 78-81 84 86 88 90 108 274 301 314 320 321 338-340 344 345 347 349 390 430-433 Regulations 317
Poor 54 55 56 64 73 83 84 86 89 107 127 129 130 158 178 307 308 311 344 348 Preferential Option 242
Popular Wisdom 143 150
Population 50 54 56 259 343 349 Density 74ff Control 133 Growth / Overpopulation 3 4 5 13 27 29 32 33 35 45 46 54-56 70-76 81 84 91 99 106-108 210 211 218 288 317 321 339 340 344 386 399 452
Positivism 277 281 282 284
Post-apartheid 257
Post-colonialism 102 Elite 102 Personality Type 14
Post-Modernism 146 158 204 213 221 222
Potential Consumption 419
Poverty 3 5 15 29 48 49 61 76 80 97 106 116 120 133 211 274 278 304 317 319 323 Gap 57 59 304 312f Absolute 60f
Power 29 40 50 56 62 80 82 86 88-90 102 116 119 130-133 149 151 160 167 170 192 194 196 208 213 278 289 302 352 Abuse of 213 388 Concentration of 179 188 438 Economic 298 314 Levels 361 Legitimate 179 Relations 133 143 Structure 159 Struggle 28 89 90 111 158 Vacuum 208
Powerlessness 169 171
Pragmatism 178 277 278
Pre-colonial Economics 123
Prebendal System 102
Presuppositions 284
Prices 308 Fixing 132
Primary Group 364-7 Cooperation 366 Solidarity 151
Private Sector 102 362 445 Ownership 289 Saving 301 Sphere 151
Private Enterprise 428
Private Interest 430
Private Transport 432
Private Spirituality 151 160 218
Privatisation 102 of Religion 151ff 160ff
Privilege 61f 177
Processing 21
Producer 32 114 285 300 317 430
Product Innovation 314
Production 21 22 24 33 101 109 110 111 114 115 116 118 121 127 192 299 300 301ff 307 319 Capital Intensive 114 Cost of 124 299 Enclave 122 127 Level of 386 Surplus 320 321 Unnecessary 431
Productive Capacity 127 299 300 301 305 306 308 312 19
Productivity 4 29 45 46 112 113 114 115 118 120 315 317 318 442 Law of Diminishing Marginal Productivity 354
Profit 86 87 131 203 219 274 282 285 299 352 Sharing 440 Short Term 352
Profitability 102 120
Progeny 198 199 201 218
Progress 187 192 201 281 353
Progressive Taxation 319 422 429 441 444
Proletariat 150
Promised Land 172 177
Propaganda 149 190
Property Rights 178 439
Prophet 174 176
Prophetic Movement 179
Prosperity 172 190 301 353
Protectionism 128 452 454
Protestant Ethic 156
Psychological Dimension 32 50 60 104 105 143
Psychology of Colonisation 227 228-32
Psychology 283
Public Consensus 150

Public Transport 423 432
Public Order 352
Public Policy 9 15 277 288
Public Wisdom 147
Purchasing Power 8 46 114 298 300ff 315ff 388 Surplus 319

Qualitative Growth 5 32 91
Quality, Enhancement of 315 321
Quality of Life 306 321 427
Quantitative Growth 5 91
Quantum Theory 282

Racial Discrimination (Racism) 56 149 205 207 212 217
Radiation 351
Radical Relativity 146
Radioactive Waste Disposal 83 87
Rain Forests 347
Rationalism 189 194 281 254
Raw Materials 21 23 24 25 112 113 117 119 120 123 124 131 189 194 314 Access to 309
Reaganomics 458
Reality 152 157 162 270 271 280 281 287 336 341 Greek Perception 187 Hebrew Perception of 187 Source of 219 Vision of 214 380
Rebellion 208
Recession 56 301ff 308 315 318 324
Reconstruction and Development Program 457
Reconstruction 241
Recycling 25 87 345 349 432
Redemption 161 171f 176 193 Action 190 Concern 163 166 168 Event 164 Resources 156 Motivations 368f
Reductionism 7 282
Reference Groups 60 149 255
Reformation 177 289
Refugee 34 89 90
Regulation: Advertising 428 Pollution 422
Relationships: Vertical 180 239 245 260
Relativism 159 193 222 282 284

Relativity of Knowledge 270 and Faith 159f
Religion 15 29 36 48 50 56 101 106 145 147 158 172 187 188 190 193-195 281 Religious Confrontation 216 Conviction 187 Fundamentalism 213 Studies 283
Religious Communities 363
Remuneration 24 25
Renaissance 187 193
Renewable Resources 13 35 79
Research 89 101 119 122 131 189 191 271 and Development 320
Resources: 82 84 88 89 91 107 192 272 Access to 320 Alternative 347 Allocation 306 Availability 20 97 133 Base 21 23 26 28 29 133 178 202 301 Control 28 102 Depletion 301 325 421 Human 201 Non-renewable 390 Scarcity 349 Utilisation 187 Wastage 307 316 320 Sharing 199 Transfer of 124
Responsibility 104 133 145 160 213 214 216 218 287 352 Communal 379 Social 289
Restlessness of God 161
Resurrection 161 165 166 168 173 220
Revelation 281
Revolution 152 158 194 209 of Rising Expectations 310f
Rich, the 80 348
Right of Existence 145 168f 199 201 216 219 229 254 286 365 382
Righteousness 165 173
Rights 36 132 of Non-humans 179
Risen Christ 168 176
Rising Expectations 310 311
Ritual 148 176 193 197 198
Robotics 317
Roman Catholic Church 193

Sabbath 171 217
Sacred Canopy 271
Sacrifice 165

Salaries, Dynamic 308 Minimum 320
Salvation 165 167 210 218
Savings 320
Scapegoating 254
Science 11 101 104 152 158 171 189f 193 200 215 270 271 281 282 287 345 445 Value-free 345 World View 151 158 189 and Religion 152ff and Technology 3 5 73 82 188 191 202 203 212 352 353 Evolution of 343
Scientists 85 167 271 284 288
Second Economy 444
Second World 84 311
Sectarianism, Fanatical 234
Secular Humanism 143 156 178
Secularisation 107 202
Secularism 151 187
Security 101 105 107 199 217 Security Apparatus 28 29 30
Selective Perception 254 259
Self-determination 211
Self-esteem 170
Self-interest / Selfishness 5 6 7 11 28 64 131 143 152 159 162 170 177 187 194 195 202 203 220 253 275 283 285 289 351 352 369 378
Self-isolation 254
Self-sufficiency 403 456
Semi-industrialisation 442
Semi-periphery 50 51 312 343
Services Sector 316 350 442
Sex / Sexuality 190 215 222 281 Sexual Relations 194 Sexual Differentiation 341
Shalom 164 172
Share Holders 440
Short Term Thinking 288 Decision Making 268
Sin 165 166 177 178
Sinks 84
Slavery 205
Slum 33 45 65 72 86 108 110 118
Small Business 130 320 Sector 441
Social: Analysts 152 159 Barriers 175 Benefits / Securities

102 105 127 Change 371
Chaos 221 Cohesion 3 105 151
196 Conflict 434 Conscience 6
Consensus 199 Context 178
Contract 207 439 Control 151
286 Convention 198 215 Cost
31 Darwinism 282 Decay 31
45 Democracy 53 56 418 439
441 Development 189 Disintegration 71 Engineering 286
Environment 7 167 168 214
287 Expectations 59 101 Factors 50 109 110 111 Formation
203 Harmony 199 Hierarchy
101 196 Instability 128 Institutions 102 166 195 272 370
Integration 101 102 150 Justice
345 Ladder 119 Mobility 101
Network 199 Order 192
Organization 47 100 101 112
Paralysis 211 Pressure 271
Problem 119 133 Processes
142 213 Prosperity 289 Pyramid 50 73 Reality 142 147 175
Relations 112 198 Sciences 9
271 283 Security 418 Services
128 306 Stability 179 Status
106 142 145 148 286 Structures 9 13 14 101 110 133
142-145 149 150 159 167 193
213 271 284 336 System 142
159 197 Unit 101 Unrest 89
106
Socialisation Process 98 142
Socialism 102 112 115 156 188
213 274 378
Socialist Parties 192
Sociology 283
Soil Erosion 29 33 71ff 78f
Solar Energy 80 82 83 345 347
Solar Radiation 344 345
Solidarity 89
Son of God 179 See also Christ.
Sorcery 199
Soul 161
South African Council of
Churches 266
South Africa 241 242 256
Space Research 306
Space 58 109 110 117 164 167
170 172 280 306 336 Outer

338 379 Social 380 Travel 188
Specialisation 25 48 101 190 283
Spending 309 Social 133
Sphere of Influence 56
Spirit 145 177 220 240 of Christ
165
Spiritual Realm / Dimension 32
145 161 World 175
Spirituality 132 307
Spiritualization 160
Squatters 440
Stalinism 147
Standard of Living 33 46 59 63
73 80 83 211 345 348
Star Wars 437
State 8 29 55 59 103 112 132
353 Apparatus 102 Bureaucracies 102 Income / Revenue
124 132 302 324 Intervention
322 Ownership 120 Spending
56 306 Subsidies 78 79 119
124 Role 372
Statistical Tools 299
Status Quo 278
Status 201 214 309
Stereotyping 255
Stewardship 179
Stimulation 307
Stone Age 346
Structural Change 143
Structural C/p-factors 98
Structural Adjustment
Programme (SAP) 128 312
Subcontracting 440
Subjugation 28 29
Subsistence Level 308
Suburb 119
Suction 314 to the Centre 117
118 119 120
Suffering 4 165 171 175 221
308
Sufficiency 279 304 305
Sun 338
Superpower 89 90 129
Supply 299f 308 318
Supply and Demand 7 14 57 115
116 117 123 124 131 297 298
333 388
Supply Side Economics 273
Supra Disciplinary 9
Supreme Being 197 198 200

Supreme Will 200
Surplus 301 Capacity 7 302 307
309 Capital 316
Survival of the Fittest 170 341
351
Survival 33 170 175 194 195
202 218 Imperative 351
Sustainability 11f 35 143 167
203 274 277ff 321 378 403
417 and Growth 7 117 121 349
Sweden 418 419
Symbiosis 196 341
Symbol 22 Symbolic Enactment
169 Symbolic Universe 5
Symbolism 148
Syncretism 174
System 11 Closed 351 of Hierarchies 56 of Meaning 144 146
149 150 152 158 174 176 188
196 198 200 202 204 207 212
214 254 271 272 284 287 351
of Relationships 170
Systems Theory 170
Systems Analysis 9 191 284

Taiwan 418 453
Tanzania 453
Tariffs 421 454 455
Taxation 59 78 117 121 123
420f 443 Negative Tax 319
Tax Concessions / Exemptions
320 431 Tax Subsidies 320
Technological advancement 27
28 82 99 103 111 112 113 114
303 314 315 318
Technological superiority 46 83
112 119 121 123 124
Technology 24 25 63 70 77 78
80 82 90 98 111 112 113 124
189 200 275 281 312 314 316
317 339 340 344 346 347 349
350 351 405 417 421 Clean
321 Labour Saving 320 Spillover 117 Technology-intensive
Production 122
Telecommunication 40 46 111
Terrorism 307
Theodicy 174
Theology 144 177 263 270 276
281 287 Theology of the Cross
175

Thermodynamics 336
Third World 12 56 63 76 78 81 82 84 86 88-90 102 103 108 111 113 119-132 196 207 211 213 311 315 317 319 322 324 325 Elite 311
This Age 172 175 177
Thought Structure 153
Throughput 419 424 433 of Resources 4
Time 46 58 109 110 115 117 164 167 170 172 175 191 280 306 336 Management 191
Tolerance 147 159
Toxic Waste 86 87 88 314 Toxic Emissions 344
Trade 24 99 116 120 125 126 131 312 International 272 Interna-tional Restrictions 320 Free Trade 116 125 126 274 386 453-455 Liberalisation 452
Trade Unions 53 56 63 115 118 192 315 440 443
Tradition / Traditionalism 14 103 143 152 160 186ff 196 197 201-204 206 215 219 Culture 6 25 48 101 104 105 106 107 148 191 210 211 217 256 310 Communalism 170 Economy 297 315 Societies 200 280 378 World view 166 212
Training 317
Transcendence 157 159 162 167ff 170ff 215f 221 287 367
Transformation 20 22 23 143 216 of Collective Mindsets 143 156 of Reality 174 187 Social 142 143 Spiritual 168 Structural 151 World 166
Translocation 118 123
Transport 29 40 44 46 110 113 114 117 120 122 124 Costs 455 Links 121

Tribal System 101
Truth 146 147 152 158 160 170 213 221f 255 270 276 281 285 Claims 145 146 148

Ultimates 146
Ultraviolet Radiation 78 85
Underdevelopment 208 318f 323
Unemployment 213 313-316 344 419 422 442 445 Benefits 320
Unification 284
United Nations 129 361
Universe 334
Universities 103
Upper Classes 50
Urbanisation 31 33 45 72 119 192
Utilitarianism 178
Utility 22 57 58 60 202 203 220 222 282 285 423 Maximisation 274 Law of Diminishing Marginal 354
Utopian 159

Value System 11 36 48 98 100 101 103 107 111 142 145 147 148 202 309
Values 149 158 196 198 271 283 284 351 352
Vegetarian Diet 77
Violence 190
Virtual Reality 288
Virtue 281
Vision 143 152 161 162 163 166 167 172 177 277 278 284 288 Comprehensive 168

Wages 308 315 317
Wants 57 58 325
Warlords 409
Wastage 309 321
Waste 4 5 21 22 25 79 83 84 86 87 91 117 301 309 340 345 347 349 Disposal 86 87 272 of

Human Potential 316
Water 46 58 83 88 105 110 Pollution 83 87 108 Scarcity 78 85 Shortage of 210
Wealth 5 22 106 117 203 Accumulation 4 194 323 Concentration 188 Creation 35 280 344 345
Weapons (see also Arms) 89 90 128 352 Trade 129 133
Welfare 8 64 312 323 Economics 273 Projects 319 Social 384 Spending 133 State 102 320
Well-being, Comprehensive 7 32 164 167 214 216 277 288 369 380 of Humankind 157
Western Civilization 145 200
Western Culture 193 Western Thought 156
Western Medicine 211
Will of God 162
Wind Energy 82 83
Wisdom 103
Women 105 106 122 150 165 264 Empowerment 445
Word of God 263
Workers 307 324 Work Force / Worker Aristocracy 64 316
Working Conditions 315
World 171 172
World Bank 127 128 322 447
World View 157 208 213 287
World Wars 192 212

Yahweh 172 173 175 176 177 178 219 220 Law 176
Year of Jubilee 27
Youth 118 122 150 201 307
Zero Sum Game 35 334
Zimbabwe 404

- Soli Deo Gloria -